Gerd Pluschke • Katharina Röltgen
Editors

Buruli Ulcer

Mycobacterium Ulcerans Disease

Editors
Gerd Pluschke
Molecular Immunology
Swiss Tropical and Public Health Institute
Basel
Switzerland

Katharina Röltgen
Department of Pathology
Stanford University
Stanford, CA
USA

https://doi.org/10.1007/978-3-030-11114-4
This book is an open access publication.

Preface

Within the 20 years after WHO has launched the Global Buruli Ulcer Initiative in 1998, considerable progress has been made toward the understanding and the control of this neglected, once mysterious, tropical skin disease. Major practical achievements were the evaluation and successful introduction of an antibiotic therapy consisting of oral rifampicin and injected streptomycin administered daily for 8 weeks, the development of a PCR-based diagnostic test to detect DNA of the causative agent *Mycobacterium ulcerans* in lesion specimens, and the establishment of national Buruli ulcer control programs in the most endemic West and Central African countries.

Genome research has shown that acquisition of the virulence plasmid pMUM has enabled *M. ulcerans* to produce mycolactone, a highly potent macrolide toxin, responsible for much of the chronic, necrotizing pathology of the disease. This evolutionary bottleneck event is considered the key step in the emergence of the species *M. ulcerans* from the environmental *M. marinum*. Genome reduction and pseudogene accumulation accompanying the further evolutionary history of *M. ulcerans* are indicative for its adaptation to a more stable ecological niche. However, despite extensive research efforts, definite reservoirs of *M. ulcerans* as well as modes of transmission to humans have remained unclear. This is in part due to the extremely slow growth rate of the pathogen, hampering attempts to obtain isolates from potential environmental sources. It appears that direct human-to-human transmission is very rare, and there may be no single, simple explanation for the acquisition of infection from other reservoirs. Proximity to stagnant or slow-flowing water bodies seems to be a feature common to all Buruli ulcer endemic regions worldwide, and man-made environmental disturbances have been implicated in the emergence of the disease in some areas. While possums and mosquitos have been proposed to play a role as reservoirs and vectors in the transmission of *M. ulcerans* in Australia, a different scenario is most likely encountered in African settings, where large chronic lesions of Buruli ulcer patients may represent a relevant reservoir. Skin injuries are considered a potential route of infection, but also contaminated biting arthropods may play a role in transmission. The development of local clonal complexes of *M. ulcerans* in individual Buruli ulcer endemic regions speaks against the existence of highly mobile reservoirs. If large, chronic Buruli ulcer lesions are a relevant reservoir, early treatment of patients in an endemic area may help to reduce transmission locally. During the course of its evolution, *M. ulcerans* has diverged into different ecovars. The classical lineage, found in Africa and Australia, is responsible for >98% of all reported Buruli ulcer cases worldwide. While the local incidence of Buruli ulcer in Africa and Australia is very high, cases are only sporadically observed in Asia and South America, where another ecovar, the ancestral *M. ulcerans* lineage, prevails and entirely different human contact patterns with the environment may lead to infection.

As there is no effective vaccine against Buruli ulcer and preventable risks for infection are not clearly identified, current disease control measures primarily rely on early case detection and rapid initiation of antibiotic treatment. Efficient implementation of this control strategy is complicated by the fact that most infections occur in remote, rural areas of Africa, where access to healthcare is limited. Logistics involved in the provision of antibiotic treatment is challenging, and laboratory facilities to perform PCR, which has become the gold standard for the laboratory diagnosis of Buruli ulcer, are centralized. Furthermore, four rounds of external quality assessment (EQA) for the molecular detection of *M. ulcerans* in clinical specimens organized since 2009 have shown that reference centers often fail to pass the expected performance indicators for PCR-based diagnosis. Daily injections, ototoxicity, and other adverse side effects of the current antibiotic treatment are causes for concern. Alternative options for a decentralized diagnosis and treatment of Buruli ulcer are urgently needed. Most of the established antibiotics used for the treatment of tuberculosis show only limited potency against *M. ulcerans.* Nevertheless, repurposing of new tuberculosis drug candidates may be the most promising avenue to develop potentially faster treatment regimens, associated with less severe adverse side effects. Whereas no substitute for rifampicin is currently available, replacement of the injectable antibiotic streptomycin with oral clarithromycin is being evaluated and may support decentralization. Moreover, thermotherapy may be considered as an option for community-based treatment. Efforts coordinated by FIND to develop a low-cost and simple rapid diagnostic test also represent a key element in the strive toward early diagnosis and treatment of Buruli ulcer at community level.

In recent years, the number of new Buruli ulcer cases reported to WHO from endemic African countries has declined. It has been hypothesized that Buruli ulcer control activities may have led to an actual decrease in incidence by early detection and treatment of previously unattended lesions, potentially fueling environmental reservoirs with the pathogen. However, a significant reduction in Buruli ulcer control program support by some key NGOs has, on the other hand, decreased active case search activities in known endemic areas and diminished efforts to identify new endemic sites. Recent focal intensification of control efforts in some endemic regions and countries has shown that Buruli ulcer is still an underreported disease. In most endemic areas, Buruli ulcer is a disease of the poor that—if untreated—can lead to permanent disabilities with very severe psychosocial and socioeconomic implications. Limited knowledge on reservoirs and vectors of the pathogen and modes of transmission constitute an unpredictable source for new infections. It is therefore of uttermost importance to mobilize funding for the maintenance of established disease control activities and healthcare infrastructures in the affected countries, for continued research, and for the validation and implementation of newly developed diagnostic tests and treatment regimens.

This book provides an overview of the history and current burden of Buruli ulcer, contains chapters on the epidemiology and control of the disease, and provides insights into current research and development activities. A steady increase in the incidence of Buruli ulcer in the past years in Victoria, Australia, shows that this neglected disease is not defeated and that its control is even in industrialized settings a difficult task.

Basel, Switzerland — Gerd Pluschke
Stanford, CA, USA — Katharina Röltgen

Contents

Buruli Ulcer: History and Disease Burden

Katharina Röltgen and Gerd Pluschke

1 History of a Mysterious Skin Disease

The first description of chronic skin ulcers consistent with the pathology of *Mycobacterium ulcerans* infection dates back to the end of the nineteenth century, when the British physician Albert Cook recorded his observations in *The Mengo Hospital Notes*, maintained in the library of the hospital in Kampala, Uganda [1]. In 1948, characteristics of similar skin ulcers were described by MacCallum and his colleagues in six patients from the Bairnsdale District in southeastern Australia [2], where the disease is therefore often referred to as Bairnsdale ulcer. The causative organism isolated from these ulcers was found to be an acid-fast mycobacterium, later named *M. ulcerans*. Noteworthy, the first isolation of the extremely slow-growing mycobacterium in culture was achieved by accidental incubation of culture plates in a faulty incubator [3], reflecting the low optimal growth temperature of the pathogen at 30–33 °C. This temperature preference is considered a major factor for the skin tropism and limited systemic dissemination of *M. ulcerans* infections. In the 1940s and 1950s a larger case series of 170 patients with necrotic skin ulcers, caused by an acid-fast mycobacterium, was recorded in the Democratic Republic of the Congo [4]. A high prevalence of *M. ulcerans* infections was noticed in the 1950s and 1960s in a geographically very limited area of the then sparsely populated Buruli County close to the Nile River in Uganda and as a consequence, the disease became more generally known as Buruli ulcer (BU) [5, 6]. In proximity to this initial infection focus, a second outbreak of the disease in Uganda was reported in Rwandans living in a refugee settlement that was opened in 1964 close to the Nile

K. Röltgen · G. Pluschke (✉)
Swiss Tropical and Public Health Institute, Basel, Switzerland

University of Basel, Basel, Switzerland
e-mail: gerd.pluschke@swisstph.ch

G. Pluschke, K. Röltgen (eds.), *Buruli Ulcer*,
https://doi.org/10.1007/978-3-030-11114-4_1

River. Intensive investigation of the 220 BU cases that occurred until 1969, when the community moved to a new locality, where case numbers declined, has provided much of the basic knowledge about the epidemiology of BU that is still valid today. Amongst other insights it became apparent that BU may affect individuals irrespective of sex and age, although the highest incidence was seen in children aged between 5 and 15 years. Furthermore, in contrast to a very low probability of person to person transmission, the involvement of an environmental reservoir in the transmission became obvious [7]. While until then clinical attention had been concentrated on ulcerative lesions, awareness of the disease in highly endemic regions has brought attention to pre-ulcerative forms of the infection including nodules, plaques and edema [8].

Until the end of the twentieth century, BU case series were reported mainly from West and Central African countries including the Congo [9], Nigeria [10], Gabon [11], Ghana [12, 13], Benin [14], and Côte d'Ivoire [15, 16] as well as from Australia [17, 18] and Papua New Guinea (PNG) [19]. Moreover, sporadic *M. ulcerans* infections occurred in a number of additional countries with tropical, subtropical and temperate climates [20], totalling up to 34 countries from which BU has hitherto been reported. However, limited awareness of the disease and the fact that it often affects poor populations in remote, rural areas has hampered the detection of new infection foci and thus the control of BU. In 1998, WHO launched the Global BU Initiative, a partnership of Member States, academic and research institutions, donors, nongovernmental organizations, WHO, and others with the intention to raise awareness and to coordinate global BU control and multidisciplinary research efforts. In addition, this Initiative was meant to strengthen BU surveillance systems and to assess the disease burden at local, national and global levels. More than 58,000 BU cases had been reported between 2002 and 2016 from 20 countries in Africa (Benin, Cameroon, Central African Republic, Congo, Côte d'Ivoire, Democratic Republic of the Congo, Equatorial Guinea, Gabon, Ghana, Guinea, Liberia, Nigeria, Sierra Leone, South Sudan, Uganda and Togo), the Americas (French Guiana), Asia (Japan), and the Western Pacific (Australia and PNG) (Table 1). Since 2008, a steady decline in the number of reported BU patients has been noticed (Fig. 1).

While this may in part be due to the establishment of effective national BU control programs, underreporting is considered likely, as there is lack of data from an increasing number of countries that have reported BU cases in the past. Moreover, surveillance activities in some of the highly endemic countries may have declined due to the decreased availability of specific funding. On the contrary, a steady increase in the number of BU patients has been reported from the BU endemic region of Victoria in southern Australia, with a peak incidence of 186 new BU infections in 2016. The geographic distribution of BU recorded by WHO in 2016 is illustrated in Fig. 2.

Table 1 Number of reported BU cases worldwide between 2002 and 2016. Data source: WHO

Country	2002	2003	2004	2005	2006	2007	2008	2009	2010	2011	2012	2013	2014	2015	2016	Total
Côte d'Ivoire	750	768	1153	1564	1872	2191	2242	2679	2533	1659	1386	1039	827	549	376	21,588
Ghana	853	737	1157	1005	1096	668	986	853	1048	971	632	550	443	275	371	11,645
Benin	565	722	925	1045	1195	1203	897	674	572	492	365	378	330	311	312	9986
Cameroon	132	223	914	265	271	230	312	323	287	256	160	133	126	133	85	3850
DRC[a]	17	119	487	51	74	340	260	172	136	209	284	214	192	234	175	2964
Togo	96	38	800	317	40	141	95	52	67	52	51	37	67	81	83	2017
Congo	102	180	235	53	370	99	126	147	107	52	38	6	Nd[b]	Nd	Nd	1515
Guinea	Nd	157	146	208	279	Nd	80	61	24	59	82	96	46	72	72	1382
Australia	32	14	34	47	72	61	40	35	42	143	101	74	89	111	186	1081
South Sudan	568	360	4	24	38	8	3	5	4	Nd	Nd	Nd	Nd	Nd	Nd	1014
Gabon	Nd	Nd	43	91	54	32	53	41	65	59	45	59	47	43	39	671
Nigeria	Nd	Nd	Nd	Nd	9	Nd	Nd	24	7	4	40	23	65	114	235	521
Uganda	117	10	7	72	5	31	24	3	Nd	Nd	Nd	Nd	Nd	Nd	Nd	269
PNG	13	18	31	Nd	Nd	26	24	8	5	8	Nd	Nd	3	11	16	163
Liberia	Nd	Nd	Nd	Nd	Nd	Nd	Nd	Nd	Nd	Nd	21	8	Nd	105	Nd	134
French Guiana	24	7	17	2	2	2	8	2	7	3	2	3	1	Nd	Nd	80
Japan	Nd	Nd	1	1	1	3	2	5	9	10	4	10	7	4	2	59
Sierra Leone	Nd	Nd	Nd	Nd	Nd	Nd	1	Nd	Nd	28	Nd	Nd	Nd	Nd	Nd	29
CAR[c]	Nd	Nd	Nd	Nd	Nd	Nd	3	Nd	Nd	Nd	Nd	Nd	Nd	Nd	Nd	3
Equ. Guinea	Nd	Nd	Nd	3	Nd	Nd	Nd	Nd	Nd	0	Nd	Nd	Nd	Nd	Nd	3
Total	3269	3353	5954	4748	5378	5035	5156	5084	4913	4005	3211	2630	2243	2043	1952	58,974

[a]Democratic Republic of the Congo
[b]No data
[c]Central African Republic

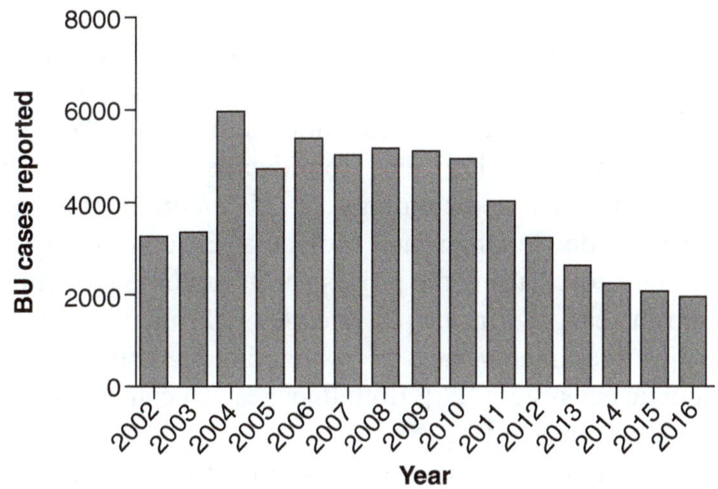

Fig. 1 Number of BU cases reported worldwide between 2002 and 2016. Graph illustrating the downward trend of new *M. ulcerans* infections reported in the past years. Data source: WHO

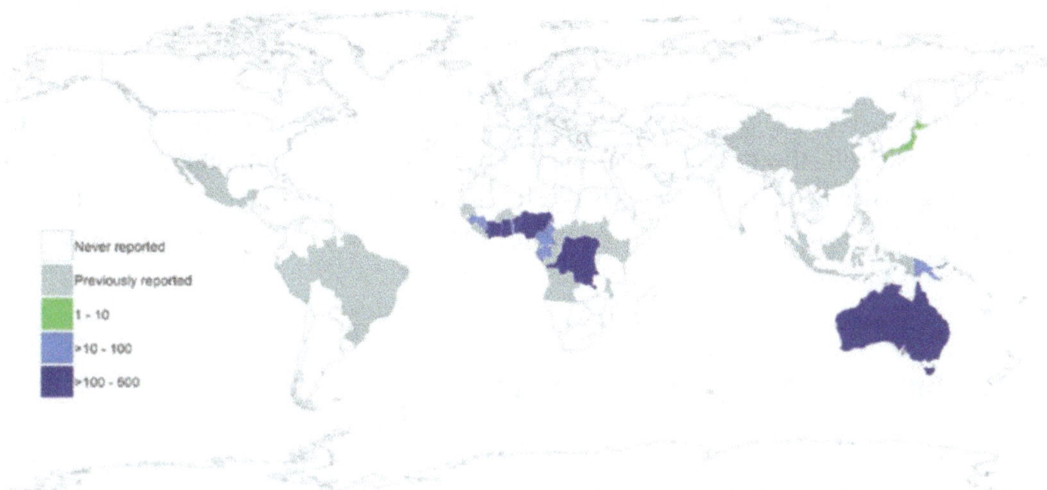

Fig. 2 Worldwide distribution of BU cases in 2016. Global map illustrating countries that have reported cases in 2016 (blue (classical lineage) and green (ancestral lineage)) and in previous years (grey). The map was kindly created by Dr. K. Ampah using 'R'. Data source: WHO

2 Evolution, Niche Adaptation, and Transmission of Distinct *M. ulcerans* Lineages

The emergence of *M. ulcerans,* which has evolved from the fish and opportunistic human pathogen *M. marinum* [21] approximately a million years ago [22], was driven by horizontal transfer of genetic material [21]. Most notably, this included a virulence plasmid, carrying genes for the synthesis of macrolide toxins, whose cytotoxic and immunosuppressive properties account for much of the typical progressive skin necrosis and chronicity of *M. ulcerans* infections. Several structural variants of these mycolactones have hitherto been described [23–25]; for more information on the toxin of *M. ulcerans* the reader is referred to chapter

"Mycolactone: More than Just a Cytotoxin" of this book. Moreover, acquisition of high-copy number insertion sequences (IS*2404* and IS*2606*) and their expansion in the *M. ulcerans* genome has resulted in extensive gene loss, pseudogene formation, and modification of gene function, coining the likely adaptation to new, restricted, ecological niches [26]. For more detailed information on the population genomics and molecular epidemiology of *M. ulcerans* the reader is referred to chapter "Population Genomics and Molecular Epidemiology of *Mycobacterium ulcerans*" of this book. During the process of its genome reduction, *M. ulcerans* has diverged into at least two principal lineages [27], the ancestral and the classical lineage, which presumably have been adapting to distinct niche environments within the aquatic ecosystem [26]. Both lineages can cause severe tissue destruction if lesions are left untreated (Fig. 3).

Until today, the exact source of the infection is not clear, in part because cultivation of *M. ulcerans* from potential environmental sources is hindered by the extremely slow growth rate of the pathogen. The marked loss of gene function as compared to the genome of *M. marinum*, including genes required for pigment biosynthesis and anaerobiosis [26], indicates that *M. ulcerans* may exist as a commensal associated with a protective organism in the ecosystem [28–30]. Despite extensive research, the mode(s) of transmission from environmental reservoirs to mammalian hosts is unclear. It is thus likely that *M. ulcerans* may be transmitted by several distinct mechanisms, each enabling the entrance of a sufficient pathogen load into the susceptible layers of subcutaneous tissue.

2.1 Ancestral *M. ulcerans* Lineage

Strains of the ancestral lineage, which are closely related to *M. marinum*, mainly cause disease in ectotherms such as fish and frogs [25, 31–33], but certain subgroups sporadically infect humans. Due to their distinct host range, some mycolactone-producing ancestral strains were initially given different species designations such as *M. marinum* for globally detected fish pathogens [25], *M. pseudoshottsii* for mycobacteria isolated from diseased Chesapeake Bay striped bass [32], or *M. liflandii* for strains isolated during a mycobacteriosis outbreak in a laboratory colony of *Xenopus tropicalis* [31]. Furthermore, ancestral strains isolated from the lesions of BU patients from Japan are often referred to as *M. ulcerans* subspecies *shinshuense* [34]. Indeed, comparative genome analysis has shown that based on the sequences of fish and frog isolates on the one hand and human disease isolates from Japan on the other hand the ancestral lineage may be subdivided into two major sub-lineages [26]. The common feature of all of these isolated pathogens is the production of mycolactone and thus it was no surprise when comparative genome analysis of ancestral and classical *M. ulcerans* strains has revealed that all mycolactone-producing mycobacteria have evolved from a common *M. marinum*-like progenitor and should therefore be considered as ecovars of a single *M. ulcerans* species [26, 35]. Ecovars of the ancestral lineage appear to

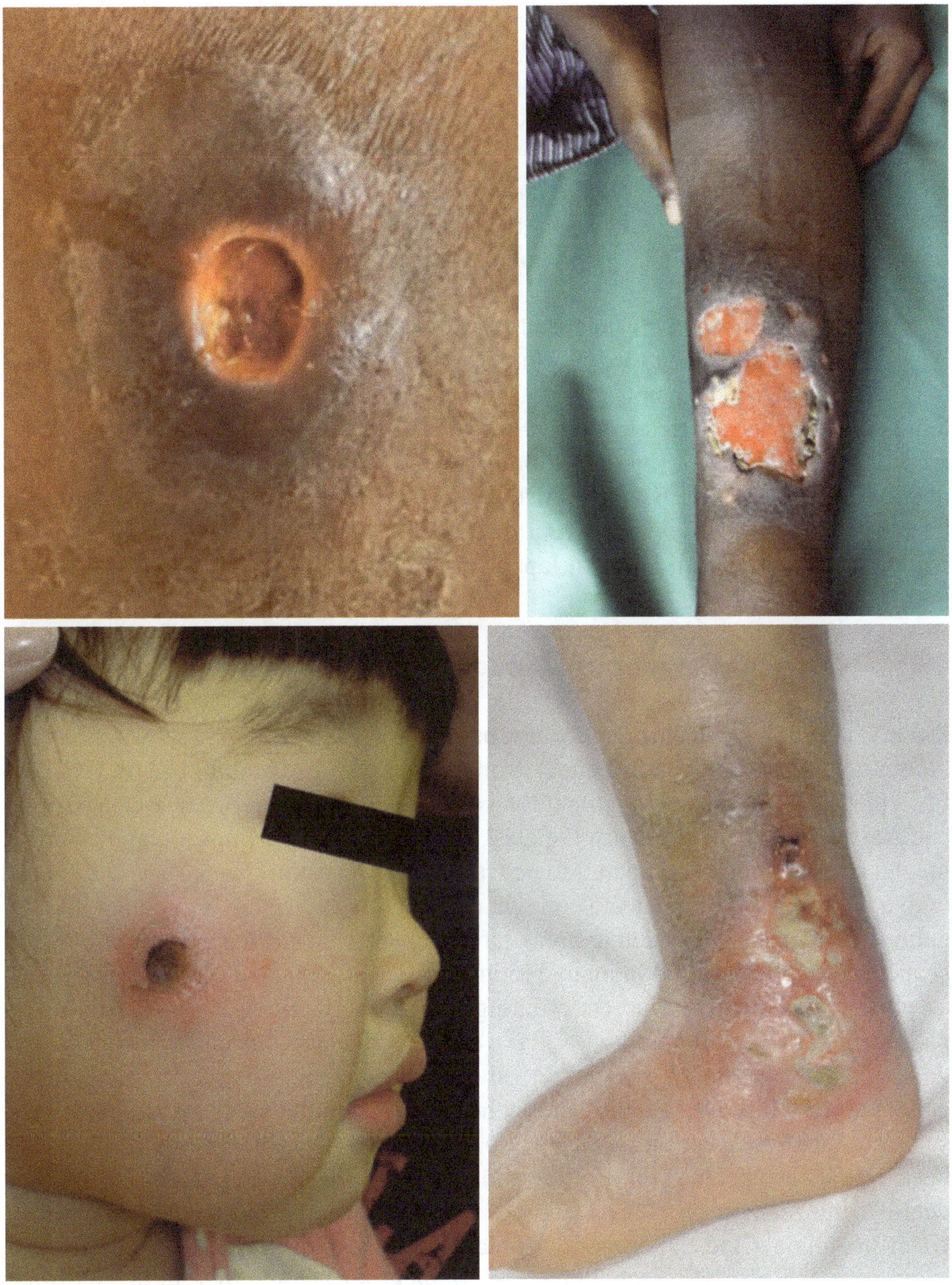

Fig. 3 Ulcerative BU lesions in patients from West Africa (upper panel) and Japan (lower panel). Photographs from Japanese patients were kindly provided by Dr. M. Ohtsuka, Fukushima Medical University

be adapting to an ecological niche environment, from which they infect humans only occasionally. Environmental studies in regions where the ancestral lineage of *M. ulcerans* prevails are still in their infancy. However, in French Guiana, *M. ulcerans*-specific DNA sequences (IS*2404* and the ketoreductase-B domain of the *M. ulcerans* mycolactone polyketide synthase genes (KR)) were found in different freshwater bodies [36]. In Japan, an outbreak of BU among several family members was linked to a stagnant water channel in the backyard of the family's house, in which a sample of a crayfish contained an IS*2404* sequence identical to that of *M. ulcerans* [37].

Sporadic human infections caused by ancestral strains were reported from the Americas (French Guiana [36], Suriname [38], Peru [39], Brazil [40] and Mexico [41]) and from Asia (Japan [42] and China [43]). Between 2002 and 2016, French Guiana and Japan reported to WHO 80 and 59 BU cases, respectively. No official records are available for the other South-American and Asian countries. For more detailed information on BU in French Guiana and Japan the reader is referred to chapters "*Mycobacterium ulcerans* Infection in French Guiana; Current State of Knowledge" and "Buruli Ulcer in Japan" of this book.

2.2 Classical *M. ulcerans* Lineage

Strains of the classical lineage are responsible for the vast majority of human *M. ulcerans* infections, reported mainly from Africa [20], Australia [17, 18, 44], and PNG [19, 45, 46]. Considering the often very high local prevalence of BU in African and Australian endemic regions, the classical lineage may be present in an environmental reservoir that—compared to ecovars of the ancestral lineage—is associated with a higher risk for humans to contract an infection. Some environmental investigations in African BU endemic areas have shown that *M. ulcerans*-specific DNA sequences can be detected in many biotic components of aquatic ecosystems, such as plants, snails, fish, or insects pointing towards the ubiquitous presence of the pathogen in these ecosystems [47]. For reasons that are currently not clear, the detection rate was much smaller in other environmental studies in African BU endemic areas [48, 49]. Studies in southeastern Australia have revealed that possums are highly susceptible to *M. ulcerans* infection. Sporadic infections have also been found in other native wildlife and domestic mammals [50, 51]. For more detailed information on BU in animals the reader is referred to chapter "Buruli Ulcer in Animals and Experimental Infection Models" of this book. Infected possums were shown to harbour high loads of *M. ulcerans* DNA in their gastrointestinal tracts. However, *M. ulcerans* cultivation from faeces failed, while being successful from skin lesions of infected possums [50, 51]. Genome comparison of classical lineage *M. ulcerans* isolates from possums and humans revealed an extremely close genetic relationship, supporting a role for these animals in the ecology of infection as a reservoir of the pathogen [26]. In Far North Queensland (FNQ), another BU endemic area of Australia, *M. ulcerans* DNA was detected in excreta of bandicoots [52]. These findings suggest that certain Australian marsupials may be highly susceptible to *M. ulcerans* infection. In contrast, until today, surveys of small mammals in African BU endemic areas have not led to the detection of an *M. ulcerans* animal reservoir [48, 53]. It has been suspected that in highly endemic areas of Africa, large chronic lesions of BU patients may contribute significantly to the dissemination of *M. ulcerans* in the environment.

Between 2002 and 2016, a total of 57,591 BU cases were reported from 16 African countries, 1081 from Australia and 163 from PNG. No official records are available for other countries, in which the presence of classical *M. ulcerans* infections was suspected, such as Kiribati [54] or Malaysia [55]. For more detailed information on BU in Africa and Australia the reader is referred to chapters "Buruli Ulcer in Africa" and "Buruli Ulcer in Australia" of this book.

3 Characteristics of BU Disease in Humans

In the following paragraphs, relevant data for BU disease in different BU endemic areas are compiled to enable a comparison of epidemiological and clinical features as well as environmental characteristics. For that purpose, PubMed was searched using the query "Buruli ulcer" OR "*Mycobacterium ulcerans*". All titles (N = ~1180) and available abstracts were screened for relevant content. Studies including more than 30 BU cases were used to extract information on demographic and geographic features of BU in these areas. In view of the multitude of publications on African BU endemic settings, representative studies conducted in eight different countries were selected. We included at least one and not more than three studies from the seven countries that have reported the vast majority of BU cases in the past decade, namely Côte d'Ivoire [56], Ghana [57–59], Benin [60–62, 76], Cameroon [63, 64], Democratic Republic of the Congo [65], Togo [66] and Republic of the Congo [67] as well as a case series from Nigeria [68], one of the few countries, which has recorded an increase in the number of reported cases over the past years. Studies were selected based on a preference for (1) quantity of BU cases (large case series), (2) quality (preferentially studies including laboratory-confirmed patients), (3) content (data on many epidemiological aspects), and (4) "up-to-datedness" (recent publications, if available and complying with other preferences). Furthermore, a very detailed study of 220 BU cases that were identified within a community of Rwandan refugees between 1964 and 1969 near the river Nile in Central Uganda was included [7] (Tables 2 and 3). BU endemic regions outside of Africa included in this comparative analysis are French Guiana (chapter "*Mycobacterium ulcerans* Infection in French Guiana; Current State of Knowledge" of this book) [69], Japan [42] (chapter "Buruli Ulcer in Japan" of this book), PNG [19, 46], FNQ, Australia [44, 70] and Victoria, Australia [71, 72] (chapter "Buruli Ulcer in Australia" of this book).

Table 2 Demographic and clinical characteristics of BU patients

Continent (lineage)	Country/site	Time span (references)	Number of BU cases (included in this table)	Age in years	Gender distribution (by age)	Overall lesion site (characteristics by gender/age)	WHO category/type of lesions
Americas (Ancestral)	French Guiana	1969–2013 [69] Chapter "*Mycobacterium ulcerans* Infection in French Guiana; Current State of Knowledge"	Total: 245, 166 confirmed[a] (all)	Median (overall): 25, Median (1969–1983): 9, Median (1984–1998): 25, Median (1999–2013): 36	51% male, 49% female	Lower limbs: 66%, upper limbs: 24%, chest: 5%, face: 1%, Nd: 4% (chest + upper limb lesions more common/lower limb lesions less common in men than in women)	Ulcers: 83%, nodules: 8%, plaques: 7%, Nd: 2%; no osteomyelitis; multiple lesions: 8%
Asia (Ancestral)	Japan	1982–2011 [42]	Total: 32, at least 23 confirmed (all)	Tendency towards middle-aged adults	37.5% male, 62.5% female	Mainly extremities	Mainly category I lesions
		1982–2016 Chapter "Buruli Ulcer in Japan"	Total: 60, at least 41 confirmed (all)	Nd[b]	Nd	Mainly upper and lower limbs followed by head	Mainly category I lesions; multiple (small) lesions: 23%

(continued)

Table 2 (continued)

Continent (lineage)	Country/site	Time span (references)	Number of BU cases (included in this table)	Age in years	Gender distribution (by age)	Overall lesion site (characteristics by gender/age)	WHO category/type of lesions
Australia/ Oceania (Classical)	FNQ	1964–2008 [70]	92 confirmed (all)	Median: 42	58% male, 42% female	Extremities: 97% → 82% of those lower limbs	Ulcers: 90%, nodules: 5%, edema: 3%, mixed: 1%
		2009–2015 [44]	95 confirmed (all)	Median: 45	57% male, 43% female	Extremities: 94% → 73% of those lower limbs	Category I: 93%, category II: 3%, category III: 4%; multiple lesions: 1%
	Victoria	1998–2011 [71]	180 confirmed (all)	Median: 61	49% male, 51% female	Lower limbs: 61%, upper limbs: 34%, over joints: 40% (upper limb lesions more common/lower limb lesions less common in men than in women)	Ulcers: 87%, nodules: 8%, edema: 5%; osteomyelitis: 1%; multiple lesions: 5%
		1998–2015 [72]	579 confirmed (all)	Median: 55	54.5% male, 45.5% female	Lower limbs: 70%, upper limbs: 27%, trunk: 2%, head/neck: 1%, over joints: 36% (upper limb lesions more common/lower limb lesions less common in men than in women)	Multiple lesions: 5.5%
	PNG	1962–1973 [46]	112 suspected[c] (all)	Peak incidence in young children; Average: 9	63% male, 37% female	Lower limbs: 69%, upper limbs: 11%, trunk: 18%, head/neck: 2%; Majority over joints	Mainly ulcers; extension below deep fascia: 20%; multiple lesions: 4%
		1979–1983 [19]	46 confirmed (all)	Majority in children <10; Mean: 11	48% male, 52% female	Mainly extremities, occasionally face and trunk	Nd

Africa (Classical)	Benin	1997–2001 [60, 76]	Total: 1700, 906 confirmed (all)	Median: 15; Highest incidence in young teenagers and individuals >50	Overall balanced	Lower limbs: 60%, upper limbs: 24%, trunk: 13%, head/neck: 3% (Lower limb lesions more common/ upper limb lesions less common in men than women) (Trunk/head/neck and upper limb lesions more common in children)	Multiple lesions: 9%; bone involvement: 13%
		2005–2011 CDTUB[d], Pobè [61]	1227 confirmed (all)	Median: 12	48% male, 52% female; patients ≤15: 57% male, >15: 33% male	Lower limbs: 60%, upper limbs: ~30%, trunk and other sites: ~10%	Category I: 22%, category II: 42%, category III: 36%; osteomyelitis: 7%; multiple lesions: 4%
		2005–2013 CDTUB, Allada [62]	476 confirmed (all)	Median: 12	51.5% male; 48.5% female; patients ≤15: 60% male, >15: 34% male	Lower limbs: 54%, upper limbs: 36%, head/neck/trunk: 10% (upper body lesions more common in younger ages)	Category I + II: 66%, category III: 34%; osteomyelitis: 1%; multiple lesions: 5%
	Cameroon	Survey in 2001 [63]	Total: 202 active, 135 confirmed (all)	Median: 14.5	57% male, 43% female	Lower limbs: 62%, upper limbs: 31%, head/neck/trunk: 7%	Ulcers: 93%, nodules: 2%, plaques: 3%, edema: 2%; bone involvement: 15%
		2010–2012 [64]	Total: 157, 88 confirmed (only confirmed)	Median: 12.5; Children <5 underrepresented; highest incidence in young teenagers and individuals >50	59% male, 41% female	Lower limbs: 56%, upper limbs: 31%, trunk: 11%, head/neck: 2%; proximity to joints: 52% (clusters around ankles and elbows) (lesions on trunk more common in men than in women)	Lesion category depending on health district; multiple lesions: 2%

(continued)

Table 2 (continued)

Continent (lineage)	Country/site	Time span (references)	Number of BU cases (included in this table)	Age in years	Gender distribution (by age)	Overall lesion site (characteristics by gender/age)	WHO category/type of lesions
	Congo	2007–2012 [67]	108 confirmed (all)	Patients <15: 56%	44% male, 56% female	Nd	Ulcers: 86%
	Côte d'Ivoire	2005–2010 [56]	1145 confirmed (all)	Patients <15: 53%, 15–49: 39%, ≥50: 8%	51% male, 49% female; patients <15: 55% male, 15–49: 46% male, ≥50: 51% male	Lower limbs: 67%, upper limbs: 26%, other sites: 7%	Category I: 29%, category II: 48%, category III: 23%; 45% ulcers, 17% nodules, 23% plaques, and 8% edema; bone involvement: 0.5%
	DRC	Survey in 2008 [65]	72 active, confirmed (all)	Patients ≤15: 38%, 16–49: 49%, >49: 14%	39% male, 61% female	Lower limbs: 53%, upper limbs: 40%, other sites: 7%	Category I: 50%, category II: 33%, category III: 17%; 76% ulcers, 18% non-ulcerated, 6% mixed
			187 active, suspected (all)	Patients ≤15: 33%, 16–49: 45%, >49: 23%	54% male, 46% female	Lower limbs: 73%, upper limbs: 17%, other sites: 10%	Category I: 48%, category II: 31%, category III: 21%; 66% ulcers, 29% non-ulcerated, 5% mixed

	Ghana	1996–2002 [57]	750 suspected (all)	Median: 12	45.5% male, 54.5% female; patients <16: 48.5% male, ≥16: 40% male	Upper limbs: 46%, lower limbs: 46%, trunk: 7%, head/neck: 5%, over joints: 55% (Trunk/head/neck lesions less common, extremities more common with increasing age)	Ulcers: 63%, nodules: 27%, plaques: 6%, edema: 4%; multiple lesions: 6%
		2001–2004 [58]	Total: 99, 94 confirmed (all)	Median: 11	42% male, 57% female	Lower limbs: 51%, upper limbs: 41%, trunk: 8%	Pre-ulcerative: 48%, early ulcers: 33%, chronic ulcers: 19%
		2008–2016 [59]	Total: 2203, 1020 confirmed (only confirmed)	Median: 20; Patients ≤15: 40%;	49% male, 51% female	Lower limbs: 60%, upper limbs: 16%	Category I: 27%, category II: 15%, category III: 35%, unknown: 23%; 73% ulcers, 9% nodules, 3% plaques, 2% edema, 12% mixed; osteomyelitis: 1%
	Nigeria	2004–2013 Treated in Benin [68]	127 confirmed (all)	Median: 10 (males), Median: 24 (females)	48% male, 52% female	Lower limbs: 57%, upper limbs: 28%, other sites: 5%, disseminated: 10%	Category I: 8%, category II: 34%, category III: 58%; osteomyelitis: 12% overall
	Togo	2007–2010 [66]	109 confirmed (all)	Median: 12	52% male, 48% female	Limbs and shoulders: 91%	Category I: 39%, category II: 38%, category III: 23%

(continued)

Table 2 (continued)

Continent (lineage)	Country/site	Time span (references)	Number of BU cases (included in this table)	Age in years	Gender distribution (by age)	Overall lesion site (characteristics by gender/age)	WHO category/type of lesions
	Uganda	1955–1969 [7]	220 suspected, partially confirmed (all)	Maximum incidence in children: 5–14	Patients <15: 59% male, 41% female, ≥15: 31% male, 69% female	(Children <5: any part of the body, 5–14 head and trunk less affected, limb lesions more frequent, ≥15: clear difference between sexes: lesions increasingly confined to lower limbs in men, in women arms and legs continue to be commonly affected)	1965 to mid-1967: majority presented with extensive, crippling ulcers, From mid-1967: people had become aware of pre-ulcerative stages, majority attended clinic at earliest sign of a lesion; multiple lesions: 5%

[a]Confirmed case: presence of a lesion clinically consistent with BU plus a positive result for any one of the available laboratory diagnostic tests (PCR, ZN, culture or histopathology)

[b]Nd = not defined

[c]Suspected case: presence of a lesion clinically consistent with BU. Inclusion of data in the table for either all suspected cases of BU or only for laboratory reconfirmed BU cases was dependent on the availability of data from the cited studies with a preference for data on only the reconfirmed BU cases

[d]CDTUB: Centre de Dépistage et de Traitement de l'Ulcère de Buruli

3.1 Demographic and Clinical Characteristics of *M. ulcerans* Infection

3.1.1 Age Distribution of BU Patients

BU has long been considered as a disease that mainly affects children under the age of 15 years living in rural areas of Africa. A comparison of the age distribution reported for various BU case series (Table 2) shows that this assumption still holds true for African BU endemic areas and for endemic sites in PNG. However, this generalized view lapsed when larger case series were reported from Victoria in southeastern Australia, where a high average age of BU patients was observed (Table 2). The peak incidence of BU among children in Africa and the elderly in Victoria may be partly related to the average age of the respective general populations in the affected areas. While the population in BU endemic areas in Africa is very young [64, 73, 74], the affected communities in Victoria are popular seaside holiday resorts, where many retired people have their homes. But even if taking the skewed age distribution of the study populations into account, the elderly were most affected in Australia [75] and a bi-modal distribution of the age-related risk of developing BU was observed in Africa with young teenagers and the elderly being overrepresented among cases [64, 74, 76]. Increasing risk of developing BU with age may be due to the gradual deterioration of the immune system in the elderly. Interestingly, a marked underrepresentation of BU patients among children below 4 years of age was found [64]. This observation was in line with subsequent sero-epidemiological studies in Ghana and Cameroon, indicating that young children are considerably less exposed to *M. ulcerans* than older children [77, 78].

In Japan, where BU disease is caused by strains of the ancestral *M. ulcerans* lineage, a tendency towards middle-aged adults was found [42]. In French Guiana, a marked transition of the most affected age groups was observed: over a period of 45 years the median age of patients increased significantly from 9 years of age in the study period between 1969 and 1983 to 36 years of age between 1999 and 2013 [69]. Taken together the age-related risk of developing BU appears to be determined both by differential exposure to *M. ulcerans* and by higher susceptibility of certain age groups.

3.1.2 Gender Distribution of BU Patients

A comparison of various BU case series (Table 2) reveals an overall balanced male:female ratio among patients, although in some studies the proportion of one of the sexes was higher than that of the other. While more female than male patients were recorded in Japan, more men than women tended to be affected in FNQ, Australia. Several studies conducted with large numbers of African BU cases revealed significant differences in the male:female ratio, when the study population was stratified by age. All of these studies have consistently reported that in children below the age of 15 years, *M. ulcerans* infection is more common in boys than in girls and conversely that in individuals above the age of 15 years the infection is more frequent in females than in males [7, 56, 61, 62] (Table 2). It appears likely that observed differences are due to different environmental contact patterns associated with the movement radius of children or different occupational exposure to environmental reservoirs.

3.1.3 Distribution of BU Lesions on the Human Body

While the mode of transmission of *M. ulcerans* remains unclear, it is commonly assumed that infection takes places via inoculation of the bacteria into the skin through direct trauma or bites of insects such as water bugs or mosquitoes [47, 51]. The site of inoculation is thought to be also the site of infection, not least because about 95% of BU patients present with a single lesion. A disseminated disease progression may only be found in those 5% of the patients presenting with multiple lesions (Table 2). Therefore, the location of BU lesions on the body of patients has been extensively studied, with the hope of detecting specific distribution patterns, such as a clustering of lesions at preferred feeding sites of biting arthropods vs. common sites for mechanical skin injuries, favouring any of the discussed mechanisms of infection. BU lesions, irrespective of whether they are caused by strains of the ancestral or the classical *M. ulcerans* lineage, occur mostly on parts of the body, which are not commonly protected by clothing (Table 2). The vast majority of studies found that lesions occurred in 90% or more of the cases on the limbs. Around two-thirds of the lesions occurred on the lower limbs, followed by the upper limbs, and less frequently the trunk, and sites on the neck/head (Table 2). While overall patterns observed in different regions, such as for example in Cameroon and Victoria, Australia appear to be surprisingly similar [64, 72], several studies have reported that the anatomical site of the lesions may vary with both gender and age. Interestingly, in studies conducted in French Guiana and Victoria upper limb lesions were more common in men than in women, who were more frequently affected on the lower limbs than were men [69, 71, 72]. Conversely, in African BU endemic areas lower limb lesions were more common and upper limb lesions were less common in men than in women [7, 76]. The most likely explanation for this differential pattern of lesion distribution is that women in Africa more commonly protect their lower limbs with clothing as compared to women in South America or Australia. Along the same line, men tend to be more often affected on the chest than women [64, 69]. Similarly, differential exposure to *M. ulcerans* may serve as an explanation for differences between age groups. It was reported that in children below 5 years of age any part of the body may be affected, while with increasing age the head and the trunk seem to be less commonly affected and limb lesions become more dominant [7, 57, 62, 76]. Failure to wear protective clothing has been identified as a risk factor for BU in several case-control studies [79–81]. In summary, the distribution of BU lesions appears to be strongly correlated with exposure of body parts to *M. ulcerans*, which may in turn be related to insect bites, skin injuries, or both.

3.1.4 WHO Categories and Clinical Manifestations of BU

M. ulcerans infections have been classified by WHO into three categories: category I is defined as a single, small lesion <5 cm in diameter, category II includes single lesions between 5 and 15 cm in diameter, plaque and edematous forms, and category III is comprised of single extensive lesions >15 cm in diameter, multiple lesions, lesions at critical sites (e.g. eye, genitalia, joints), and osteomyelitis [82]. There are marked differences in the proportion of BU patients presenting at health facilities with lesions of one of these categories for different geographical regions (Table 2). In Japan and Australia, where healthcare is readily available, the vast majority of BU patients present with category I lesions, for which standard national antibiotic regimens are highly effective. On the contrary, a significant proportion of BU patients in Africa and in particular those detected by active case search, present with more advanced stages of the disease, for which time required for healing is prolonged and healing is often accompanied by permanent disability. The percentage of patients with category II and III lesions is strongly dependent on the degree of remoteness of a population and consequently access to healthcare as well as with awareness of the disease among populations at risk. A high degree of awareness among the Rwandan population in the Kinyara refugee settlement as well as the establishment of a BU treatment center, has for example led to reporting at early stages of the disease [7]. On the contrary, due to a lack of adequate health infrastructure for the diagnosis and treatment of BU, a large cohort of patients from southwestern Nigeria presented to a health center in Benin with 58% category III and only 8% category I lesions [68]. Osteomyelitis is mainly observed in African populations [61, 63, 68] and only occasionally in Australia [71]. The fact that ulcers are the most common form of BU lesions in all of the affected geographical areas is related to the relatively unspecific, nodular onset and the often reported painlessness [84] of *M. ulcerans* disease.

3.2 Geographical Features of *M. ulcerans* Infection Foci

3.2.1 Environmental and Climatic Characteristics of BU Endemic Areas

Environmental and climatic factors are highly diverse in different *M. ulcerans* infection foci, not only if areas are compared where either the ancestral or the classical lineage prevails, but also among areas affected by the same lineage (Table 3).

French Guiana is located at the Atlantic Ocean in northern South America between Suriname and northern Brazil. Most of the inhabitants reside along a 50 km wide coastline, which is mainly composed of marshy savannah and mangroves, whereas the rest of the country consists of dense, barely accessible rainforest. Located between latitudes 2° and 6° N and rising only to modest elevations, the climate in French Guiana is tropical (hot and humid) all year round. While rainfall may be heavy from January to June/July, the period between August and December is considered the main dry season [85]. The highest mean incidence of BU was detected in the western coastal area around the towns of Sinnamary and Mana adjacent to the Sinnamary and La Mana rivers, respectively [69]. On the contrary, Japan extends through a wide latitude from 30° to 45° N, with climate conditions ranging from humid subtropical to humid continental. BU cases are sporadically distributed throughout the large, mountainous Honshu Island, which separates the Sea of Japan

Table 3 Geographical characteristics of BU endemic areas

Continent (lineage)	Country/ site	Refs.	Location/ environment	Climate	Seasons	Seasonality of infections	Linkage to environmental factors
Americas (Ancestral)	French Guiana	[69] Chapter "*Mycobacterium ulcerans* Infection in French Guiana; Current State of Knowledge"	Coastal, surrounded by marshy savannah and mangroves	Tropical (hot and humid)	Dry season: August–December, rainy season: January–July	After peak in rainfall increase in the number of BU cases during a subsequent dry period (short-term and long-term rainfall patterns)	Construction of a dam (Petit-Saut Dam) on the Sinnamary river: decline in case numbers; Deforestation and creation of rice fields in Mana: increase in incidence
Asia (Ancestral)	Japan	[42] Chapter "Buruli Ulcer in Japan"	Sporadically distributed throughout the large Honshu island	Temperate (humid, subtropical in most of Honshu; humid, continental in northern Honshu)	Four distinct seasons including a period with higher rainfall between mid of June and end of July	80% of BU cases diagnosed during autumn/winter (indicating infection during/shortly after the hot rainy season in summer)	Due to wide distribution of cases no general associations; one case of familial occurrence connected with water channel in the backyard of their house

Australia/ Oceania (Classical)	FNQ	[44, 70]	Daintree River catchment and adjacent coastal lowlands	Tropical (hot and humid)	Dry season during winter: May–November, rainy season during summer: November–April	Cases presented in 2011 during the late dry season after an unusually wet rainy season	Flooding after cyclones or exceptionally long and wet rainy seasons associated with BU cases
	Victoria	Chapter "Buruli Ulcer in Australia"	Low-lying, coastal towns, focal distribution	Temperate	Four distinct seasons; rainfall spread throughout the year	Diagnosis mainly during cooler months (April–September) between autumn and spring, with a peak in winter (indicating infection during warmer summer months)	Outbreak on Phillip island was suspected to be connected with golf course irrigation system; strong association of recent cases on Bellarine and Mornington Peninsulas with animal reservoir (possums)
	PNG	[46]	Coastal Oro Province, endemic villages around Kumusi River[a]	Tropical (hot and humid)	Rainfall high throughout the year, relatively drier months: May–November	Higher incidence in drier months (May–November)	Great flood following the eruption of nearby volcano Mount Lamington associated with BU cases

(continued)

Table 3 (continued)

Continent (lineage)	Country/ site	Refs.	Location/ environment	Climate	Seasons	Seasonality of infections	Linkage to environmental factors
Africa (Classical)	Different African countries	[56–60, 63–68] Chapter "Buruli Ulcer in Africa"	Highly focal distribution in river valleys, often affecting rural villages and therefore exact distribution within affected countries is often unkown	Tropical (hot and humid)	Region-specific dry and rainy seasons	Analysis of seasonal patterns of *M. ulcerans* infections complicated by delayed reporting of patients and combined analyses of country-wide data, which may obscure local climate variation; necessity of future investigations on seasonal patterns of the infection in areas with high quality surveillance systems	Proximity to rivers and tributaries
	Uganda	[7]	Community of refugees close to Nile River		Rainy seasons: March–May and September–November	Diagnosis mainly from September–November and to lesser extent January and February	Proximity to the Nile River (but not actual contact)

[a]BU has been reported from many parts of PNG (e.g. one large focus along Sepik River); information is mainly available for BU focus in villages along the Kumusi River ("Kumusi ulcer")

from the North Pacific Ocean. Much of Honshu belongs to the temperate zone with humid, subtropical climate characterized by four distinct seasons including a period with higher rainfall between mid of June and end of July. Humid, continental climate may be encountered in northern Honshu, with cold winters commonly experiencing snowfall [86]. The 60 sporadic BU cases reported between 2003 and 2016 in Japan were distributed over 17 of the 47 prefectures of the country (chapter "Buruli Ulcer in Japan" of this book). Interestingly, in some of the northern regions reporting BU cases, the temperature can be below zero degrees Celsius during the coldest season of the year. In Australia, *M. ulcerans* infections mainly occur in two geographically distinct areas in a temperate climatic zone in the south-eastern state of Victoria, and in tropical FNQ. In Victoria, *M. ulcerans* infections are consistently associated with low-lying coastal areas, where rainfall is spread throughout the year [75]. After first BU cases were described in the Bairnsdale region [2], the pathogen appeared to migrate westwards [87] causing larger outbreaks in the 1990s in Phillip Island [18] and Frankston/Langwarrin [17]. Since 1998 new foci have continuously appeared on the Bellarine and Mornington Peninsulas [71, 80, 88]. The second significant BU endemic area of Australia in FNQ comprises a rim of coastal valleys and lowlands surrounding the Dagmar Range, and extending from Daintree in the North to Mossman in the South [44]. The climate in FNQ is characterized by two distinct seasons with a warm, dry season during the winter period in May–October/November and a warm, wet season during the summer period in November/December–April [52]. Information on the distribution of BU in PNG is sparse, although infections have sporadically occurred in different parts of the country. One main BU focus was seen in the coastal Oro province in villages along the Kumusi River, where the climate is hot and humid and rainfall is high throughout the year with a relatively drier period from May to November [46]. BU endemic areas of West and Central Africa are mainly located in remote, rural areas and are characterized by a focal distribution with often high local prevalence rates. Affected regions have a tropical climate, typically with distinct dry and rainy seasons and are often associated with slow flowing and stagnant water bodies in proximity to large rivers. In West Africa the dry period lasts generally from November to February and is followed by region-specific rainy seasons. The southern areas of West Africa in the Guinea Coast region, where most of the BU cases occur, commonly experience two rainy seasons, one from April to July and a shorter one in September/October. In contrast, only one rainy period between July and September is observed in the North [89]. Distinct, region-specific dry and rainy seasons are also observed in Central African regions reporting BU cases [90].

Taken together *M. ulcerans* infection foci may be located in tropical or temperate climatic zones with or without distinct seasons, are connected with coastal lowlands in French Guiana, Australia and PNG, and with water bodies in remote, rural inland or coastal areas of Africa (Table 3).

3.2.2 Seasonality of BU Transmission

The detection and interpretation of seasonal fluctuations in the incidence of *M. ulcerans* disease is complicated by a combination of different factors, including the extremely long incubation period of the infection and the lack of knowledge on the mode of transmission and accordingly the timing of the infection, unless the patient has visited an endemic area only once and for a short time. Furthermore, features of early lesions are relatively uncharacteristic, causing delays in detection of the disease and reporting to health facilities. Well-documented records of BU cases among short-term visitors to BU endemic sites represent a unique source for the estimation of the time from exposure to disease onset. This approach was applied in south-eastern Australia by systematic identification of BU patients with a single visit exposure to one of the well-known focal BU endemic areas. The mean incubation period determined for patients infected in Victoria was 4.5 months with a wide variation from 32 to 264 days [91, 92]. In the Daintree region of FNQ, a large spike of BU cases was observed in September and October 2011, 7–8 month after an exceptionally long and very wet rainy season with peak rainfall in February 2011 that was suspected to be connected with the occurrence of the infections [44]. While this may speak for a longer average incubation period in FNQ as compared to Victoria, in the Kinyara refugee camp in Uganda, the period between short stays of visitors and the development of BU was estimated to be between 1 and 3 months [7]. These apparent differences in the mean incubation period may be due to mode of transmission-related differences in the inoculation dose of *M. ulcerans*. All of these studies have shown that the time interval between exposure and onset of symptoms and finally to the diagnosis of BU can vary substantially on an individual basis, which may obscure potentially distinct seasonal variations in transmission intensity. Nevertheless, seasonal patterns of BU incidence have been reported for infection foci caused by both ancestral and classical *M. ulcerans* strains (Table 3). In French Guiana, *M. ulcerans* was found to be correlated with short-term (6 months) and long-term (a decade) rainfall patterns and the El Niño-Southern oscillation [85], (chapter "*Mycobacterium ulcerans* Infection in French Guiana; Current State of Knowledge" of this book). In Japan, approximately 80% of the reported BU cases were diagnosed during autumn and winter [42] (chapter "Buruli Ulcer in Japan" of this book), indicating contraction of the infection during or shortly after the hot and rainy season in summer. Between 2009 and 2015, most BU cases in FNQ were diagnosed during the dry season from July to November, which is likely explained by a surge in transmission during the wet season [44]. For the BU endemic focus in villages along the Kumusi River in PNG, a consistently higher incidence of cases in the drier months (May–November) was recorded [46].

Analyses of seasonal patterns of *M. ulcerans* infection are particularly difficult for remote BU endemic regions in Africa, where very long delays in reporting of patients to the public health system have to be taken into account. In addition, several studies assessing seasonal patterns of the disease in Africa have used country-wide data, ignoring local differences in rainfall patterns. In a study assessing seasonal patterns of BU in an endemic area of Cameroon, case incidences between 2002 and 2012 peaked in March. Assuming a delay between infection and diagnosis of 5–6 months, this suggested that the risk of infection is highest during the high rainy season from August to October [93].

Similarly, transmission in the Kinyara refugee camp, was estimated to be greatest from July to September [7].

Seasonal patterns of BU incidence may result from differences in the environmental presence of the pathogen or potential vectors, as well as from behaviour-associated variations in exposure of populations to *M. ulcerans*. These variations may be triggered by climatic conditions and connected landscape dynamics as well as by seasonal changes in agricultural and other activities intensifying environmental exposure. Considering the general association of BU endemicity with aquatic ecosystems, one of the main drivers of increased transmission may be flooding and the seasonal appearance of stagnant temporary water bodies and swamps, providing a breeding ground for potential reservoirs/vectors and/or the pathogen itself.

3.2.3 Linkage of the Emergence of BU Foci to Environmental Changes

Altered environmental conditions may be favourable for the growth and dissemination of pathogens and may lead to increased contact of populations to pathogens, i.e. their vector(s) and/or environmental reservoir(s) [94, 95]. Therefore changes in land use including intensification and changes in the nature of agricultural activities, water management projects, deforestation and urbanization as well as natural phenomena, such as cyclones, flooding, or volcanic eruption can lead to ecological edge effects that promote disease emergence. It is suspected, that the emergence of some of the *M. ulcerans* infection foci was related to different types of ecological and environmental disturbances (Table 3).

In French Guiana, an increase in the incidence of BU in Mana between 1984 and 1988 was observed just after the creation of rice fields [69]. On the other hand, construction of the hydroelectric Petit-Saut Dam on the Sinnamary River in 1994 upstream of an adjacent BU endemic area has been associated with a significant decline in case numbers, possibly due to a better regulation of water flows. In Japan, a BU outbreak affecting several members of a family was linked to a stagnant agricultural water channel [37] and in Victoria an outbreak in Phillip Island in the 1990s was suspected to be triggered by a newly formed lake and/or a golf course irrigation system [96]. The population in BU endemic areas along the Kumusi River in PNG has claimed that infections had only occurred after a great flood following the eruption of the nearby volcano Mount Lamington. In FNQ, a spike of cases seen in 2011 was connected with the aforementioned exceptionally long and very wet rainy season. The subsequent decrease in the number of cases since 2011, may be due to much drier conditions in the following years [44]. In African BU endemic areas associations have been suspected between BU incidence and damming of rivers or streams leading to the creation of artificial lakes; e.g. a small stream on the University campus in Ibadan, Nigeria [97], the Mapé River in Cameroon [98], the Densu River in Ghana, or the Bandama River in Côte d'Ivoire [99]. High risk zones for the contraction of BU were also connected with land cover changes; e.g. establishment of irrigated rice and banana fields in Côte d'Ivoire [100] or areas of mining and agricultural activity in Ghana [101].

4 Risk/Protective Factors for BU

In order to identify risk factors for BU, a number of case-control studies were conducted in areas endemic for the classical lineage of *M. ulcerans*. In view of a limited reliability of the clinical diagnosis, laboratory confirmation of patients enrolled in these studies is a crucial prerequisite for the quality of obtained results. Due to the multitude of studies conducted on behavioural risk factors for BU, only those with laboratory-confirmed patients were included here. On the other hand, considering that only very limited information is available to date on potential host genetic factors and on the impact of BCG vaccination and HIV co-infection, also studies with only partly reconfirmed cases are discussed in the corresponding paragraphs. The BU case confirmation status of all of these studies is listed in Table 4.

Table 4 Assessment of risk factors for the contraction of BU caused by classical *M. ulcerans* strains identified by comparative case-control studies

Country/ site	Time span of study (Ref.)	No of confirmed[a] BU cases	No of controls (matching factors)	Assessed risk factors	Risk factors identified	Protective factors identified
Behavioural risk factors						
Australia (Victoria)	1998–2005 [80]	49	609 (community-based)	Lifestyle and insect exposure	Mosquito bites on lower legs and lower arms	Use of insect repellent, wearing protective clothing, washing of wounds
Benin	2006–2008 [105]	104	312 (community-based; matched by age, sex and village of residence)	Water sources, family relationships	BU history in the family, contact with natural water sources	
Côte d'Ivoire	2012 [102]	51	102 (hospital/ health center-based; matched by age (±5 years), sex, and type of residency)	Socio-sanitary, environment, and behaviour	Regular contact with unprotected surface water and absence of protective equipment during agricultural activities	Good knowledge about the risks that may result in BU and perception about the disease causes

Table 4 (continued)

Country/site	Time span of study (Ref.)	No of confirmed[a] BU cases	No of controls (matching factors)	Assessed risk factors	Risk factors identified	Protective factors identified
Ghana	1999 [104]	51	51 (hospital-based; matched by age group, sex, BCG)	Water-related	Swimming in rivers on a habitual basis	
	2000 [79]	116	116 (community-based; matched by age and village)	Environment and behavior	Wading in a river or stream	Wearing a shirt while farming, sharing indoor living space with livestock, bathing with toilet soap
	2010–2011 [81]	113	113 (community-based; matched by age, sex and village)	Demography, socio-economy, health and hygiene as well as environment	Presence of wetland, insect bites in water, use of adhesive when injured, and washing in river	Use of alcohol for injuries, covering limbs during farming
	2013–2015 [103]	176	176 (community-based; matched by age (±5 years), sex, and place of residence)	Demography, environment, and behaviour	Farming in swampy areas, farming while wearing short clothing, insect bites, and application of leaves on wounds	Farming in long clothing, washing wounds with water, and application of adhesive bandage on wounds

(continued)

Table 4 (continued)

Country/site	Time span of study (Ref.)	No of confirmed[a] BU cases	No of controls (matching factors)	Assessed risk factors	Risk factors identified	Protective factors identified
Togo	2014–2015 [106]	83	128 (community-based; matched by sex and place of residence)	Socio-demography, environment and behaviour	Bathing with water from open borehole, frequently crossing or swimming in a river, and receiving cuts, scratches or insect bites near a river	Using detergents for washing clothes or dishes
Host genetic factors						
Benin	2005–2013 [111]	208	300 (community-based; matched by age, sex, water contact habits, and ethnic background)	Autophagy-related genes	rs1333955 SNP in PARK2	rs2241880 SNP in ATG16L1
Ghana	2006 [110]	102 of 182	191 (community-based; matched by age)	Polymorphisms in the natural resistance-associated macrophage protein gene (SLC11A1)	D543N in SLC11A1	
	2011 [109]	96	384 (community-based; matched by age, sex, ethnicity and home village)	Polymorphisms in genes known to be associated with susceptibility to Tuberculosis and Leprosy	rs9282799 SNP in iNOS and rs2069705 in IFNG	

Table 4 (continued)

Country/ site	Time span of study (Ref.)	No of confirmed[a] BU cases	No of controls (matching factors)	Assessed risk factors	Risk factors identified	Protective factors identified
BCG vaccination						
Benin	2002–2003 [117]	134 of 279	988 (community-based; matched by age and sex)	BCG vaccination	None	None
Côte d'Ivoire	2001 [119]	56 of 116	116 (community-based; matched by age, sex, and village)	BCG vaccination (and other risk factors not discussed here)	No history of BCG vaccination	
DRC, Ghana and Togo	2010–2013 [118]	401	826 (mostly family members and neighbours)	BCG vaccination	None	None
Ghana	2000 [79]	116	116 (community-based; matched by age and village)	BCG vaccination (and other risk factors discussed above)	None	None
HIV co-infection						
Benin	2002–2003 [120]	258 of 426	613 (community)	HIV infection	HIV infection	
Ghana	2000 [79]	116	116 (community-based; matched by age and village)	HIV infection (and other risk factors discussed above)	None	None

[a]Laboratory confirmation of clinically diagnosed patients by at least one diagnostic test

4.1 Behavioural Risk Factors

Failure to wear protective clothing [102, 103] and activities close to or in certain unprotected water sources [79, 81, 102–106] have repeatedly been connected with an increased risk of contracting BU. Conversely, wearing protective clothing [79–81, 103], wound care [80, 81, 103] and good hygiene [79, 106] have been identified as factors conferring some degree of protection (Table 4). In southeastern Australia, where an involvement of mosquitoes in the transmission of *M. ulcerans* infections as either biological or purely mechanical vectors of the pathogen has been suggested, mosquito bites were identified as a risk factor. In addition to using insect repellent, immediate cleansing of wounds has also been associated with a decreased risk of contracting an *M. ulcerans* infection [80]. In African settings, insect bites, but also cuts and scratches near or in water bodies have been linked with an increased risk of infection [106]. One of the case-control studies came to the conclusion that good knowledge about BU and the risk factors involved in infection had a protective effect [102].

4.2 Host Genetic Factors

With the availability of a large number of human genome sequences, millions of polymorphic markers have been identified [107]. For case-control studies aiming at uncovering genetic factors that influence disease susceptibility, there are two main principal approaches; one based on testing candidate genes and the other based on genome-wide association studies. While candidate gene studies have higher statistical power, they cannot discover new markers or marker combinations. Even with several thousands of samples, genome-wide association studies can on the other hand be underpowered [108] and in the case of infectious diseases the element of exposure adds complexity and can contribute to statistical noise.

In view of the limited number of BU patients that can be enrolled for geographically restricted case-control studies, it is not surprising that only candidate gene studies have so far been conducted for BU. To date, single nucleotide polymorphisms (SNPs) in the inducible nitric oxide synthase gene *iNOS* and in the interferon gamma gene *IFNG* [109], the natural resistance-associated macrophage protein gene *SLC11A1* (*NRAMP1*) [110], and the autophagy-related E3 ubiquitin protein ligase gene *PARK2* [111] have been linked to susceptibility to BU (Table 4). However, these initial findings should be reconfirmed in future studies enrolling larger BU patient cohorts.

The fact that SNPs, which reduce the promoter activity of *iNOS* and *IFNG* implicated in macrophage activation, were found to increase susceptibility to BU, support the hypothesis that macrophages may be of crucial importance for the containment of *M. ulcerans* infections. Also, the influence of polymorphisms in SLC11A1 implicated in the transport of divalent cations to late endosomes/lysosomes [112] and in the ubiquitin ligase PARK2 implicated in resistance to intracellular pathogens [113], hint to a key role of macrophages in the early stage of an *M. ulcerans* infection, when low mycolactone levels may still permit elimination of the inoculated mycobacteria. For more detailed information on the immunology of BU the reader is referred to chapter "The Immunology of Buruli Ulcer" of this book. It should be noted however, that selection of candidate genes investigated so far was strongly biased towards host polymorphisms that have previously shown significant associations with susceptibility to intracellular mycobacteria, such as *M.*

tuberculosis and *M. leprae*. Genome-wide association studies for susceptibility to BU could thus potentially reveal even more robust gene-disease association data for genes that are relevant for other immune effector mechanisms.

4.3 BCG Vaccination

No specific vaccine is currently available to prevent *M. ulcerans* disease, but the protective effectiveness of *Mycobacterium bovis* Bacillus Calmette-Guérin (BCG) vaccination has been controversially discussed [114]. Two randomized controlled trials, one at the settlement of Rwandan refugees in Kinyara between 1967 and 1969 [115] and a second in another BU endemic area of Uganda in the 1970s [116] indicated a certain protective effect of BCG vaccination, although protection appeared to be short-lived [116]. Results of three case-control studies showed no evidence of a protective effect of BCG vaccination on the risk of developing BU [79, 117, 118], whereas no history of BCG vaccination appeared to be associated with BU in another study [119] (Table 4). Uncertainty about the BCG vaccination status among study populations with missing vaccination records should be considered as a potential confounder in these investigations.

4.4 BU-HIV Co-infection

BU-HIV co-infection is not uncommon due to the high HIV prevalence in many BU endemic areas. But until today, there is paucity of information on epidemiological and clinical relationships of the two infections. A case-control study conducted in Benin found that the prevalence of HIV in BU patients (2.6%) was significantly higher ($P = 0.003$) than that of the local control population (0.3%) [120], indicating that HIV increases the risk of BU. Similar percentages were reported in another case-control study performed in Ghana, with 5% of the BU patients and 0.9% of the control individuals testing positive for HIV, although—probably due to the small number of study participants—this association was not statistically significant [79] (Table 4). Furthermore, the prevalence of HIV among BU patients was significantly higher than that of the regional estimated prevalence [121] or that of other patients or pregnant women attending the same health facilities [122]. HIV infection weakens the immune system and seems to also affect the clinical presentation of BU, as co-infected patients tend to have more severe and more often multifocal lesions than HIV-negative patients [121, 122]. This is also indicated by a number of BU-HIV case reports describing aggressive, multifocal BU disease progression [123–128]. Management of BU-HIV co-infection is challenging [129]. It is recommended that all BU patients should be tested for HIV co-infection and that HIV-positive individuals should receive early antiretroviral treatment to reduce mortality (for more information on the management of BU-HIV co-infection see chapter "Management of BU-HIV Co-infection" of this book), which seems to be higher in BU-HIV co-infected than in HIV-negative patients [121, 122, 126, 128].

5 Diagnosis and Treatment of BU in Different Geographical Settings

The diagnosis of *M. ulcerans* infections is not trivial. The disease has a wide spectrum of clinical manifestations including ulcerative lesions as well as non-ulcerative forms such as nodules, plaques, and edema. As a consequence, the differential diagnosis of BU is broad, particularly in tropical endemic areas, where the prevalence of skin conditions with similar presentations is high [130]. Findings from a recent retrospective assessment of the diagnosis of BU in Ghana between 2008 and 2016 has revealed that in over 50% of all cases, a clinical suspicion of BU could not be confirmed by laboratory testing [59]. While surgical excision of lesions had long been the only treatment option for BU, pre-treatment laboratory confirmation of *M. ulcerans* infections has gained further in importance after the introduction of antibiotic therapy in 2004. In recent years, PCR targeting the IS*2404* element of *M. ulcerans* has become a gold standard for the diagnosis of BU in reference laboratories of the endemic countries. PCR-based laboratory confirmation of suspected BU cases is routinely performed in countries with resource rich healthcare systems and good laboratory infrastructure such as Japan [131] and Australia [132]. In French Guiana, the proportion of laboratory-confirmed BU cases increased from 52% before PCR became available in 2000 to 92% after 2000 [69]. However, the main drawback of PCR-based diagnosis for rural African endemic areas is its limited accessibility. To date, the only laboratory test locally available is microscopy of Ziehl-Neelsen stained smears from lesion specimens, a method with low sensitivity. In-country PCR reconfirmation in Africa can only be performed at a few reference laboratories. Samples are often stored for bulk shipments to these laboratories, leading to delayed delivery of results. Initiation of antibiotic therapy is thus often based on clinical diagnosis only, bearing the risk that some of the BU patients (and also misdiagnosed non-BU patients) do not receive appropriate treatment. A low-tech, accurate, rapid diagnostic test for the diagnosis of BU is of urgent need. For more information on the laboratory diagnosis of BU see chapter "Laboratory Diagnosis of Buruli Ulcer: Challenges and Future Perspectives" of this book.

WHO treatment recommendations include an eight-week course of a rifampicin-based combination of antibiotics, and—if needed—adjunct debridement and skin grafting. Until recently, rifampicin—the most effective drug against *M. ulcerans*—was prescribed in combination with intramuscular doses of streptomycin. However, administration of streptomycin through injections has a major impact on patient acceptance and adherence, and is often not practical in rural BU endemic areas. Moreover, prolonged use of streptomycin has been shown to cause permanent ototoxicity and transient nephrotoxicity in BU patients [133]. Therefore, the WHO Technical Advisory Group on Buruli ulcer recommended in 2017 to replace streptomycin with oral clarithromycin for BU treatment. Current first-line eight-week treatment regimens include rifampicin combined with clarithromycin, moxifloxacin or ciprofloxacin in Australia [134], a triple combination of rifampicin, levofloxacin, and clarithromycin in Japan (chapter "Buruli Ulcer in Japan" of this book) and rifampicin combined with amikacin or clarithromycin in French Guiana (chapter "*Mycobacterium ulcerans* Infection in French Guiana; Current State of Knowledge" of this book). WHO has developed a policy of free supplies of antibiotics to the affected countries on request from national BU control programs, which should in turn ensure that BU treatment facilities have an uninterrupted supply of the antibiotics [82]. However, the logistics involved in the provision of medication to remote, rural health facilities is challenging and hence antibiotics are not always available to patients in these areas. Alternative options for the treatment of BU, in particular the potential of new tuberculosis drug candidates are currently being investigated. Furthermore, local thermotherapy at the site of the lesions with simple phase change material devices showed promising results in a phase II clinical trial [135] and may be suitable for treatment at community level. For more information on the treatment of BU see chapters "Antimicrobial Treatment of *Mycobacterium ulcerans* Infection" and "Thermotherapy of Buruli Ulcer" of this book. Another important component to improve the healing process of lesions and to prevent secondary infections is adequate wound management. Interested readers are referred to chapter "Secondary Infection of Buruli Ulcer Lesions" of this book dealing with secondary infection and management of BU lesions.

6 Socio-Economic Burden of BU for Patients and their Families

The diagnosis and treatment of BU patients in Australia, Japan, and French Guiana is secured by well-resourced universal healthcare systems. In contrast, health services in remote BU endemic communities of Africa and PNG are often very limited. Although in many BU endemic countries, antibiotic treatment of BU is free of charge, other expenditures and required efforts may prevent patients from seeking care at the formal health system. The economic burden arising from transport, accommodation and food for patients and caregivers, lost earnings and work force may be too high for the affected families [136, 137], in particular if long hospital stays are required. Cultural beliefs, perceptions regarding the effectiveness of BU treatment, and stigma derived from the perceived mystical origin of the disease further aggravate the situation [138, 139] (see chapter "Social Science Contributions to BU Focused Health Service Research in West-Africa" of this book for more information on social science aspects of BU). In many BU endemic areas, patients prefer to first consult traditional healers and only refer to hospitals as a last resort [137]. Although BU is a slowly progressing disease, delays in receiving adequate medical care is detrimental to the treatment outcome. Chemotherapy is highly effective for early stages of BU, whereas the management of advanced stages is often complicated by delayed wound healing requiring prolonged hospitalization. Moreover, treatment of advanced BU may not prevent permanent functional disabilities.

Neglected tropical diseases (NTDs) have a substantial, cumulative effect on the economy and the health of affected populations, imposing a devastating social and economic burden on affected individuals, households and communities. A number of studies in BU endemic African countries have revealed that BU can push households into poverty, whether patients were hospitalized or not. For instance, the median cost burden of hospitalization of BU patients in Central Cameroon was reported to correspond to 25% of the annual earnings of a household due to high non-medical costs and productivity loss. More than half of the families were forced to withdraw financial and hence social support to the patient, resulting in the patient's isolation at the hospital. Indeed, social isolation of in-patients has been mentioned as the principal cause for abandonment of biomedical treatment [140]. On the other hand, for non-hospitalized BU patients in Ghana, transportation and other costs added up to 45% of the annual income of households. For these outpatients, social isolation was also an issue, particularly for children, who were often not accompanied for treatment [141]. In a study conducted in Nigeria it was revealed that costs before a definite diagnosis of BU was established (including costs for medications, drugs, wound care, hospitalization, transportation, food, and others), accounted for the brunt of the total costs for the treatment of BU. Costs were catastrophic for 50% of all affected households [142].

Taken together, new, socially more compatible intervention strategies, such as a more decentralized system of diagnosis and care as well as improved community mobilization and education of populations concerning BU to reduce care-seeking delays are of urgent need. Results of a recent pilot BU outreach campaign and decentralized care program in one of the most endemic districts of Benin demonstrated the great value of such interventions, reflected by a strong community support and a dramatic increase in the detection of BU cases that was associated with immediate, free, and accessible care [143]. However, a major challenge will be the mobilization of both financial and human (organization, training, health staff etc.) resources to implement and sustain decentralized care at a larger scale and on a long-term basis.

7 Outlook

Launching of the Global BU Initiative in 1998 and adding BU to the WHO list of NTDs, has helped to increase awareness of the disease among affected populations, health staff, potential donors, and researchers. As a result, national BU control programs were established in the affected countries, healthcare provision to BU patients was improved, and new diagnostic tools and treatment modalities have been developed. The decline in the number of new infections reported to WHO in the past years might on the one hand be a result of the progress made in the fight against the disease. Active case finding and treatment of patients with chronic ulcerative BU lesions, which may represent significant reservoirs of the pathogen, may be an important corner stone for the control of the disease. But on the other hand, the declining number of reported BU patients has also led to a decrease in awareness of and interest in the disease. The reduction in the number of reported cases may therefore be in part related to a decline in active case search activities in known BU endemic regions and also to a lack of efforts to identify new BU endemic areas. The large BU burden in Nigeria only became apparent when Nigerian patients presented to established BU health facilities in Benin [68, 144], indicating that the true number of BU cases in regions where no fully functional BU control programs are in place, may be vastly underestimated. While the detection and treatment of BU in Australia, Japan, and French Guiana will continue to be covered by their universal healthcare systems, a major task for the next years will be the sustained control of BU in Africa. For that purpose, funding for control activities and research has to be mobilized to maintain established healthcare infrastructure in the affected countries and to implement newly developed diagnostic tests and treatment regimens. Support by NGOs has played in many settings an important role in the development of BU control activities and treatment centers and was crucial for BU research funding (an example for funding of a BU research consortium by an NGO is provided in Chapter "Transdisciplinary Research and Action to Stop Buruli Ulcer: A Case Study from Philanthropy"). This has major implications for sustainability as reductions in program support by major partners may lead to severe drops in performance of BU control activities [145] and a loss in research output and capacity development.

Fragmentation of health interventions and services is a widely discussed issue, since narrowly targeted interventions can generate in particular in low resource settings inequity and substantial extra costs. Therefore, it has been suggested to integrate BU control into the broader context of care for skin NTDs [146]. The co-endemicity of several of these diseases, such as BU, yaws, scabies, mycetoma, and leprosy may allow maximizing financial and human resources by the implementation of integrated approaches, including training of health workers in differential diagnosis of skin NTDs, basic dermatology and wound management [147].

References

1. Cook A (1897) The Mengo hospital notes. Makere College Medical School Library, Kampala
2. MacCallum P, Tolhurst JC et al (1948) A new mycobacterial infection in man. J Pathol Bacteriol 60(1):93–122
3. Parson W (1999) Mycobacterium ulcerans. Lancet 354(9196):2171
4. Janssens PG, Quertinmont MJ, Sieniawski J, Gatti F (1959) Necrotic tropical ulcers and mycobacterial causative agents. Trop Geogr Med 11:293–312
5. Clancey JK, Dodge OG, Lunn HF, Oduori ML (1961) Mycobacterial skin ulcers in Uganda. Lancet 2(7209):951–954
6. Lunn HF, Connor DH, Wilks NE, Barnley GR, Kamunvi F, Clancey JK et al (1965) Buruli (Mycobacterial) ulceration in Uganda. (a new focus of Buruli ulcer in Madi District, Uganda): report of a field study. East Afr Med J 42:275–288
7. The Uganda Buruli Group (1971) Epidemiology of Mycobacterium ulcerans infection (Buruli ulcer) at Kinyara, Uganda. Trans R Soc Trop Med Hyg 65(6):763–775
8. The Uganda Buruli Group (1970) Clinical features and treatment of pre-ulcerative Buruli lesions (Mycobacterium ulcerans infection). Report II of the Uganda Buruli group. Br Med J 2(5706):390–393
9. Smith JH (1970) Epidemiologic observations on cases of Buruli ulcer seen in a hospital in the lower Congo. Am J Trop Med Hyg. 19(4):657–663

10. Oluwasanmi JO, Solankee TF, Olurin EO, Itayemi SO, Alabi GO, Lucas AO (1976) Mycobacterium ulcerans (Buruli) skin ulceration in Nigeria. Am J Trop Med Hyg 25(1):122–128
11. Burchard GD, Bierther M (1986) Buruli ulcer: clinical pathological study of 23 patients in Lambarene, Gabon. Trop Med Parasitol 37(1):1–8
12. van der Werf TS, van der Graaf WT, Groothuis DG, Knell AJ (1989) Mycobacterium ulcerans infection in Ashanti region, Ghana. Trans R Soc Trop Med Hyg 83(3):410–413
13. Amofah GK, Sagoe-Moses C, Adjei-Acquah C, Frimpong EH (1993) Epidemiology of Buruli ulcer in Amansie west district, Ghana. Trans R Soc Trop Med Hyg 87(6):644–645
14. Muelder K, Nourou A (1990) Buruli ulcer in Benin. Lancet 336(8723):1109–1111
15. Darie H, Le Guyadec T, Touze JE (1993) Epidemiological and clinical aspects of Buruli ulcer in Ivory Coast. 124 recent cases. Bull Soc Pathol Exot 86(4):272–276
16. Marston BJ, Diallo MO, Horsburgh CR Jr, Diomande I, Saki MZ, Kanga JM et al (1995) Emergence of Buruli ulcer disease in the Daloa region of cote d'Ivoire. Am J Trop Med Hyg 52(3):219–224
17. Johnson PD, Veitch MG, Leslie DE, Flood PE, Hayman JA (1996) The emergence of Mycobacterium ulcerans infection near Melbourne. Med J Aust 164(2):76–78
18. Veitch MG, Johnson PD, Flood PE, Leslie DE, Street AC, Hayman JA (1997) A large localized outbreak of Mycobacterium ulcerans infection on a temperate southern Australian island. Epidemiol Infect 119(3):313–318
19. Igo JD, Murthy DP (1988) Mycobacterium ulcerans infections in Papua New Guinea: correlation of clinical, histological, and microbiologic features. Am J Trop Med Hyg. 38(2):391–392
20. Röltgen K, Pluschke G (2015) Epidemiology and disease burden of Buruli ulcer: a review. Res Rep Trop Med 2015(6):59–73
21. Stinear TP, Seemann T, Pidot S, Frigui W, Reysset G, Garnier T et al (2007) Reductive evolution and niche adaptation inferred from the genome of Mycobacterium ulcerans, the causative agent of Buruli ulcer. Genome Res 17(2):192–200
22. Qi W, Kaser M, Roltgen K, Yeboah-Manu D, Pluschke G (2009) Genomic diversity and evolution of Mycobacterium ulcerans revealed by next-generation sequencing. PLoS Pathog 5(9):e1000580
23. Mve-Obiang A, Lee RE, Portaels F, Small PL (2003) Heterogeneity of mycolactones produced by clinical isolates of Mycobacterium ulcerans: implications for virulence. Infect Immun 71(2):774–783
24. Mve-Obiang A, Lee RE, Umstot ES, Trott KA, Grammer TC, Parker JM et al (2005) A newly discovered mycobacterial pathogen isolated from laboratory colonies of Xenopus species with lethal infections produces a novel form of mycolactone, the Mycobacterium ulcerans macrolide toxin. Infect Immun 73(6):3307–3312
25. Ranger BS, Mahrous EA, Mosi L, Adusumilli S, Lee RE, Colorni A et al (2006) Globally distributed mycobacterial fish pathogens produce a novel plasmid-encoded toxic macrolide, mycolactone F. Infect Immun 74(11):6037–6045
26. Doig KD, Holt KE, Fyfe JA, Lavender CJ, Eddyani M, Portaels F et al (2012) On the origin of Mycobacterium ulcerans, the causative agent of Buruli ulcer. BMC Genomics 13:258
27. Kaser M, Rondini S, Naegeli M, Stinear T, Portaels F, Certa U et al (2007) Evolution of two distinct phylogenetic lineages of the emerging human pathogen Mycobacterium ulcerans. BMC Evol Biol 7:177
28. Eddyani M, De Jonckheere JF, Durnez L, Suykerbuyk P, Leirs H, Portaels F (2008) Occurrence of free-living amoebae in communities of low and high endemicity for Buruli ulcer in southern Benin. Appl Environ Microbiol 74(21):6547–6553
29. Gryseels S, Amissah D, Durnez L, Vandelannoote K, Leirs H, De Jonckheere J et al (2012) Amoebae as potential environmental hosts for Mycobacterium ulcerans and other mycobacteria, but doubtful actors in Buruli ulcer epidemiology. PLoS Negl Trop Dis 6(8):e1764
30. Amissah NA, Gryseels S, Tobias NJ, Ravadgar B, Suzuki M, Vandelannoote K et al (2014) Investigating the role of free-living amoebae as a reservoir for Mycobacterium ulcerans. PLoS Negl Trop Dis 8(9):e3148

31. Trott KA, Stacy BA, Lifland BD, Diggs HE, Harland RM, Khokha MK et al (2004) Characterization of a Mycobacterium ulcerans-like infection in a colony of African tropical clawed frogs (Xenopus tropicalis). Comp Med 54(3):309–317
32. Rhodes MW, Kator H, McNabb A, Deshayes C, Reyrat JM, Brown-Elliott BA et al (2005) Mycobacterium pseudoshottsii sp. nov., a slowly growing chromogenic species isolated from Chesapeake Bay striped bass (Morone saxatilis). Int J Syst Evol Microbiol 55(Pt 3):1139–1147
33. Stragier P, Hermans K, Stinear T, Portaels F (2008) First report of a mycolactone-producing Mycobacterium infection in fish agriculture in Belgium. FEMS Microbiol Lett 286(1):93–95
34. Yoshida M, Nakanaga K, Ogura Y, Toyoda A, Ooka T, Kazumi Y et al (2016) Complete genome sequence of Mycobacterium ulcerans subsp. shinshuense. Genome Announc 4(5):e01050–e01016
35. Yip MJ, Porter JL, Fyfe JA, Lavender CJ, Portaels F, Rhodes M et al (2007) Evolution of Mycobacterium ulcerans and other mycolactone-producing mycobacteria from a common Mycobacterium marinum progenitor. J Bacteriol 189(5):2021–2029
36. Morris A, Gozlan R, Marion E, Marsollier L, Andreou D, Sanhueza D et al (2014) First detection of Mycobacterium ulcerans DNA in environmental samples from South America. PLoS Negl Trop Dis 8(1):e2660
37. Ohtsuka M, Kikuchi N, Yamamoto T, Suzutani T, Nakanaga K, Suzuki K et al (2014) Buruli ulcer caused by Mycobacterium ulcerans subsp shinshuense: a rare case of familial concurrent occurrence and detection of insertion sequence 2404 in Japan. JAMA Dermatol 150(1): 64–67
38. Boleira M, Lupi O, Lehman L, Asiedu KB, Kiszewski AE (2010) Buruli ulcer. An Bras Dermatol 85(3):281–298. quiz 99-301
39. Guerra H, Palomino JC, Falconi E, Bravo F, Donaires N, Van Marck E et al (2008) Mycobacterium ulcerans disease, Peru. Emerg Infect Dis 14(3):373–377
40. dos Santos VM, Noronha FL, Vicentina EC, Lima CC (2007) Mycobacterium ulcerans infection in Brazil. Med J Aust 187(1):63–64
41. Coloma JN, Navarrete-Franco G, Iribe P, Lopez-Cepeda LD (2005) Ulcerative cutaneous mycobacteriosis due to Mycobacterium ulcerans: report of two Mexican cases. Int J Lepr Other Mycobact Dis 73(1):5–12
42. Yotsu RR, Nakanaga K, Hoshino Y, Suzuki K, Ishii N (2012) Buruli ulcer and current situation in Japan: a new emerging cutaneous Mycobacterium infection. J Dermatol 39(7):587–593
43. Faber WR, Arias-Bouda LM, Zeegelaar JE, Kolk AH, Fonteyne PA, Toonstra J et al (2000) First reported case of Mycobacterium ulcerans infection in a patient from China. Trans R Soc Trop Med Hyg 94(3):277–279
44. Steffen CM, Freeborn H (2016) Mycobacterium ulcerans in the Daintree 2009-2015 and the mini-epidemic of 2011. ANZ J Surg 88(4):E289–E293
45. Reid IS (1967) Mycobacterium ulcerans infection: a report of 13 cases at the Port Moresby general hospital, Papua. Med J Aust 1(9):427–431
46. Radford AJ (1974) Mycobacterium Ulcerans infections in Papua New Guinea. P N G Med J 17(2):145–149
47. Merritt RW, Walker ED, Small PL, Wallace JR, Johnson PD, Benbow ME et al (2010) Ecology and transmission of Buruli ulcer disease: a systematic review. PLoS Negl Trop Dis 4(12):e911
48. Vandelannoote K, Durnez L, Amissah D, Gryseels S, Dodoo A, Yeboah S et al (2010) Application of real-time PCR in Ghana, a Buruli ulcer-endemic country, confirms the presence of Mycobacterium ulcerans in the environment. FEMS Microbiol Lett 304(2):191–194
49. Bratschi MW, Ruf MT, Andreoli A, Minyem JC, Kerber S, Wantong FG et al (2014) Mycobacterium ulcerans persistence at a village water source of Buruli ulcer patients. PLoS Negl Trop Dis 8(3):e2756
50. Fyfe JA, Lavender CJ, Handasyde KA, Legione AR, O'Brien CR, Stinear TP et al (2010) A major role for mammals in the ecology of Mycobacterium ulcerans. PLoS Negl Trop Dis 4(8):e791
51. Röltgen K, Pluschke G (2015) Mycobacterium ulcerans disease (Buruli ulcer): potential reservoirs and vectors. Curr Clin Microbiol Rep 2(1):35–43

52. Roltgen K, Pluschke G, Johnson PDR, Fyfe J (2017) Mycobacterium ulcerans DNA in bandicoot excreta in Buruli ulcer-endemic area, northern Queensland, Australia. Emerg Infect Dis 23(12):2042–2045
53. Durnez L, Suykerbuyk P, Nicolas V, Barriere P, Verheyen E, Johnson CR et al (2010) Terrestrial small mammals as reservoirs of Mycobacterium ulcerans in Benin. Appl Environ Microbiol 76(13):4574–4577
54. Christie M (1987) Suspected Mycobacterium ulcerans disease in Kiribati. Med J Aust 146(11):600–604
55. Pettit JH, Marchette NJ, Rees RJ (1966) Mycobacterium ulcerans infection. Clinical and bacteriological study of the first cases recognized in South East Asia. Br J Dermatol 78(4): 187–197
56. N'krumah RT, Kone B, Cisse G, Tanner M, Utzinger J, Pluschke G et al (2016) Characteristics and epidemiological profile of Buruli ulcer in the district of Tiassale, south Cote d'Ivoire. Acta Trop 175:138–144
57. Hospers IC, Wiersma IC, Dijkstra PU, Stienstra Y, Etuaful S, Ampadu EO et al (2005) Distribution of Buruli ulcer lesions over body surface area in a large case series in Ghana: uncovering clues for mode of transmission. Trans R Soc Trop Med Hyg 99(3):196–201
58. Mensah-Quainoo E, Yeboah-Manu D, Asebi C, Patafuor F, Ofori-Adjei D, Junghanss T et al (2008) Diagnosis of Mycobacterium ulcerans infection (Buruli ulcer) at a treatment centre in Ghana: a retrospective analysis of laboratory results of clinically diagnosed cases. Tropical Med Int Health 13(2):191–198
59. Yeboah-Manu D, Aboagye SY, Asare P, Asante-Poku A, Ampah K, Danso E et al (2018) Laboratory confirmation of Buruli ulcer cases in Ghana, 2008-2016. PLoS Negl Trop Dis 12(6):e0006560
60. Debacker M, Aguiar J, Steunou C, Zinsou C, Meyers WM, Guedenon A et al (2004) Mycobacterium ulcerans disease (Buruli ulcer) in rural hospital, southern Benin, 1997-2001. Emerg Infect Dis 10(8):1391–1398
61. Vincent QB, Ardant MF, Adeye A, Goundote A, Saint-Andre JP, Cottin J et al (2014) Clinical epidemiology of laboratory-confirmed Buruli ulcer in Benin: a cohort study. Lancet Glob Health 2(7):e422–e430
62. Capela C, Sopoh GE, Houezo JG, Fiodessihoue R, Dossou AD, Costa P et al (2015) Clinical epidemiology of Buruli ulcer from Benin (2005-2013): effect of time-delay to diagnosis on clinical forms and severe phenotypes. PLoS Negl Trop Dis 9(9):e0004005
63. Noeske J, Kuaban C, Rondini S, Sorlin P, Ciaffi L, Mbuagbaw J et al (2004) Buruli ulcer disease in Cameroon rediscovered. Am J Trop Med Hyg 70(5):520–526
64. Bratschi MW, Bolz M, Minyem JC, Grize L, Wantong FG, Kerber S et al (2013) Geographic distribution, age pattern and sites of lesions in a cohort of Buruli ulcer patients from the Mape Basin of Cameroon. PLoS Negl Trop Dis 7(6):e2252
65. Mavinga Phanzu D, Suykerbuyk P, Saunderson P, Ngwala Lukanu P, Masamba Minuku JB, Imposo DB et al (2013) Burden of Mycobacterium ulcerans disease (Buruli ulcer) and the underreporting ratio in the territory of Songololo, Democratic Republic of Congo. PLoS Negl Trop Dis 7(12):e2563
66. Bretzel G, Huber KL, Kobara B, Beissner M, Piten E, Herbinger KH et al (2011) Laboratory confirmation of Buruli ulcer disease in Togo, 2007-2010. PLoS Negl Trop Dis 5(7):e1228
67. Marion E, Obvala D, Babonneau J, Kempf M, Asiedu KB, Marsollier L (2014) Buruli ulcer disease in Republic of the Congo. Emerg Infect Dis 20(6):1070–1072
68. Marion E, Carolan K, Adeye A, Kempf M, Chauty A, Marsollier L (2015) Buruli ulcer in South Western Nigeria: a retrospective cohort study of patients treated in Benin. PLoS Negl Trop Dis 9(1):e3443
69. Douine M, Gozlan R, Nacher M, Dufour J, Reynaud Y, Elguero E et al (2017) Mycobacterium ulcerans infection (Buruli ulcer) in French Guiana, South America, 1969–2013: an epidemiological study. Lancet Planet Health 1(2):e65–e73

70. Steffen CM, Smith M, McBride WJ (2010) Mycobacterium ulcerans infection in North Queensland: the 'Daintree ulcer'. ANZ J Surg 80(10):732–736
71. Boyd SC, Athan E, Friedman ND, Hughes A, Walton A, Callan P et al (2012) Epidemiology, clinical features and diagnosis of Mycobacterium ulcerans in an Australian population. Med J Aust 196(5):341–344
72. Yerramilli A, Tay EL, Stewardson AJ, Kelley PG, Bishop E, Jenkin GA et al (2017) The location of Australian Buruli ulcer lesions-implications for unravelling disease transmission. PLoS Negl Trop Dis 11(8):e0005800
73. Ampah KA, Asare P, Binnah DD, Maccaulley S, Opare W, Roltgen K et al (2016) Burden and historical trend of Buruli ulcer prevalence in selected communities along the Offin River of Ghana. PLoS Negl Trop Dis 10(4):e0004603
74. Debacker M, Portaels F, Aguiar J, Steunou C, Zinsou C, Meyers W et al (2006) Risk factors for Buruli ulcer, Benin. Emerg Infect Dis 12(9):1325–1331
75. Johnson PD, Azuolas J, Lavender CJ, Wishart E, Stinear TP, Hayman JA et al (2007) Mycobacterium ulcerans in mosquitoes captured during outbreak of Buruli ulcer, southeastern Australia. Emerg Infect Dis 13(11):1653–1660
76. Debacker M, Aguiar J, Steunou C, Zinsou C, Meyers WM, Scott JT et al (2004) Mycobacterium ulcerans disease: role of age and gender in incidence and morbidity. Trop. Med Int Health 9(12):1297–1304
77. Roltgen K, Bratschi MW, Ross A, Aboagye SY, Ampah KA, Bolz M et al (2014) Late onset of the serological response against the 18 kDa small heat shock protein of Mycobacterium ulcerans in children. PLoS Negl Trop Dis 8(5):e2904
78. Ampah KA, Nickel B, Asare P, Ross A, De-Graft D, Kerber S et al (2016) A Seroepidemiological approach to explore transmission of mycobacterium ulcerans. PLoS Negl Trop Dis 10(1):e0004387
79. Raghunathan PL, Whitney EA, Asamoa K, Stienstra Y, Taylor TH Jr, Amofah GK et al (2005) Risk factors for Buruli ulcer disease (Mycobacterium ulcerans infection): results from a case-control study in Ghana. Clin Infect Dis 40(10):1445–1453
80. Quek TY, Athan E, Henry MJ, Pasco JA, Redden-Hoare J, Hughes A et al (2007) Risk factors for Mycobacterium ulcerans infection, southeastern Australia. Emerg Infect Dis 13(11):1661–1666
81. Kenu E, Nyarko KM, Seefeld L, Ganu V, Kaser M, Lartey M et al (2014) Risk factors for buruli ulcer in Ghana-a case control study in the Suhum-Kraboa-Coaltar and Akuapem south districts of the eastern region. PLoS Negl Trop Dis 8(11):e3279
82. WHO (2012) Treatment of Mycobacterium ulcerans disease (Buruli ulcer): guidance for health workers. World Health Organization, Geneva
83. Chauty A, Ardant MF, Adeye A, Euverte H, Guedenon A, Johnson C et al (2007) Promising clinical efficacy of streptomycin-rifampin combination for treatment of buruli ulcer (Mycobacterium ulcerans disease). Antimicrob Agents Chemother 51(11): 4029–4035
84. Marion E, Song OR, Christophe T, Babonneau J, Fenistein D, Eyer J et al (2014) Mycobacterial toxin induces analgesia in buruli ulcer by targeting the angiotensin pathways. Cell 157(7):1565–1576
85. Morris A, Gozlan RE, Hassani H, Andreou D, Couppie P, Guegan JF (2014) Complex temporal climate signals drive the emergence of human water-borne disease. Emerg Microbes Infec 3:e56
86. Hudman LE, Jackson RH (2003) Geography of travel & tourism, 4th edn. Thomson/Delmar Learning, Clifton Park. xi, 534 p., 8 p. of plates p
87. Buultjens AH, Vandelannoote K, Meehan CJ, Eddyani M, de Jong BC, Fyfe JAM et al (2018) Comparative genomics shows Mycobacterium ulcerans migration and expansion has preceded the rise of Buruli ulcer in South-Eastern Australia. Appl Environ Microbiol 84(8):e02612–e02617

88. Carson C, Lavender CJ, Handasyde KA, O'Brien CR, Hewitt N, Johnson PD et al (2014) Potential wildlife sentinels for monitoring the endemic spread of human buruli ulcer in south-East Australia. PLoS Negl Trop Dis 8(1):e2668
89. Froidurot S, Diedhiou A (2017) Characteristics of wet and dry spells in the west African monsoon system. Atmos Sci Lett 18(3):125–131
90. Guenang GM, Kamga FM (2012) Onset, retreat and length of the rainy season over Cameroon. Atmos Sci Lett 13(2):120–127
91. Trubiano JA, Lavender CJ, Fyfe JA, Bittmann S, Johnson PD (2013) The incubation period of Buruli ulcer (Mycobacterium ulcerans infection). PLoS Negl Trop Dis 7(10):e2463
92. Loftus MJ, Trubiano JA, Tay EL, Lavender CJ, Globan M, Fyfe JAM et al (2018) The incubation period of Buruli ulcer (Mycobacterium ulcerans infection) in Victoria, Australia - remains similar despite changing geographic distribution of disease. PLoS Negl Trop Dis 12(3):e0006323
93. Landier J, Constantin de Magny G, Garchitorena A, Guegan JF, Gaudart J, Marsollier L et al (2015) Seasonal patterns of Buruli ulcer incidence, Central Africa, 2002-2012. Emerg Infect Dis 21(8):1414–1417
94. Morse SS (1995) Factors in the emergence of infectious-diseases. Emerg Infect Dis 1(1): 7–15
95. Patz JA, Olson SH, Uejio CK, Gibbs HK (2008) Disease emergence from global climate and land use change. Med Clin N Am 92(6):1473
96. Johnson PD, Veitch MG, Flood PE, Hayman JA (1995) Mycobacterium ulcerans infection on Phillip Island. Victoria Med J Aust 162(4):221–222
97. Oluwasanmi JO, Itayemi SO, Alabi GO (1975) Buruli (mycobacterial) ulcers in Caucasians in Nigeria. Brit J Plast Surg 28(2):111–113
98. Marion E, Landier J, Boisier P, Marsollier L, Fontanet A, Le Gall P et al (2011) Geographic expansion of Buruli ulcer disease, Cameroon. Emerg Infect Dis 17(3):551–553
99. N'Krumah RTAS, Kone B, Cisse G, Tanner M, Utzinger J, Pluschke G et al (2017) Characteristics and epidemiological profile of Buruli ulcer in the district of Tiassale, south cote d'Ivoire. Acta Trop 175:138–144
100. Brou T, Broutin H, Elguero E, Asse H, Guegan JF (2008) Landscape diversity related to Buruli ulcer disease in cote d'Ivoire. Plos Neglect Trop D 2(7):e271
101. Wu J, Tschakert P, Klutse E, Ferring D, Ricciardi V, Hausermann H et al (2015) Buruli ulcer disease and its association with land cover in southwestern Ghana. PLoS Negl Trop Dis 9(6):e0003840
102. N'Krumah RT, Kone B, Tiembre I, Cisse G, Pluschke G, Tanner M et al (2016) Socio-environmental factors associated with the risk of contracting Buruli ulcer in Tiassale, south cote d'Ivoire: a case-control study. PLoS Negl Trop Dis 10(1):e0004327
103. Aboagye SY, Asare P, Otchere ID, Koka E, Mensah GE, Yirenya-Tawiah D et al (2017) Environmental and behavioral drivers of Buruli ulcer disease in selected communities along the Densu River basin of Ghana: a case-control study. Am J Trop Med Hyg 96(5): 1076–1083
104. Aiga H, Amano T, Cairncross S, Adomako J, Nanas OK, Coleman S (2004) Assessing water-related risk factors for Buruli ulcer: a case-control study in Ghana. Am J Trop Med Hyg 71(4):387–392
105. Sopoh GE, Barogui YT, Johnson RC, Dossou AD, Makoutode M, Anagonou SY et al (2010) Family relationship, water contact and occurrence of Buruli ulcer in Benin. PLoS Negl Trop Dis 4(7):e746
106. Maman I, Tchacondo T, Kere AB, Piten E, Beissner M, Kobara Y et al (2018) Risk factors for Mycobacterium ulcerans infection (Buruli ulcer) in Togo horizontal line a case-control study in Zio and Yoto districts of the maritime region. BMC Infect Dis 18(1):48
107. Hinds DA, Stuve LL, Nilsen GB, Halperin E, Eskin E, Ballinger DG et al (2005) Whole-genome patterns of common DNA variation in three human populations. Science 307(5712):1072–1079

108. Menashe I, Rosenberg PS, Chen BE (2008) PGA: power calculator for case-control genetic association analyses. BMC Genet 9:36
109. Bibert S, Bratschi MW, Aboagye SY, Collinet E, Scherr N, Yeboah-Manu D et al (2017) Susceptibility to Mycobacterium ulcerans disease (Buruli ulcer) is associated with IFNG and iNOS gene polymorphisms. Front Microbiol 8:1903
110. Stienstra Y, van der Werf TS, Oosterom E, Nolte IM, van der Graaf WT, Etuaful S et al (2006) Susceptibility to Buruli ulcer is associated with the SLC11A1 (NRAMP1) D543N polymorphism. Genes Immun 7(3):185–189
111. Capela C, Dossou AD, Silva-Gomes R, Sopoh GE, Makoutode M, Menino JF et al (2016) Genetic variation in autophagy-related genes influences the risk and phenotype of Buruli ulcer. PLoS Negl Trop Dis 10(4):e0004671
112. Goswami T, Bhattacharjee A, Babal P, Searle S, Moore E, Li M et al (2001) Natural-resistance-associated macrophage protein 1 is an H+/bivalent cation antiporter. Biochem J 354(Pt 3):511–519
113. Manzanillo PS, Ayres JS, Watson RO, Collins AC, Souza G, Rae CS et al (2013) The ubiquitin ligase parkin mediates resistance to intracellular pathogens. Nature 501(7468): 512–516
114. Zimmermann P, Finn A, Curtis N (2018) Does BCG vaccination protect against non-Tuberculous mycobacterial infection? A systematic review and meta-analysis. J Infect Dis 218(5):679–687
115. The Uganda Buruli group (1969) BCG vaccination against mycobacterium ulcerans infection (Buruli ulcer). First results of a trial in Uganda. Lancet 1(7586):111–5
116. Smith PG, Revill WD, Lukwago E, Rykushin YP (1976) The protective effect of BCG against Mycobacterium ulcerans disease: a controlled trial in an endemic area of Uganda. Trans R Soc Trop Med Hyg 70(5–6):449–457
117. Nackers F, Dramaix M, Johnson RC, Zinsou C, Robert A (2006) E DEBB, et al. BCG vaccine effectiveness against Buruli ulcer: a case-control study in Benin. Am J Trop Med Hyg 75(4):768–774
118. Phillips RO, Phanzu DM, Beissner M, Badziklou K, Luzolo EK, Sarfo FS et al (2015) Effectiveness of routine BCG vaccination on buruli ulcer disease: a case-control study in the Democratic Republic of Congo, Ghana and Togo. PLoS Negl Trop Dis 9(1):e3457
119. Ahoua L, Guetta AN, Ekaza E, Bouzid S, N'Guessan R, Dosso M (2009) Risk factors for Buruli ulcer in cote d'Ivoire: results of a case-control study, august 2001. Afr J Biotechnol 8(4):536–546
120. Johnson RC, Nackers F, Glynn JR, de Biurrun Bakedano E, Zinsou C, Aguiar J et al (2008) Association of HIV infection and Mycobacterium ulcerans disease in Benin. AIDS 22(7):901–903
121. Christinet V, Comte E, Ciaffi L, Odermatt P, Serafini M, Antierens A et al (2014) Impact of human immunodeficiency virus on the severity of buruli ulcer disease: results of a retrospective study in Cameroon. Open Forum Infect Dis 1(1):ofu021
122. Tuffour J, Owusu-Mireku E, Ruf MT, Aboagye S, Kpeli G, Akuoku V et al (2015) Challenges associated with Management of Buruli Ulcer/human immunodeficiency virus Coinfection in a treatment Center in Ghana: a case series study. Am J Trop Med Hyg 93(2): 216–223
123. Bayonne Manou LS, Portaels F, Eddyani M, Book AU, Vandelannoote K, de Jong BC (2013) Mycobacterium ulcerans disease (Buruli ulcer) in Gabon: 2005-2011. Med Sante Trop 23(4):450–457
124. Toll A, Gallardo F, Ferran M, Gilaberte M, Iglesias M, Gimeno JL et al (2005) Aggressive multifocal Buruli ulcer with associated osteomyelitis in an HIV-positive patient. Clin Exp Dermatol 30(6):649–651
125. Komenan K, Elidje EJ, Ildevert GP, Yao KI, Kanga K, Kouame KA et al (2013) Multifocal Buruli ulcer associated with secondary infection in HIV positive patient. Case Rep Med 2013:348628

126. Johnson RC, Ifebe D, Hans-Moevi A, Kestens L, Houessou R, Guedenon A et al (2002) Disseminated Mycobacterium ulcerans disease in an HIV-positive patient: a case study. AIDS 16(12):1704–1705
127. Bafende AE, Lukanu NP, Numbi AN (2002) Buruli ulcer in an AIDS patient. S Afr Med J 92(6):437
128. Kibadi K, Colebunders R, Muyembe-Tamfum JJ, Meyers WM, Portaels F (2010) Buruli ulcer lesions in HIV-positive patient. Emerg Infect Dis 16(4):738–739
129. O'Brien DP, Ford N, Vitoria M, Christinet V, Comte E, Calmy A et al (2014) Management of BU-HIV co-infection. Tropical Med Int Health 19(9):1040–1047
130. Junghanss T, Johnson RC, Pluschke G (2014) Mycobacterium ulcerans disease. In: Farrar J, Hotez PJ, Junghanss T, Kang G, Lalloo D, White NJ (eds) Manson's tropical diseases, 23rd edn. Saunders, Edinburgh, pp 519–531
131. Nakanaga K, Hoshino Y, Yotsu RR, Makino M, Ishii N (2011) Nineteen cases of Buruli ulcer diagnosed in Japan from 1980 to 2010. J Clin Microbiol 49(11):3829–3836
132. Fyfe JA, Lavender CJ, Johnson PD, Globan M, Sievers A, Azuolas J et al (2007) Development and application of two multiplex real-time PCR assays for the detection of Mycobacterium ulcerans in clinical and environmental samples. Appl Environ Microbiol 73(15):4733–4740
133. Klis S, Stienstra Y, Phillips RO, Abass KM, Tuah W, van der Werf TS (2014) Long term streptomycin toxicity in the treatment of Buruli ulcer: follow-up of participants in the BURULICO drug trial. PLoS Negl Trop Dis 8(3):e2739
134. O'Brien DP, Jenkin G, Buntine J, Steffen CM, McDonald A, Horne S et al (2014) Treatment and prevention of Mycobacterium ulcerans infection (Buruli ulcer) in Australia: guideline update. Med J Aust 200(5):267–270
135. Vogel M, Bayi PF, Ruf MT, Bratschi MW, Bolz M, Um Boock A et al (2016) Local heat application for the treatment of Buruli ulcer: results of a phase II open label single center non comparative clinical trial. Clin Infect Dis 62(3):342–350
136. Renzaho AM, Woods PV, Ackumey MM, Harvey SK, Kotin J (2007) Community-based study on knowledge, attitude and practice on the mode of transmission, prevention and treatment of the Buruli ulcer in Ga West District, Ghana. Tropical Med Int Health 12(3): 445–458
137. Mulder AA, Boerma RP, Barogui Y, Zinsou C, Johnson RC, Gbovi J et al (2008) Healthcare seeking behaviour for Buruli ulcer in Benin: a model to capture therapy choice of patients and healthy community members. Trans R Soc Trop Med Hyg 102(9):912–920
138. Koka E, Yeboah-Manu D, Okyere D, Adongo PB, Ahorlu CK (2016) Cultural understanding of wounds, Buruli ulcers and their Management at the Obom sub-district of the Ga south municipality of the Greater Accra region of Ghana. PLoS Negl Trop Dis 10(7):e0004825
139. Peeters Grietens K, Toomer E, Um Boock A, Hausmann-Muela S, Peeters H, Kanobana K et al (2012) What role do traditional beliefs play in treatment seeking and delay for Buruli ulcer disease?--insights from a mixed methods study in Cameroon. PLoS One 7(5): e36954
140. Grietens KP, Boock AU, Peeters H, Hausmann-Muela S, Toomer E, Ribera JM (2008) "It is me who endures but my family that suffers": social isolation as a consequence of the household cost burden of Buruli ulcer free of charge hospital treatment. PLoS Negl Trop Dis 2(10):e321
141. Amoakoh HB, Aikins M (2013) Household cost of out-patient treatment of Buruli ulcer in Ghana: a case study of Obom in Ga south municipality. BMC Health Serv Res 13:507
142. Chukwu JN, Meka AO, Nwafor CC, Oshi DC, Madichie NO, Ekeke N et al (2017) Financial burden of health care for Buruli ulcer patients in Nigeria: the patients' perspective. Int Health 9(1):36–43
143. Amoussouhoui AS, Sopoh GE, Wadagni AC, Johnson RC, Aoulou P, Agbo IE et al (2018) Implementation of a decentralized community-based treatment program to improve the management of Buruli ulcer in the Ouinhi district of Benin, West Africa. PLoS Negl Trop Dis 12(3):e0006291

144. Ayelo GA, Anagonou E, Wadagni AC, Barogui YT, Dossou AD, Houezo JG et al (2018) Report of a series of 82 cases of Buruli ulcer from Nigeria treated in Benin, from 2006 to 2016. PLoS Negl Trop Dis 12(3):e0006358
145. Tabah EN, Nsagha DS, Bissek AC, Njamnshi AK, Bratschi MW, Pluschke G et al (2016) Buruli ulcer in Cameroon: the development and impact of the national control Programme. PLoS Negl Trop Dis 10(1):e0004224
146. Mitja O, Marks M, Bertran L, Kollie K, Argaw D, Fahal AH et al (2017) Integrated control and Management of Neglected Tropical Skin Diseases. PLoS Negl Trop Dis 11(1):e0005136
147. Barogui YT, Diez G, Anagonou E, Johnson RC, Gomido IC, Amoukpo H et al (2018) Integrated approach in the control and management of skin neglected tropical diseases in Lalo, Benin. PLoS Negl Trop Dis 12(6):e0006584

Buruli Ulcer in Africa

Earnest Njih Tabah, Christian R. Johnson,
Horace Degnonvi, Gerd Pluschke, and Katharina Röltgen

1 Introduction

The first description of skin lesions resembling those caused by *Mycobacterium ulcerans*, dates back to the late nineteenth century, when the missionary physician Albert Cook [1] recorded a range of chronic, necrotizing skin ulcers in patients in Uganda. In the 1950s and 1960s a larger case series of patients with similar ulcers was detected in today's Nakasongola district in Uganda [2, 3], formerly known as Buruli County. Since then, cases of the disease, henceforth designated "Buruli ulcer" (BU), were reported from 20 additional African countries, where the major burden commonly falls on children aged five to 15 years. BU in Africa is characterized by a patchy geographical distribution, affecting mainly rural communities with often very high local prevalence rates. Access to the formal health sector in these regions is limited and as a result knowledge on the actual distribution and frequency of infections is scanty [4]. The occurrence of *M. ulcerans* infections in Africa is closely linked to areas of land drained by rivers and their tributaries. While the probability of person-to-person transmission is thought to be very low, the nature of relevant environmental reservoirs is highly controversial and the mode by which the pathogen is transmitted from

E. N. Tabah (✉)
National Yaws, Leishmaniasis, Leprosy and Buruli Ulcer Control Programme, Ministry of Public Health, Yaounde, Cameroon

C. R. Johnson
Fondation Raoul Follereau, Paris, France

University of Abomey-Calavi, Abomey-Calavi, Benin

H. Degnonvi
University of Abomey-Calavi, Abomey-Calavi, Benin

G. Pluschke · K. Röltgen
Swiss Tropical and Public Health Institute, Basel, Switzerland

University of Basel, Basel, Switzerland
e-mail: gerd.pluschke@swisstph.ch

G. Pluschke, K. Röltgen (eds.), *Buruli Ulcer*,
https://doi.org/10.1007/978-3-030-11114-4_2

environmental sources to humans is not clarified [5, 6]. Several routes for the introduction of *M. ulcerans* into the susceptible layers of the skin are discussed.

Causes of BU disease are commonly perceived by the local population as somewhat mysterious and are often associated with witchcraft or sorcery [7–9]. Also insect bites, contamination of skin lesions, and contact with swamps and water bodies often connected with changes in ecology are considered risk factors for contracting BU and a concept of dual causality is frequently encountered, particularly among affected populations in West African countries [9]. As a consequence, patients may first consult traditional healers or prayer camps to deal with witchcraft before seeking biomedical treatment at hospitals or health centers. Other patients may consider care seeking at the formal health sector only as a last resort [8, 10, 11]. Findings from a biosocial analysis of BU among fishermen in northwestern Uganda revealed that late presentation for biomedical treatment resulted from a perceived lack of its efficacy and a perceived efficacy of herbalists' treatment, which was sought promptly after first signs of lesions appeared [12]. These insights explain why in many BU endemic regions a high proportion of patients present to formal health facilities with large lesions, which require extended periods for healing and often result in permanent disabilities. According to the WHO classification system, BU lesions fall into one of three categories. Category I includes single, small lesions (nodules or ulcers) below five cm in diameter, Category II comprises single lesions between five and 15 cm in diameter as well as plaque and edematous forms, and Category III includes single lesions above 15 cm in diameter, multiple lesions, lesions at critical sites such as eyes, genitalia, and joints, as well as osteomyelitis [13]. Category II and III lesions are particularly prevalent in remote areas, where access to healthcare is limited and awareness of the disease is low. Surveillance and reporting of cases supported by community health workers, teachers, and other community volunteers are important elements for the control of BU. As long as preventable risks are not clearly identified and no vaccine is available, the main goal is to diagnose and treat patients in an early disease stage, when most lesions heal fast and without adjunct surgical treatment so that long-term sequelae and other complications can be avoided.

A momentum for the establishment of organized National BU Control Programs (NBUCPs) in the most affected countries was created by the Yamoussoukro Declaration and the global BU Initiative, launched by WHO in 1998 [14]. The three main pillars of global and national BU control strategies included (1) the strengthening of health systems by the development of infrastructure and provision of training for health workers, (2) sensitization and involvement of communities by information and education campaigns to facilitate early case detection and reporting, and (3) standardized case management in terms of diagnosis, treatment, and prevention of disability.

BU has been reported from five of the six WHO regions except the European region [4]. However, Africa bears the brunt of the disease burden with around 57,500 cases reported to WHO from 16 African countries between 2002 and 2016, representing 98% of all BU cases recorded worldwide during that time. The decreasing numbers of cases reported to WHO in the past couple of years play their part in further neglect of BU. Major efforts are required in the coming years to mobilize resources for the establishment, maintenance, and expansion of BU control activities.

2 Management of BU in Africa

2.1 Structure of Health Systems in Africa and Implications for the Management of BU

Health systems are organized in many African countries in three levels, each composed of administrative structures and care facilities [15, 16]. The central (tertiary) level constituted by the technical departments of the Ministry of Public Health is in charge of the development of national health strategies, national hospitals as well as reference laboratories. In the case of BU, control activities at this level are organized by NBUCPs, which define control strategies, plan the implementation of interventions, supervise and evaluate these interventions. National hospitals are usually not directly involved in the management of BU patients, whereas laboratory confirmation of suspected BU patients by PCR analysis is performed by national reference laboratories. The intermediate (secondary) level is composed of regional delegations of public health and regional hospitals, supervising control activities for BU. The peripheral (primary) level, where the implementation of BU control activities is based, is the health district, comprising the district health service, district hospitals, primary health centers, and communities. While the district health service organizes community sensitization and BU case-finding in collaboration with the primary health centers and community stakeholders, the district hospitals act as BU diagnostic and treatment centers (BU-DTCs), supervise BU care in primary health centers, and assist the district health service in community-based activities.

BU-DTC facilities have in the past often been built or rehabilitated and equipped by support partners to provide adequate infrastructure for BU diagnosis and treatment, including surgical theatres, wound dressing rooms, laboratories, physiotherapy units, and admission wards. Within the framework of national BU surveillance, all BU-DTCs are provided with BU case-definition, diagnostic and treatment guidelines and other documentation on the disease for use by health workers. Standard WHO BU case record files and registers referred to as BU01 and BU02 forms, respectively, are used at the BU-DTCs to document information on each patient. The BU-DTC health staff receives specific training through workshops organized by the NBUCPs and facilitated by BU experts, using the WHO training modules and guidelines on BU care. The data recording and reporting process at BU-DTCs is regularly monitored and supervised by the NBUCPs to ensure correctness and completeness [17, 18].

Main activities of the BU-DTCs include early case detection at the community level, training of village health workers and strengthening of the community-based surveillance system, information, education, and communication campaigns in communities and schools, strengthening of the health system and infrastructure, management of equipment, transport and logistics, standardized recording and reporting using WHO BU01 and BU02 forms, standardized case management (diagnosis and treatment), prevention of disability/rehabilitation, supportive activities, advocacy, social mobilization, partnerships, and operational research.

Box 1: Benin

The first case of BU in Benin was registered in 1977 at the Saint Camille Hospital in Dogbo (Couffo Department). A first focus of the disease was detected in the Ouinhi district (Zou Department) in 1988 [19], cases of which were reported by Muelder and Nourou in 1990 [20]. The NBUCP in Benin was created in 1997 and surveillance using the WHO BU01 and BU02 forms started in 2003. In Benin, there are four peripheral BU-DTCs, which are locally referred to as 'Centre de Dépistage et de Traitement de l'Ulcère de Buruli (CDTUB)'. These centers, which are responsible for the implementation of BU control activities, are distributed across the main endemic regions and located in Allada, Lalo, Pobe, and Zangnanado (Fig. 1). Health workers at these facilities have considerable experience in diagnosing and treating BU patients. Early case detection and referral of patients to the CDTUBs is supervised by health workers of the nearest health post, but relies strongly on community-based surveillance teams, comprising village volunteers ('relais communautaires') and teachers ('focal points'). These teams are also responsible for the follow-up of patients after treatment. In order to facilitate reporting at each CDTUB, BU cases are registered on a BU02 form, which is sent out each quarter to regional authorities and the NBUCP, where data are analyzed and mapped. Feedback is provided annually by the NBUCP at a review meeting attended by the heads of all CDTUBs and other partners involved in BU activities in Benin. The CDTUBs perform quarterly data analyses, which they feed back to the teams in each center. Refresher workshops for the teams are conducted and new team members are trained [17]. This system allows to (1) conduct permanent active surveys for BU, (2) determine the burden of the disease, (3) provide the most adapted care for each patient. With this system in place, the number of reported BU cases increased steadily from 2003, reached a peak of 1203 cases in 2007 and decreased continuously thereafter, to 312 in 2016. The geographical distribution of BU cases in 2016 is shown in Fig. 1.

In contrast to the decrease in the number of new BU cases reported in recent years in Benin, the percentage of patients diagnosed with WHO Category III lesions has increased (Fig. 2). Between 2007 and 2011 the number of new BU cases with Category I and II lesions detected was much higher than those with Category III lesions, whereas almost half of the newly detected BU cases between 2012 and 2016 were diagnosed with Category III lesions.

Therefore, the current challenge for the Ministry of Health through the NBUCP is to effectively address the issue of severe, chronic BU lesions. According to national health statistics, 267 new BU cases were detected in 2017. Of these, 46% presented with ulcers, 30% with mixed forms, 13% with plaques, and only a minority with nodules and edema. For a significant proportion of cases (23%) in 2017 BU-related disability was registered at entry.

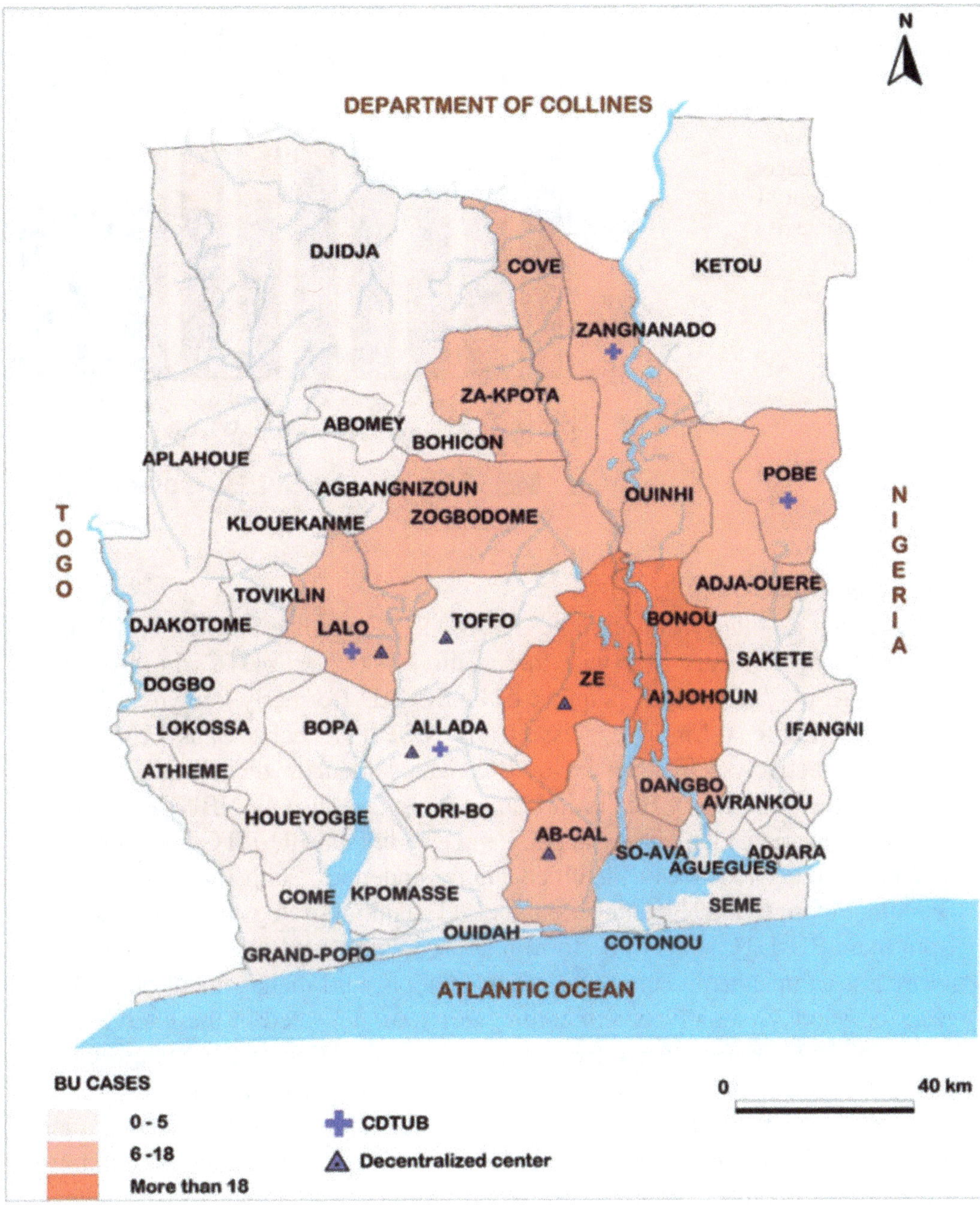

Fig. 1 Geographical distribution of BU in Southern Benin. The map illustrates the number of reported BU cases in 2016 as well as the location of the CDTUBs in the main BU endemic region of Benin

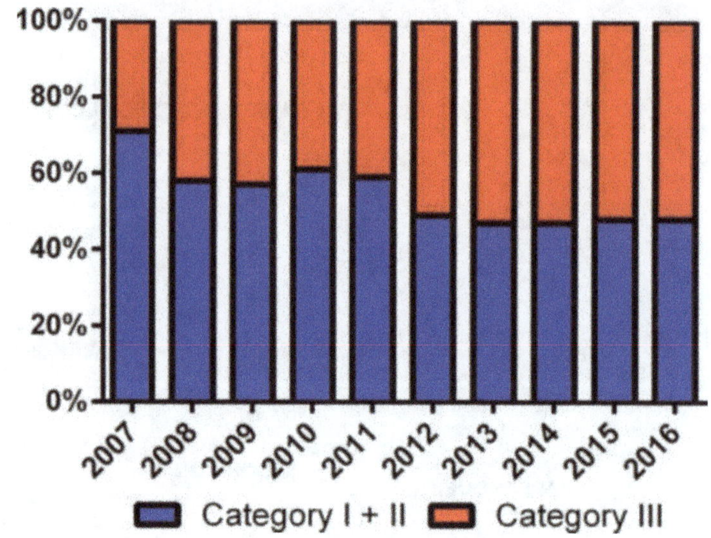

Fig. 2 Evolution of new BU cases detected in Benin according to the WHO classification system. The percentage of patients diagnosed with Category III (red) and Category I and II (blue) lesions is shown

Box 2 Cameroon

BU was first reported in Cameroon in 1969 [21]. Organized BU control in the country began 33 years later with the creation of two BU-DTCs in Ayos and Akonolinga in 2002 with the support of FAIRMED (formerly known as ALES (Aide Aux Lépreux Emmaüs-Suisse)) and MSF (Médecins Sans Frontières) Suisse, respectively. Effectiveness of BU management in these initial centers prompted the Ministry of Health in Cameroon to establish an NBUCP in 2004. Following a national BU survey in 2004, three additional BU-DTCs were created in 2006 in the three newly identified highly endemic health districts, namely Bankim, Mbonge, and Ngoantet-Mbalmayo. The initial strategy of medical teams from these BU-DTCs was to perform mass screening for BU, sensitization and awareness campaigns in communities and schools, and management of the BU cases detected. Gradually, health committees were activated in the intervention areas, and became responsible for community sensitization and awareness campaigns for BU, case-finding, and referral. Community participation in referral of suspected cases to the BU-DTCs for confirmation and treatment increased after many volunteers were trained to recognize BU in their communities. As communities became more aware of BU, active case-finding gradually gave way to passive case-finding with cases coming on their own or being referred to BU-DTCs.

In Cameroon, the WHO BU02 forms together with activity reports are sent by BU-DTCs to the NBUCP on a monthly basis. Between 2002 and 2016, 3850 BU cases were reported in the country, with a peak of 914 cases seen in 2004, explained by the national survey for BU in that year. Considerably lower numbers of cases were reported in recent years with a trough of 85 cases in 2016. Between 2002 and 2014 the number of health districts identified to be endemic for BU rose from 2 to 64 (Fig. 3). While endemic regions were mainly detected in the South and Central part of Cameroon, a national survey is required to confirm the suspected presence of BU in the northern part of the country. However, it has been shown that a reduction in program resources and activities by the major support partners over the past years has led to decreased surveillance activities, which has negatively impacted performance indicators in Cameroon [18].

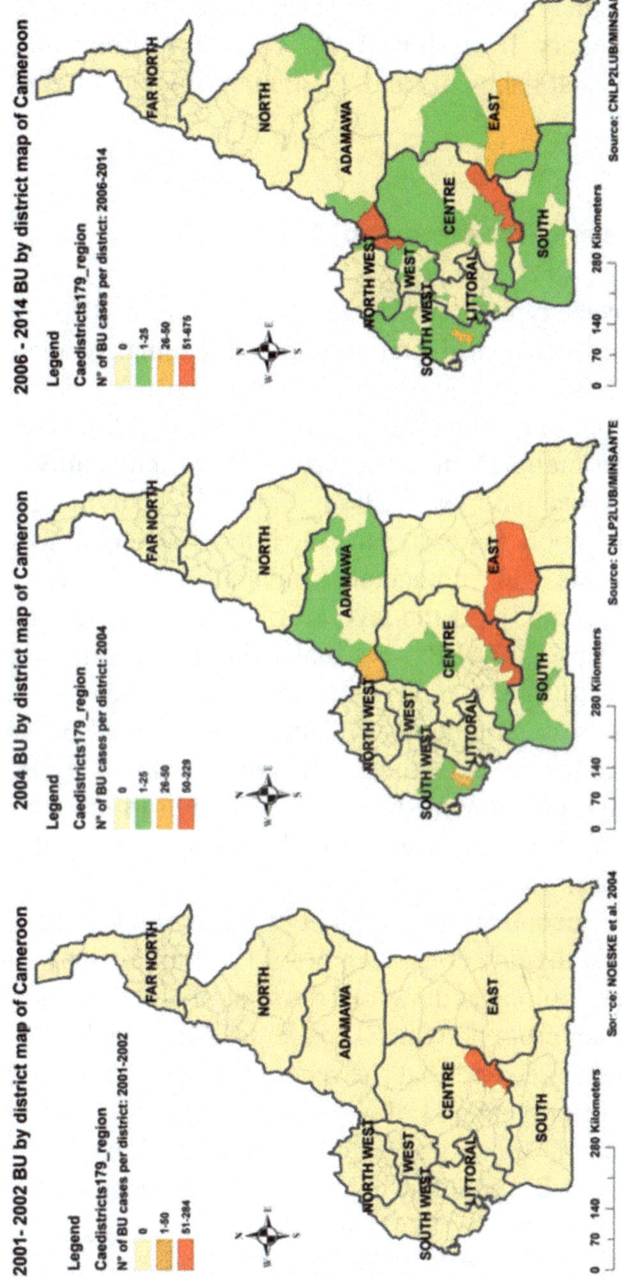

Fig. 3 Increase in the number of health districts identified to be BU endemic between 2001 and 2014 [18]

In many African BU endemic areas, government expenditure on health care is extremely low. Low per capita income and limited potential for domestic revenue mobilization hamper the ability of governments to respond effectively to national health problems, such as BU. Many NBUCPs depend strongly on external partner funding, which has decreased considerably in the past few years in some of the BU endemic countries, leading to a decrease in BU surveillance and laboratory confirmation of cases. This may in turn lead to a rise in the proportion of patients reporting with WHO Category III lesions [18]. Therefore, financing remains a key challenge for the establishment and sustainability of effective BU control programs.

2.2 Diagnosis and Treatment of BU

In addition to relatively typical ulcerative disease stages, BU may present in the form of rather unspecific, non-ulcerative nodules, edema, or plaque lesions. The clinical diagnosis of BU in Africa is complicated by many skin conditions with similar presentation ranging from cysts, lipoma, psoriasis, skin lymphomas, tropical ulcers, ulcerated skin malignancies, and venous or vascular ulcers, over bacterial skin infections such as actinomycosis, boils, cellulitis, ecthyma, folliculitis, furuncle, impetigo, noma, treponematosis, parasitic infections including cutaneous leishmaniasis and myiasis, to cutaneous tuberculosis, leprosy, and atypical mycobacteriosis [17, 22, 23]. At the level of district health facilities, microscopic detection of acid-fast bacilli (AFBs) is usually the only available laboratory diagnostic test for BU. As this method has limited sensitivity and specificity, specimens from suspected lesions are commonly collected and shipped in bulk to diagnostic reference laboratories for the detection of *M. ulcerans* DNA by PCR analysis, the current gold standard for the diagnosis of BU. To avoid further delay and dropout of patients, treatment of clinically suspected BU cases is often started before PCR results are reported back to the health centers.

Since 2004, WHO recommends treatment of BU with an 8-week course of daily combination antibiotic therapy consisting of oral rifampicin and injectable streptomycin [13]. However, prolonged duration of streptomycin therapy can cause persistent hearing loss and nephrotoxicity in BU patients [24]. In 2017 the WHO Technical Advisory Group on BU decided that the recommendation for treatment should be changed to oral clarithromycin and rifampicin, pending the full results of a clinical trial for the new regimen. At national level, this combination has already been introduced in several African BU endemic countries. Free provision of antibiotics to BU treatment centers is managed by the NBUCPs. However, logistics involved in the supply of the drugs to remote, rural health facilities is complicated and access to the required antibiotics is not always secured. Other direct and indirect treatment costs, such as for transport and stay at health facilities, wound management, and loss of labor of patients and caretakers have to be covered by patients and their families. The main source of funding for these costs is direct household spending, mostly through out-of-pocket payments, which may prevent patients from seeking care or may have catastrophic consequences for the household economies [25, 26]. The situation is further aggravated by the fact that many patients report with advanced stages of the disease, for which treatment is both much more demanding and more expensive. Although antibiotic treatment is effective, extensive destruction of tissue complicates healing and often leads to contractures and deformities. For such lesions, surgical debridement, skin grafting, and physical re-education may be required after antibiotic therapy. Lifelong functional limitations are a common outcome of large ulcers and are associated with loss of workforce, school abandonment [27], stigma, and social exclusion [28].

Early detection of BU, the development of an inexpensive, sensitive and specific point-of-care diagnostic test suitable for pre-treatment diagnosis, as well as the development of shorter treatment regimens suitable for decentralized care in rural endemic areas are key priorities for future efforts to improve the control of BU in Africa. For the diagnosis at field sites, the application of *M. ulcerans* antigen capture-based approaches are currently being evaluated [29]. Local thermotherapy of BU lesions using heat packs filled with phase change material [30] may be developed into an alternative treatment option if antibiotic treatment is not indicated, not tolerated or not readily available. Considering the continued preference of many patients to first seek care from traditional healers, sustained collaborations between community health workers, BU-DTC health staff, and traditional healers is essential for the implementation of BU control activities [10].

3 Geographical Distribution of BU: Reporting of Cases Versus Actual Situation in Africa

The geographical distribution of BU cases in Africa by country reported to WHO between 2002 and 2016 is illustrated in Fig. 4, with Côte d'Ivoire, Ghana, and Benin being the most affected.

These three countries were among the first with well-established BU surveillance systems [31–34], after their governments have signed the Yamoussoukro declaration on BU in 1998, an agreement to mobilize resources for the establishment of NBUCPs with technical support from WHO. In Cameroon, which is ranked fourth among the most affected countries, a wide-spread distribution of endemic regions was detected upon increased case-finding activities (Fig. 3) [18], strongly suggesting that the number of BU cases is vastly underestimated in some other BU endemic countries, where organized BU control programs have only more recently been set in place. This is for example illustrated by large series of BU patients from Nigeria, who presented to established health facilities in Benin over the past few years [35, 36], demonstrating an urgent need for the improvement of BU control activities in Nigeria [37].

Reasons for a marked reduction of new infections reported from Africa in the past few years are not entirely understood, but may at least in part be related to a decline in international support for BU control programs [18]. In Ghana, the annual BU case confirmation rates gradually decreased over the past years, from a high

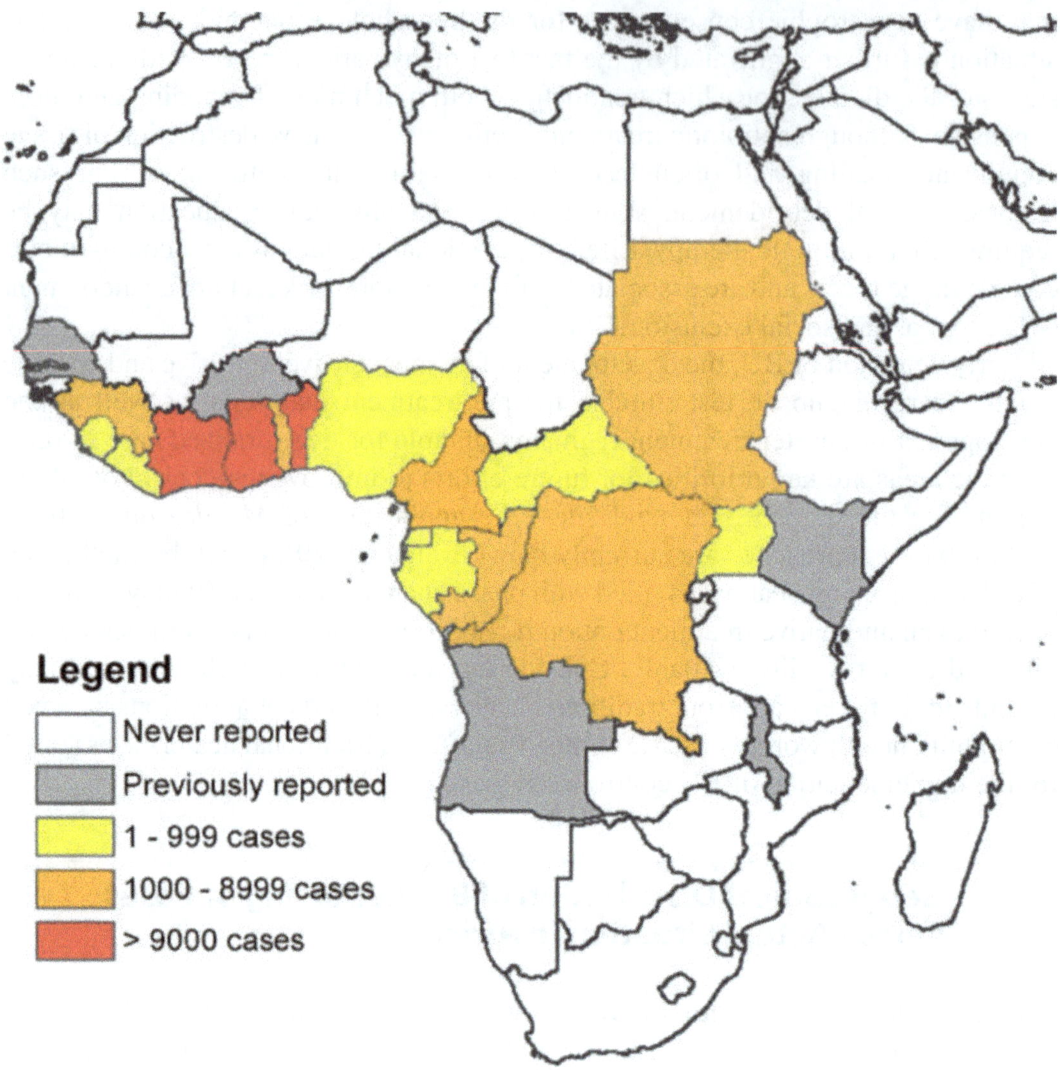

Fig. 4 Geographical distribution of BU in Africa by country. The map shows an accumulated number of cases reported to WHO between 2002 and 2016. Data source: WHO

proportion of 76% in 2009 to a trough of only 15% in 2016. High confirmation rates in earlier years may be attributed to prior training activities on case detection and proper specimen collection provided by the NBUCP to healthcare givers within the facilities of the Ghana Health Service as well as to quarterly early case search activities conducted within the framework of the Stop BU project by researchers of the Noguchi Memorial Institute for Medical Research. The downward trend in confirmation rates may be a reflection of both an actual reduction in BU incidence and ceasing outreach activities by the NBUCP in recent years [38]. On the other hand, initiation of surveillance activities in the endemic countries may have led to the detection of many patients with long-standing infections and the current number of reported cases may reflect more the true incidence of BU. As the shedding of the bacteria from chronic BU lesions is hypothesized to fuel potential reservoirs of the pathogen, early identification and treatment of BU patients may lead to a reduction of the bacterial burden in the environment, which may in turn reduce transmission. This view is supported by the fact that no major animal reservoir has been detected to date in Africa, as opposed to BU endemic areas of southern Australia, where large numbers of possums in BU endemic settings were shown to be infected with *M. ulcerans* [39]. Another aspect to be considered is the potential cyclical occurrence of the disease associated with environmental or climatic factors. Also, a number of African countries, where BU cases had been detected in the past, do no longer report cases to WHO. These include countries where political instability or overburdened health systems may prevent efforts to control BU. On the other hand, also the possibility of over-reporting in areas where access to reliable laboratory confirmation is limited has to be taken into consideration, when estimating the actual burden of BU in Africa.

4 Distribution of BU Among Affected Populations in Africa

The distribution of BU within endemic countries is highly focal and local prevalence rates may vary from one village to another. Affected populations commonly live in remote, rural areas, which are sometimes largely isolated from the world around. Families typically live on subsistence-level agriculture and small-scale husbandry and obtain their water from nearby rivers or tributaries. Stagnant water bodies, which are used for washing clothes or bathing, appear to be strongly associated with the occurrence of BU [40] (Fig. 5).

In many areas the emergence of BU or an increase in the incidence of BU have been ascribed to environmental disturbances such as damming of rivers, establishment of permanent wet agricultural areas like rice fields, deforestation, sand digging or mining activities with remaining water holes and seasonal ponds. For example, on the campus of the University of Ibadan, Nigeria, damming of a small stream flowing through the campus was associated with the emergence of BU among

Fig. 5 Typical BU endemic settings in Africa. Left photograph: Health post in Cameroon. Right photograph: Water site in Cameroon used for doing the laundry or bathing

Caucasians living on site [41]. In northern Liberia, BU cases emerged when the Manor River was dammed and wetlands extended for swamp rice farming [42]. Other examples include the damming of the Mapé River in Cameroon [43], the Densu River in Ghana [44], and the Bandama River in Côte d'Ivoire [45]. Apart from man-made environmental changes, heavy rainfall may also lead to an upsurge of BU; in Uganda, an outbreak in the Busoga District was for example related to unprecedented flooding of the lakes of Uganda between 1962 and 1964 [46].

All age groups may be affected with an often equal gender distribution. However, children between five and 15 years of age are vastly overrepresented among cases in Africa, even if the age distribution of the typically very young general population is taken into account [47, 48]. In contrast, very young children seem to be underrepresented [47], and sero-epidemiological studies of populations living in BU endemic areas of Ghana and Cameroon have shown that children below the age of four years are also less exposed to *M. ulcerans* than older individuals [49, 50]. This suggests that exposure to *M. ulcerans* intensifies at an age, when children start to have more intense contact with the environment, outside their hitherto confined movement range. While the risk of developing BU seems to drop in young adults, the population age adjusted cumulative incidence of BU in the elderly was reported to be similar to that in older children [47, 48]. This may be related to immunosenescence, the gradual deterioration of the immune system associated with natural age advancement. The fact that in Japan and Australia middle-aged or elderly individuals are most commonly affected may at least in part reflect the much higher average age of the population living in the BU endemic areas.

Mounting evidence exists that infection with HIV increases the risk of BU [51]. Moreover, a number of case studies report a more aggressive progression of BU in HIV-positive individuals [52–54]. Although helminth infections elicit an immune response potentially enhancing susceptibility to mycobacterial diseases, no association between BU and schistosomiasis was found in one study [55]. Potential associations of BU with other co-infections prevalent in African BU endemic areas or with malnutrition have not been systematically investigated so far. Studies on a potential link between susceptibility to BU and host genetic factors are in their infancy [56, 57] and are hampered by the limited number of BU cases that can be enrolled in the studies.

5 The Etiology of BU in Africa

M. ulcerans is an acid-fast mycobacterium that has evolved from an *M. marinum*-like progenitor through an evolutionary bottleneck event; the acquisition of a virulence plasmid, encoding the enzymatic machinery for the synthesis of the unique macrolide toxin mycolactone [58]. Subsequent genome reduction and pseudogene development is indicative of an adaptation of *M. ulcerans* to a more stable ecological niche [58]. Definite identification of an environmental niche is complicated by the extraordinarily long generation time of *M. ulcerans*. This hampers the isolation of the bacterium from potential environmental sources, as *M. ulcerans* is readily overgrown by other, less fastidious organisms, including other environmental mycobacteria that are also resistant to decontamination methods developed for primary isolation [59, 60]. Considering that the occurrence of BU is commonly associated with stagnant water bodies in river basins, a habitat in aquatic ecosystems seems likely. Indeed, *M. ulcerans*-specific DNA sequences were detected by PCR analyses in environmental samples, such as biofilms, water, soil and plants, as well as in various aquatic animals [5]. For unclear reasons, vastly different PCR positivity rates have been reported in various studies conducted in BU endemic regions of Africa [40, 61, 62]. In contrast to endemic sites in Australia, where terrestrial mammals may be implicated as reservoirs in the disease ecology of *M. ulcerans* [39, 63], no similar reservoir has so far been detected in Africa [6, 64]. Therefore, it has been hypothesized that shedding bacilli from large chronic lesions of BU patients into the environment may play a role in *M. ulcerans* transmission in African BU endemic areas [4]. While mosquitoes have been proposed as vectors of *M. ulcerans* in Australia, several modes of transmission including vector-mediated, but also skin trauma-induced may be involved in African settings [5]. Direct person-to-person transmission is considered unlikely.

All *M. ulcerans* isolates derived from lesions of BU patients from Africa have been found to belong to the classical *M. ulcerans* lineage. Genetic diversity of these isolates is very low and only comparative whole genome sequencing allowed resolving the population structure and evolutionary history of African *M. ulcerans* disease isolates. These genome analyses have identified in many BU endemic areas, local clonal complexes of *M. ulcerans* that show limited diversification by the accumulation of point mutations and are associated with particular hydrological drainage areas [44, 65–67]. After introduction of *M. ulcerans* in a particular area, the local clone seems to remain isolated, which allows some point mutations to become fixed in that population. Spread of these locally confined clonal complexes between endemic areas seems to occur only rarely [68, 69], speaking against the existence of a highly mobile animal reservoir. While the dominating African *M. ulcerans* sublineage MU_A1 has been endemic in Africa for several hundred years, another less common and geographically more restricted African sublineage (MU_A2) has been recently identified, which seems to have been introduced into the African continent in the late nineteenth and early twentieth century [66]. Sequence analyses of the virulence plasmid pMUM encoding genes for the biosynthesis of mycolactones have shown that all mycolactone producing mycobacteria have evolved from the same progenitor which has acquired the plasmid [70]. Different *M. ulcerans* lineages produce different species of mycolactone; the *M. ulcerans* disease isolates from Africa mainly produce mycolactone A/B, the most potent form of the toxin [71, 72]. There is no doubt that the ability to produce mycolactone is critical for the evolution and persistence of *M. ulcerans* as a human pathogen. While the combination of cytotoxic and immunosuppressive properties of mycolactone is thought to confer a fitness advantage for *M. ulcerans* in mammalian hosts by preventing an immune system-mediated elimination of the bacteria, it is not yet clear how the bacteria might benefit from mycolactone production in aquatic niche environments.

6 Future Perspectives

If human BU lesions prove to be a relevant maintenance reservoir for *M. ulcerans* transmission, active case-finding programs, improved disease surveillance, early diagnosis, and adequate treatment may lead to a reduction in disease transmission. This hypothesis is supported by the recent decline in the number of new infections reported to WHO from many African countries and particularly from areas, where effective BU control programs have been implemented. A major challenge for the coming years will be the maintenance of established control strategies and the implementation of improved diagnostic tools and treatment approaches.

In many regions of Africa a number of tropical skin diseases, such as cutaneous leishmaniasis, leprosy, lymphatic filariasis, mycetoma, onchocerciasis, and yaws are co-endemic with BU. Screening of individuals for the presence of skin conditions in communities or schools offers the opportunity to detect these neglected diseases using a common rather than a disease-specific approach. Development of locally adapted triage criteria and diagnostic algorithms for the recognition of changes in the appearance of the skin can furthermore allow health workers at the primary health care level making appropriate decisions on either treatment or referral of patients. After specific treatment, repair of tissue damage often requires similar wound management approaches. Therefore, integrated strategies for the control and management of these diseases are now strongly promoted by the WHO Department of control of neglected tropical diseases (WHO/NTD) [73].

References

1. Billington WR (1970) Albert Cook 1870-1951: Uganda pioneer. Br Med J 4(5737):738–740
2. Clancey JK, Dodge OG, Lunn HF, Oduori ML (1961) Mycobacterial skin ulcers in Uganda. Lancet 2(7209):951–954
3. Clancey JK (1964) Mycobacterial skin ulcers in Uganda: description of a new Mycobacterium (Mycobacterium Buruli). J Pathol Bacteriol 88:175–187
4. Röltgen K, Pluschke G (2015) Epidemiology and disease burden of Buruli ulcer: a review. Res Rep Trop Med 2015(6):59–73
5. Merritt RW, Walker ED, Small PL, Wallace JR, Johnson PD, Benbow ME et al (2010) Ecology and transmission of Buruli ulcer disease: a systematic review. PLoS Negl Trop Dis 4(12):e911
6. Röltgen K, Pluschke G (2015) Mycobacterium ulcerans disease (Buruli ulcer): potential reservoirs and vectors. Curr Clin Microbiol Rep 2(1):35–43
7. Mulder AA, Boerma RP, Barogui Y, Zinsou C, Johnson RC, Gbovi J et al (2008) Healthcare seeking behaviour for Buruli ulcer in Benin: a model to capture therapy choice of patients and healthy community members. Trans R Soc Trop Med Hyg 102(9):912–920
8. Aujoulat I, Johnson C, Zinsou C, Guedenon A, Portaels F (2003) Psychosocial aspects of health seeking behaviours of patients with Buruli ulcer in southern Benin. Trop Med Int Health 8(8):750–759
9. Peeters Grietens K, Toomer E, Um Boock A, Hausmann-Muela S, Peeters H, Kanobana K et al (2012) What role do traditional beliefs play in treatment seeking and delay for Buruli ulcer disease?--insights from a mixed methods study in Cameroon. PLoS One 7(5):e36954
10. Awah PK, Boock AU, Mou F, Koin JT, Anye EM, Noumen D et al (2018) Developing a Buruli ulcer community of practice in Bankim, Cameroon: a model for Buruli ulcer outreach in Africa. PLoS Negl Trop Dis 12(3):e0006238

11. Renzaho AM, Woods PV, Ackumey MM, Harvey SK, Kotin J (2007) Community-based study on knowledge, attitude and practice on the mode of transmission, prevention and treatment of the Buruli ulcer in Ga West District. Ghana Trop Med Int Health 12(3):445–458
12. Pearson G (2018) Understanding perceptions on 'Buruli' in northwestern Uganda: a biosocial investigation. PLoS Negl Trop Dis 12(7):e0006689
13. WHO (2012) Treatment of Mycobacterium ulcerans disease (Buruli ulcer): guidance for health workers. World Health Organization, Geneva
14. WHO (2000) Buruli ulcer: mycobacterium ulcerans infection. World Health Organization, Geneva
15. Hayriye IŞik HN (2013) Health care system and health financing structure, the case of Cameroon. Int Anatol Acad Online J Health Sci 1(2):24–44
16. Bertrand NAS (2012) Analysis of determinants of public hospitals efficiency in Cameroon. Int J Econ Commer Res 2(2):31–65
17. Junghanss T, Johnson RC, Pluschke G (2014) Mycobacterium ulcerans disease. In: Farrar J, Hotez PJ, Junghanss T, Kang G, Lalloo D, White NJ (eds) Manson's tropical diseases, 23rd edn. Saunders, Edinburgh, pp 519–531
18. Tabah EN, Nsagha DS, Bissek AC, Njamnshi AK, Bratschi MW, Pluschke G et al (2016) Buruli ulcer in Cameroon: the development and impact of the national control programme. PLoS Negl Trop Dis 10(1):e0004224
19. Muelder K (1988) Buruli ulcer in Benin. Trop Dr 18(2):53
20. Muelder K, Nourou A (1990) Buruli ulcer in Benin. Lancet 336(8723):1109–1111
21. Ravisse P (1977) Skin ulcer caused by Mycobacterium ulcerans in Cameroon. I. Clinical, epidemiological and histological study. Bull Soc Pathol Exot Fil 70(2):109–124
22. Bratschi MW, Njih Tabah E, Bolz M, Stucki D, Borrell S, Gagneux S et al (2012) A case of cutaneous tuberculosis in a Buruli ulcer-endemic area. PLoS Negl Trop Dis 6(8):e1751
23. Toutous Trellu L, Nkemenang P, Comte E, Ehounou G, Atangana P, Mboua DJ et al (2016) Differential diagnosis of skin ulcers in a Mycobacterium ulcerans endemic area: data from a prospective study in Cameroon. PLoS Negl Trop Dis 10(4):e0004385
24. Klis S, Stienstra Y, Phillips RO, Abass KM, Tuah W, van der Werf TS (2014) Long term streptomycin toxicity in the treatment of Buruli ulcer: follow-up of participants in the BURULICO drug trial. PLoS Negl Trop Dis 8(3):e2739
25. Chukwu JN, Meka AO, Nwafor CC, Oshi DC, Madichie NO, Ekeke N et al (2017) Financial burden of health care for Buruli ulcer patients in Nigeria: the patients' perspective. Int Health 9(1):36–43
26. Grietens KP, Boock AU, Peeters H, Hausmann-Muela S, Toomer E, Ribera JM (2008) "It is me who endures but my family that suffers": social isolation as a consequence of the household cost burden of Buruli ulcer free of charge hospital treatment. PLoS Negl Trop Dis 2(10):e321
27. Stienstra Y, van Roest MH, van Wezel MJ, Wiersma IC, Hospers IC, Dijkstra PU et al (2005) Factors associated with functional limitations and subsequent employment or schooling in Buruli ulcer patients. Tropical Med Int Health 10(12):1251–1257
28. Owusu AaA C (2012) The socioeconomic burden of Buruli ulcer disease in the GA West District of Ghana. Ghana J Dev Stud 9(1):5–20
29. Dreyer A, Roltgen K, Dangy JP, Ruf MT, Scherr N, Bolz M et al (2015) Identification of the Mycobacterium ulcerans protein MUL_3720 as a promising target for the development of a diagnostic test for Buruli ulcer. PLoS Negl Trop Dis 9(2):e0003477
30. Vogel M, Bayi PF, Ruf MT, Bratschi MW, Bolz M, Um Boock A et al (2016) Local heat application for the treatment of Buruli ulcer: results of a phase II open label single center non comparative clinical trial. Clin Infect Dis 62(3):342–350
31. Johnson RC, Sopoh GE, Barogui Y, Dossou A, Fourn L, Zohoun T (2008) Surveillance system for Buruli ulcer in Benin: results after four years. Sante 18(1):9–13
32. Kanga JM, Kacou ED, Kouame K, Kassi K, Kaloga M, Yao JK et al (2006) Fighting against Buruli ulcer: the Cote-d'Ivoire experience. Bull Soc Pathol Exot 99(1):34–38
33. Amofah G, Bonsu F, Tetteh C, Okrah J, Asamoa K, Asiedu K et al (2002) Buruli ulcer in Ghana: results of a national case search. Emerg Infect Dis 8(2):167–170

34. Ackumey MM, Kwakye-Maclean C, Ampadu EO, de Savigny D, Weiss MG (2011) Health services for Buruli ulcer control: lessons from a field study in Ghana. PLoS Negl Trop Dis 5(6):e1187
35. Marion E, Carolan K, Adeye A, Kempf M, Chauty A, Marsollier L (2015) Buruli ulcer in South Western Nigeria: a retrospective cohort study of patients treated in Benin. PLoS Negl Trop Dis 9(1):e3443
36. Ayelo GA, Anagonou E, Wadagni AC, Barogui YT, Dossou AD, Houezo JG et al (2018) Report of a series of 82 cases of Buruli ulcer from Nigeria treated in Benin, from 2006 to 2016. PLoS Negl Trop Dis 12(3):e0006358
37. Otuh PI, Soyinka FO, Ogunro BN, Akinseye V, Nwezza EE, Iseoluwa-Adelokiki AO et al (2018) Perception and incidence of Buruli ulcer in Ogun State, South West Nigeria: intensive epidemiological survey and public health intervention recommended. Pan Afr Med J 29:166
38. Yeboah-Manu D, Aboagye SY, Asare P, Asante-Poku A, Ampah K, Danso E et al (2018) Laboratory confirmation of Buruli ulcer cases in Ghana, 2008-2016. PLoS Negl Trop Dis 12(6):e0006560
39. Fyfe JA, Lavender CJ, Handasyde KA, Legione AR, O'Brien CR, Stinear TP et al (2010) A major role for mammals in the ecology of Mycobacterium ulcerans. PLoS Negl Trop Dis 4(8):e791
40. Bratschi MW, Ruf MT, Andreoli A, Minyem JC, Kerber S, Wantong FG et al (2014) Mycobacterium ulcerans persistence at a village water source of Buruli ulcer patients. PLoS Negl Trop Dis 8(3):e2756
41. Oluwasanmi JO, Itayemi SO, Alabi GO (1975) Buruli (Mycobacterial) Ulcers in Caucasians in Nigeria. Brit J Plast Surg 28(2):111–113
42. Monson MH, Gibson DW, Connor DH, Kappes R, Hienz HA (1984) Mycobacterium ulcerans in Liberia: a clinicopathologic study of 6 patients with Buruli ulcer. Acta Trop 41(2): 165–172
43. Marion E, Landier J, Boisier P, Marsollier L, Fontanet A, Le Gall P et al (2011) Geographic expansion of Buruli ulcer disease. Cameroon Emerg Infect Dis 17(3):551–553
44. Röltgen K, Pluschke G (2010) Single nucleotide polymorphism typing of Mycobacterium ulcerans reveals focal transmission of Buruli ulcer in a highly endemic region of Ghana. PLoS Negl Trop Dis 4(7):e751
45. N'Krumah RTAS, Kone B, Cisse G, Tanner M, Utzinger J, Pluschke G et al (2017) Characteristics and epidemiological profile of Buruli ulcer in the district of Tiassale, south Cote d'Ivoire. Acta Trop 175:138–144
46. Barker DJ (1971) Buruli disease in a district of Uganda. J Trop Med Hyg 74(12):260–264
47. Bratschi MW, Bolz M, Minyem JC, Grize L, Wantong FG, Kerber S et al (2013) Geographic distribution, age pattern and sites of lesions in a cohort of Buruli ulcer patients from the Mape Basin of Cameroon. PLoS Negl Trop Dis 7(6):e2252
48. Debacker M, Aguiar J, Steunou C, Zinsou C, Meyers WM, Scott JT et al (2004) Mycobacterium ulcerans disease: role of age and gender in incidence and morbidity. Tropical Med Int Health 9(12):1297–1304
49. Röltgen K, Bratschi MW, Pluschke G (2014) Late onset of the serological response against the 18 kDa small heat shock protein of Mycobacterium ulcerans in children. PLoS Negl Trop Dis 8(5):e2904
50. Ampah KA, Nickel B, Asare P, Ross A, De-Graft D, Kerber S et al (2016) A sero-epidemiological approach to explore transmission of Mycobacterium ulcerans. PLoS Negl Trop Dis 10(1):e0004387
51. Johnson RC, Nackers F, Glynn JR, de Biurrun Bakedano E, Zinsou C, Aguiar J et al (2008) Association of HIV infection and Mycobacterium ulcerans disease in Benin. AIDS 22(7):901–903
52. Toll A, Gallardo F, Ferran M, Gilaberte M, Iglesias M, Gimeno JL et al (2005) Aggressive multifocal Buruli ulcer with associated osteomyelitis in an HIV-positive patient. Clin Exp Dermatol 30(6):649–651

53. Kibadi K, Colebunders R, Muyembe-Tamfum JJ, Meyers WM, Portaels F (2010) Buruli ulcer lesions in HIV-positive patient. Emerg Infect Dis 16(4):738–739
54. Komenan K, Elidje EJ, Ildevert GP, Yao KI, Kanga K, Kouame KA et al (2013) Multifocal Buruli Ulcer associated with secondary infection in HIV positive patient. Case Rep Med 2013:348628
55. Stienstra Y, van der Werf TS, van der Graaf WT, Secor WE, Kihlstrom SL, Dobos KM et al (2004) Buruli ulcer and schistosomiasis: no association found. Am J Trop Med Hyg 71(3):318–321
56. Capela C, Dossou AD, Silva-Gomes R, Sopoh GE, Makoutode M, Menino JF et al (2016) Genetic variation in autophagy-related genes influences the risk and phenotype of Buruli ulcer. PLoS Negl Trop Dis 10(4):e0004671
57. Bibert S, Bratschi MW, Aboagye SY, Collinet E, Scherr N, Yeboah-Manu D et al (2017) Susceptibility to Mycobacterium ulcerans disease (Buruli ulcer) is associated with IFNG and iNOS gene polymorphisms. Front Microbiol 8:1903
58. Stinear TP, Seemann T, Pidot S, Frigui W, Reysset G, Garnier T et al (2007) Reductive evolution and niche adaptation inferred from the genome of Mycobacterium ulcerans, the causative agent of Buruli ulcer. Genome Res 17(2):192–200
59. Yeboah-Manu D, Bodmer T, Mensah-Quainoo E, Owusu S, Ofori-Adjei D, Pluschke G (2004) Evaluation of decontamination methods and growth media for primary isolation of Mycobacterium ulcerans from surgical specimens. J Clin Microbiol 42(12):5875–5876
60. Yeboah-Manu D, Danso E, Ampah K, Asante-Poku A, Nakobu Z, Pluschke G (2011) Isolation of Mycobacterium ulcerans from swab and fine-needle-aspiration specimens. J Clin Microbiol 49(5):1997–1999
61. Williamson HR, Benbow ME, Campbell LP, Johnson CR, Sopoh G, Barogui Y et al (2012) Detection of Mycobacterium ulcerans in the environment predicts prevalence of Buruli ulcer in Benin. PLoS Negl Trop Dis 6(1):e1506
62. Vandelannoote K, Durnez L, Amissah D, Gryseels S, Dodoo A, Yeboah S et al (2010) Application of real-time PCR in Ghana, a Buruli ulcer-endemic country, confirms the presence of Mycobacterium ulcerans in the environment. FEMS Microbiol Lett 304(2):191–194
63. Roltgen K, Pluschke G, Johnson PDR, Fyfe J (2017) Mycobacterium ulcerans DNA in bandicoot excreta in Buruli ulcer-endemic area, Northern Queensland. Aust Emerg Infect Dis 23(12):2042–2045
64. Durnez L, Suykerbuyk P, Nicolas V, Barriere P, Verheyen E, Johnson CR et al (2010) Terrestrial small mammals as reservoirs of Mycobacterium ulcerans in benin. Appl Environ Microbiol 76(13):4574–4577
65. Vandelannoote K, Jordaens K, Bomans P, Leirs H, Durnez L, Affolabi D et al (2014) Insertion sequence element single nucleotide polymorphism typing provides insights into the population structure and evolution of Mycobacterium ulcerans across Africa. Appl Environ Microbiol 80(3):1197–1209
66. Vandelannoote K, Meehan CJ, Eddyani M, Affolabi D, Phanzu DM, Eyangoh S et al (2017) Multiple introductions and recent spread of the emerging human pathogen Mycobacterium ulcerans across Africa. Genome Biol Evol 9(3):414–426
67. Bolz M, Bratschi MW, Kerber S, Minyem JC, Um Boock A, Vogel M et al (2015) Locally confined clonal complexes of mycobacterium ulcerans in two Buruli ulcer endemic regions of Cameroon. PLoS Negl Trop Dis 9(6):e0003802
68. Ablordey AS, Vandelannoote K, Frimpong IA, Ahortor EK, Amissah NA, Eddyani M et al (2015) Whole genome comparisons suggest random distribution of Mycobacterium ulcerans genotypes in a Buruli ulcer endemic region of Ghana. PLoS Negl Trop Dis 9(3):e0003681
69. Lamelas A, Ampah KA, Aboagye S, Kerber S, Danso E, Asante-Poku A et al (2016) Spatiotemporal co-existence of two Mycobacterium ulcerans clonal complexes in the Offin River Valley of Ghana. PLoS Negl Trop Dis 10(7):e0004856
70. Doig KD, Holt KE, Fyfe JA, Lavender CJ, Eddyani M, Portaels F et al (2012) On the origin of Mycobacterium ulcerans, the causative agent of Buruli ulcer. BMC Genomics 13:258

71. Mve-Obiang A, Lee RE, Portaels F, Small PL (2003) Heterogeneity of mycolactones produced by clinical isolates of Mycobacterium ulcerans: implications for virulence. Infect Immun 71(2):774–783
72. Scherr N, Gersbach P, Dangy JP, Bomio C, Li J, Altmann KH et al (2013) Structure-activity relationship studies on the macrolide exotoxin mycolactone of Mycobacterium ulcerans. PLoS Negl Trop Dis 7(3):e2143
73. Mitja O, Marks M, Bertran L, Kollie K, Argaw D, Fahal AH et al (2017) Integrated control and management of neglected tropical skin diseases. PLoS Negl Trop Dis 11(1):e0005136

Buruli Ulcer in Australia

Paul D. R. Johnson

1 Bairnsdale Ulcer and the Discovery of *Mycobacterium ulcerans*

"Bairnsdale ulcer" (synonymous with Buruli ulcer (BU)) was first recognised and named as a distinct clinical entity in the late 1930s by General Practitioners working in Bairnsdale—a regional town in east Gippsland, Victoria, Australia [1, 2] (Fig. 1). They suspected their patients were suffering from a new type of infection caused by an acid-fast bacillus, but with clinical features distinct from tuberculosis and leprosy. Pathology specimens forwarded to the University of Melbourne and the transfer of patients from Bairnsdale to the Alfred Hospital in Melbourne provided researchers with clinical samples from which the microbiology and pathology of *Mycobacterium ulcerans* infection were first definitively described [3]. Although not listed as authors on the original 1948 publication "A New Mycobacterial Infection in Man" the contribution of these doctors was pivotal in the discovery of *M. ulcerans*. Glen Buckle, who with Jean Tolhurst first isolated *M. ulcerans* in pure culture, later drew special attention to the important contribution of Drs D. S. Alsop, L. E. Clay and J. R. Searls of Bairnsdale, and Dr. K. E. Torode of Colac [1].

In their initial research, Buckle and Tolhurst showed that *M. ulcerans* was able to grow on media that supported other pathogenic mycobacteria provided the incubation temperature was kept at 30–33 °C. They noted that growth slowed above 35 °C and that cells started to slowly die above 37 °C. They went on to establish experimental infections in mice, rats and rabbits and observed that while local lesions were produced at the site of subcutaneous inoculations, when inoculated into the

P. D. R. Johnson (✉)
Infectious Diseases Department, Austin Health and University of Melbourne, Heidelberg, VIC, Australia

WHO Collaborating Centre for Mycobacterium ulcerans, VIDRL, Peter Doherty Institute for Infection and Immunity, Melbourne, VIC, Australia
e-mail: Paul.Johnson@austin.org.au

G. Pluschke, K. Röltgen (eds.), *Buruli Ulcer*,
https://doi.org/10.1007/978-3-030-11114-4_3

Fig. 1 Australia with locations of confirmed BU acquisition since 1937 shown in red

peritoneal cavity, peripheral lesions developed on distant extremities in cooler body areas (scrota, tails). Sir Frank Fenner, a renowned Australian scientist who later worked on the eradication of smallpox, studied *M. ulcerans* in the 1950s. Fenner demonstrated that an inoculum of only 5–10 cells would reliably produce lesions in mice that appeared after at least 150 days (5 months) and that BCG was protective in an animal model at low but not higher inoculums [2]. Interestingly, the very low inoculum required to produce an experimental infection has recently been confirmed in new research that demonstrated a mosquito could initiate an *M. ulcerans* infection in a mouse-tail coated with *M. ulcerans* cells. In these experiments, the calculated effective inoculum size was in the range of just 2–3 cells [4].

Between 1948 and 1975 Radford reported 39 human cases of BU in Australia: 12 from Victoria including those in the original report, 4 from the Northern Territory (three from Croker Island and one from the coastal mainland near Darwin) and at least 20 from near Rockhampton and near Cairns in Queensland. A single case report from New South Wales (NSW) was published in 1954 but was thought likely to be an imported infection as the patient had epidemiological links to Papua New Guinea [2].

The Bairnsdale region in Victoria (latitude 38°S) is approximately 2000 km by road from Rockhampton in Queensland (latitude 23°S) and almost 3000 km from Cairns (latitude 17°S) (Fig. 1). There are large continuous human populations along the NSW Coast between the Victorian and Queensland endemic regions, yet locally acquired BU remains almost unknown there with the exception of one published confirmed case [5] and a small number of others from southern coastal NSW very close to the Victorian border. There has also been only one confirmed case from the northern tropical coast of Western Australia but no local transmission in the more populated temperate areas further south [6]. BU does not occur in South Australia or Tasmania, western Victoria or anywhere in the dry Australian interior. The reason for this patchwork focal distribution within and between states is unknown, but the geographic pattern has remained stable from the 1950s, except in Victoria.

All Australian isolates of *M. ulcerans* belong to the virulent so called "Classical Lineage" that also causes BU in Africa [7, 8], but there are small differences at the genomic level that allow location of Australian clinical isolates to specific geographic regions. Variable number of tandem repeat (VNTR) typing has been used to successfully attribute region of acquisition when newly diagnosed patients report multiple potential exposures [9]. More recently a detailed investigation using whole genome sequencing of 178 *M. ulcerans* isolates from different regions of Australia spanning 70 years has been conducted [10]. The results strongly support single introduction events into each of the major Australian endemic areas with subsequent local evolution. It also appears that *M. ulcerans* first reached Australia in the north, possibly from Papua New Guinea and has moved in large skip-steps down the east coast of Australia and along the Southern coast of Victoria as far as Melbourne. These introductions are historically recent and may have been assisted by coastal shipping activities down the east coast since European settlement. In Victoria movement from east to west along the southern mainland coast is continuing with the most recent recognised introduction event occurring in about 2003 to Rye on the Mornington peninsula [10] (Figs. 1 and 2).

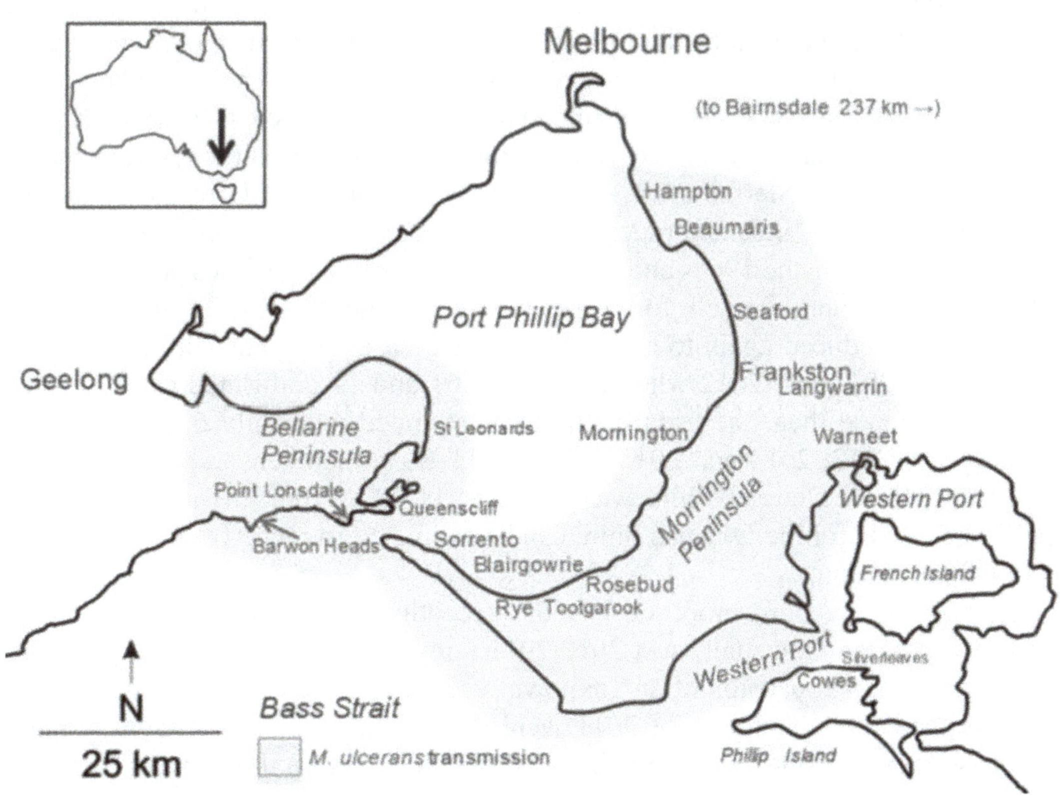

Fig. 2 Sketch map showing of southern approaches to the city of Melbourne including Port Phillip Bay, Western Port, the Mornington and Bellarine Peninsulas and towns mentioned in the text

2 Buruli Ulcer in Queensland

Buruli ulcer was recognised in Queensland from the early 1950s and possibly even earlier [3, 11]. In Queensland, the local names Daintree ulcer or Mossman ulcer have been used in preference to Bairnsdale ulcer [12]. To avoid confusion in this chapter, the term "Buruli ulcer" refers to all cases of *M. ulcerans* infection acquired in Australia, past and present.

Two separate endemic areas were recognised in Queensland from the early 1950s, one in the Douglas Shire just north of Cairns, particularly between Mossman and the Daintree River [13, 14], and a second on the Capricorn Coast near Rockhampton and Yeppoon [2, 11]. There have also been single cases or small clusters linked to Nambour near Brisbane, the Glass House Mountains (Sunshine Coast), Maryborough (Fraser Coast), Townsville [15, 16], and Port Douglas near Cairns and Mossman [2, 11, 16, 17]. However, due to the long incubation period of BU [18] it is difficult to confirm that any one location is endemic based on a single case unless a detailed travel history is obtained that excludes contact with all other regions or there is other evidence such as strain typing of the isolate to support geographic attribution. For most historical cases, these details are no longer available.

The most active recent Queensland focus of *M. ulcerans* transmission is in the far north of the state, from Mossman to the Daintree River in the Douglas Shire. The great majority of infections associated with this region have occurred in permanent residents rather than visitors [12]. In her Master of Science thesis published in 1996, May Smith, for many years Director of Nursing at Mossman Hospital, reported 41 cases of BU acquired in this small geographic region [13]. The first likely case occurred in 1952 and the first confirmed case in 1964. However, local aboriginal people had been aware of a disease resembling BU in this region for several generations. Subsequently, Steffen *et al.* reviewed five decades of BU in the far North Queensland endemic area and reported a gradual increase from approximately one case every 2 years in the 1960s and 1970s to approximately two cases per year in the 1980s rising to approximately four cases per year in the 2000s [12]. Subsequently the incidence reduced again to an average of two cases per year with the notable exception of 2011 and 2012 when there were 61 and 11 confirmed cases respectively [17]. Since then, the usual pattern has resumed with only 6, 2 and 1 new diagnoses in 2013, 2014 and 2015. In the 2011 Australian census, the permanent population of the Douglas Shire was just over 11,000. From this a crude annual incidence of BU for the Douglas Shire can be estimated to range from 9 to 550 per 100,000 population.

Steffen and Freeborn proposed that the exceptional year of 2011 followed an unusually long and unusually wet 2010/2011 rainy season that may have led to an abrupt temporary expansion of an unknown reservoir or vector. Interestingly, most cases in the exceptional year of 2011 were diagnosed in the dry season (winter) which is likely explained by a surge in transmission during their unusual wet season (summer) followed by a delay of several months attributable to the long incubation period of BU and an additional pre-diagnostic interval [18]. Despite the very different climate and geography, exactly the same temporal pattern is observed in temperate Victoria with peak transmission likely occurring in summer but most new diagnoses made in winter and early spring [19, 20].

3 Buruli Ulcer in Victoria

In the original definitive description of *Mycobacterium ulcerans* published by MacCallum et al. in 1948, 6 patients were described, 5 from "in and around" the town of Bairnsdale and a sixth case from Colac over 400 km to the west of Bairnsdale by road (Fig. 2). It was noted that the Bairnsdale cases were from separate households unknown to each other, suggesting from the outset chance exposure to a dispersed environmental pathogen rather than exposure to single point source or person to person transmission [3]. Also noted at the time was the wide age distribution with the disease affecting children and adults and both genders equally which has remained a consistent pattern in Victoria since then [21–23]. John Hayman, a pathologist with a long-standing interest in BU, reported 42 cases in Victoria, all from the general Bairnsdale region, including the original 5 from the 1930s to 1990. Notably only 14 of these were directly linked to the town of Bairnsdale itself, 8 were from the nearby lakeside hamlet of Loch Sport but others were from up to 50 km to the east of Bairnsdale. In seeking to understand the distribution he observed, Professor Hayman proposed proximity of cases to relict rain forest and suggested this was intermittently disturbed through fire or flooding leading to temporary seeding of lower lying areas near human habitation with *M. ulcerans*. He also proposed that *M. ulcerans* may be distributed through aerosols generated by wind action over water bodies that were supporting temporary blooms of *M. ulcerans* recently dislodged from its usual rainforest habitat upstream [24].

From 1990 there was a notable change in the epidemiology of BU in Victoria as *M. ulcerans* appeared to switch its behaviour from low level endemicity in a fixed geographic region to high-level transmission in completely new regions not previously endemic. Initially, a handful of new cases were identified by John Hayman around Tooradin and Warneet on the shores of Western Port, 240 km to the west of Bairnsdale [25] (Fig. 2). Next, at East Cowes (known as "Silverleaves") on Phillip Island a local General Practitioner, Dr. Paul Flood, recognised a cluster of unusual ulcers in patients in his practice from late 1992 which were subsequently confirmed by histology and culture to be caused by *M. ulcerans* [26]. From 1992–1995 there were at least 25 new cases, almost all within a very small region surrounding a golf course and a newly formed shallow lake that had developed following the construction of a fire access track [27]. The golf course was irrigated with treated recycled water purchased from the Phillip Island sewerage facility which was delivered to a permanent dam on the golf course and allowed to mix there with natural ground water before the dam contents were pumped to sprinklers on the greens and fairways. Victorian public health authorities arranged for improved drainage of the newly formed lake behind the access track in 1993 and altered the way recycled water was used on the golf course from 1995. This was followed by a sustained reduction in new cases linked to Phillip Island in subsequent years [21] and east Cowes/Silverleaves is currently disease free.

We attempted to support the hypothesis that the golf course irrigation system and/or shallow lake that formed behind the fire access track close to houses where people had acquired BU had become contaminated with *M. ulcerans* initially with direct culture of environmental samples and later by direct detection with PCR in water samples collected from the outbreak region. In a pilot study aimed at developing a PCR method to detect *M. ulcerans* in environmental samples we discovered the insertion sequence (IS) *2404* and developed an IS*2404*-based PCR assay in 1995 [28, 29]. We then investigated whether *M. ulcerans* was present in stored samples obtained from the dam and golf course irrigation system. In the first ever example of environmental detection of *M. ulcerans*, we identified PCR-positive water samples from the Cowes golf course pumping system and ground water that had collected behind the fire access track the previous year [28, 29]. Our attempts to directly culture *M. ulcerans* from environmental samples were unsuccessful.[1]

Clinical samples supplied by Professor John Hayman were crucial in the initial validation of the new PCR assay and it was quickly appreciated that IS*2404* PCR would be an excellent rapid diagnostic test for BU and soon become the diagnostic test of choice in Australia, due to its quick turn around and exceptional sensitivity and specificity. PCR based on the IS*2404* target is now also the gold standard for BU diagnosis in reference laboratories worldwide [16, 28, 30, 31].

The sudden appearance of a new focus of transmission at Cowes, Phillip Island, the high attack rate (up to 6% of the permanent population of east Cowes were affected) [21] and the recognition of infection in visitors who may have spent only brief periods in the outbreak area has become a hallmark of the new epidemiology of BU in Victoria since 1990. At the same time as the Phillip Island outbreak, there was a similar less intense outbreak over a slightly longer period (1990–1996) at Frankston/Langwarrin on the Mornington peninsula [25]. Dr. Mark Veitch, an epidemiological intelligence officer with the Department of Human Services in Victoria described the detailed epidemiology of the Cowes outbreak, and made an interesting link between the two outbreaks (Frankston/Langwarrin from 1990, East Cowes from 1992) noting that sand mined near Langwarrin had been used in the construction of a new road system in East Cowes (Silverleaves) just prior to the outbreak there [21].

Between 1995 and early 2000s there were very few cases of BU in humans in Victoria and the disease returned to its former obscurity. However, the situation changed again abruptly with large new outbreaks on the Bellarine Peninsula and most recently on the Mornington Peninsula (Fig. 2). At the time of writing (late 2017) there has been an exponential increase in cases in Victoria for the past 4 years (Fig. 3), and several new endemic areas have become established. The re-emergence began in 1998 when cases of BU were linked to St Leonards. St Leonards is a small town on the Bellarine Peninsula with a permanent population of 2480 (2016 Census) and like Cowes on Phillip Island also a popular summer holiday destination. The outbreak at

[1] This work was performed by scientists at the Mycobacterium Reference Laboratory, then located at Fairfield Infectious Diseases Hospital.

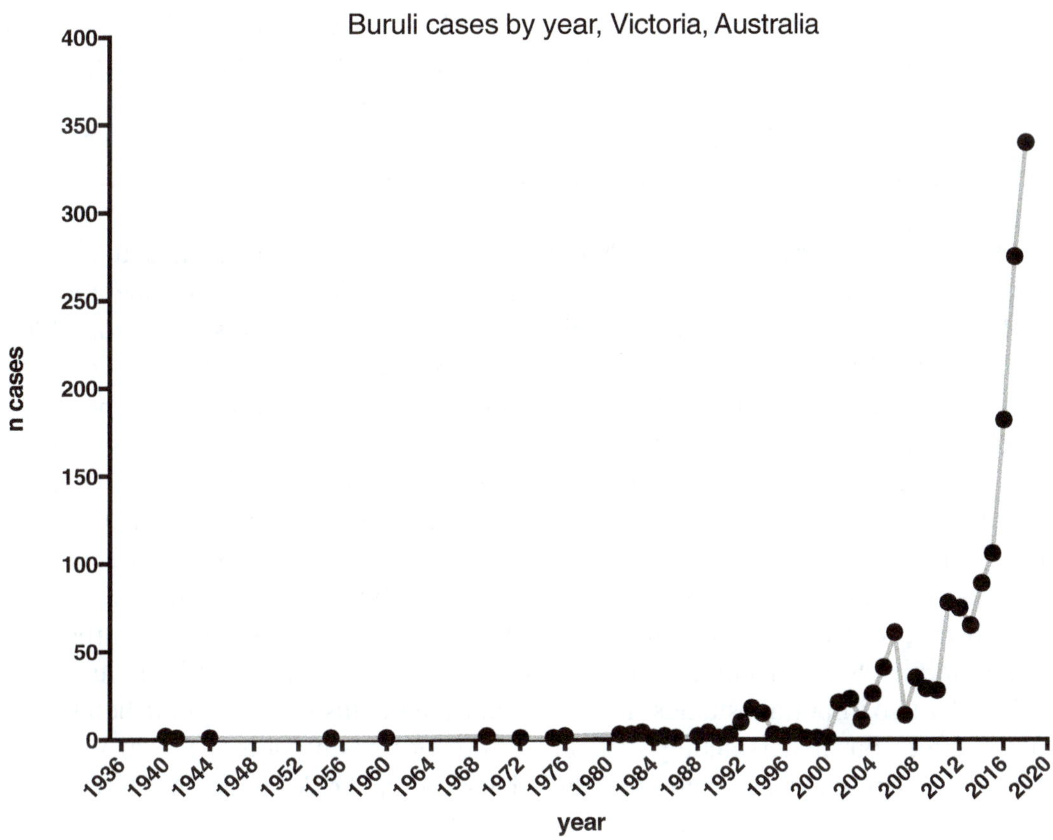

Fig. 3 Cases (n) of confirmed BU in Victoria by year 1940 to November 2017. Cases prior to 1992 courtesy private database created by Professor John Hayman

St. Leonards was investigated by the Victorian Department of Human Services Communicable Diseases Branch. There were two cases in 1998, one more in 2000 and then 11 diagnosed in 2001. Environmental PCR was attempted following the success at Phillip Island but no positive results were obtained from low-lying water sources or from the local golf course irrigation system (recycled water not used).

Cases have continued to occur at St Leonards since 1998 albeit at low intensity after 2001. Possible explanations to explain the outbreak at the time included much higher than usual rainfall and local reports of high mosquito numbers. A small proportion of mosquitoes trapped in CO_2 traps at St. Leonards were subsequently shown to harbour *M. ulcerans* DNA [32].

In 2002 new cases of BU abruptly appeared at Point Lonsdale, 20 km around the coast to the south of St Leonards (Fig. 2) heralding the onset of an intense, sustained outbreak that peaked in 2011 and is slowly abating now, but has been an important local public health issue at Point Lonsdale for 15 years [23]. Point Lonsdale is a small seaside resort town with a permanent population of around 2684 (2016 census) which increases significantly during the summer. The crude incidence in 2011 was estimated as 770/100,000 of the permanent population (C. Lavender *personal communication*). Many of the local population are retirees and a feature of the Point Lonsdale outbreak was the high attack rate in older people with up to 3.7% of all residents aged over 75 requiring treatment for BU [22]. A second notable feature at Point Lonsdale and elsewhere in Victoria since 1990 has been the high proportion of visitors who developed BU [22]. At Point Lonsdale in the first 2 years of the outbreak only local residents were affected but after 2004 this changed with both an increase in outbreak intensity and the appearance of disease in visitors and residents in almost equal proportions [22].

The changing epidemiology of BU in Victoria and concern from local councils and influential residents prompted the creation of a series of Victorian Government Public Health Grants to investigate the new epidemiological patterns of disease. From 2001 diagnostic PCR for BU became a standard diagnostic test performed at the Victorian Infectious Diseases Reference Laboratory, and from January 2004 in response to increased local concern BU became a legally notifiable infection in Victoria which has greatly aided mapping of cases and new endemic areas. In 2005 a new BU focus appeared at Barwon Heads (and the adjacent town of Ocean Grove) a further 10 km around the coast from Point Lonsdale (Fig. 2). From 2012 onwards new endemic foci have appeared along the Mornington Peninsula affecting towns including Sorrento, Blairgowrie, Rye Tootgarook, Mornington, Frankston and surrounding suburbs, and further north at Seaford and Beaumaris, just 24 km from the centre of Melbourne.

In 2007 two important studies from the Bellarine Peninsula were published which have changed our understanding of likely modes of transmission of BU in Victoria. Quek et al. conducted a case control study that examined risk factors in 49 cases and 609 controls. The key new finding of this research was a statistically significant association between mosquito bites and risk of BU. In their final multivariate model being bitten by mosquitoes was found to increase risk (odds ratio 2.60 [95% c.i. 1.22–5.53]) and use of insect repellent to reduce risk (odds ratio 0.37 [95% c.i. 0.19–0.69]) [33]. In the second study conducted at Point Lonsdale between 2004 and 2007, 11,500 mosquitoes were trapped of which approximately 4/1000 were PCR-positive for IS*2404* and in a subset of samples there was molecular evidence that we were detecting the human outbreak strain of *M. ulcerans* [22, 30]. At the time of writing the detection of *M. ulcerans* in several species of mosquitoes has been repeatedly confirmed in Victoria [32] but similar studies have so far been negative in Africa [34].

4 Buruli Ulcer in Animals in Victoria

During the 1980s, in the Bairnsdale region of Victoria, *M. ulcerans* infection was identified in 11 Koalas (arboreal marsupials) from a population of about 200 animals on Raymond Island, just a few km southeast of Bairnsdale [35–38]. This was the first recognition of naturally occurring BU in any species other than humans anywhere. The affected animals were mature, usually male and it was suggested that lesion distribution was consistent with wounds acquired during social behaviour such as fighting.

In the late 1990s after the decline in human cases at Phillip Island, several sick adult possums (also native arboreal marsupials) were detected at Phillip Island with ulcerative disease, at least three of which were confirmed to have *M. ulcerans* infection by histology and PCR [35]. Although brushtail possums had previously been shown to be susceptible under experimental conditions to *M. ulcerans* infection and to transmit the infection to other co-housed individuals [39], this was the first recognition of natural infection in possums. There is a koala reserve at Phillip Island just 2 km from the endemic area in east Cowes, yet no cases of BU in this koala population were observed despite notification to local wild life officers of the presence of the disease in local humans and possums.

Subsequently, while seeking possible environmental sources of mosquito exposure to *M. ulcerans* we identified positive PCR signals in environmental samples at Point Lonsdale, with by far the strongest signals arising from possum excreta. This led to a systematic investigation of the possible role that possums may play as a reservoir and environmental amplifier of *M. ulcerans* in Victoria. We performed environmental surveys and trapped and screened 63 possums (42 ringtail, 21 brushtail). During the surveys, we discovered that 42% of possum excreta samples collected at Point Lonsdale were strongly PCR-positive for *M. ulcerans* compared with <1% in non-endemic areas. Further, 9/42 trapped ringtail possums had clinical BU lesions confirmed by PCR as did 1/21 brushtail possums. Additional trapped animals without clinical lesions were found to be excreting high levels of *M. ulcerans* DNA in their faeces. This research confirmed the validity of using possum excreta surveys as a proxy for mapping the occurrence of BU in possum populations, and for the first time suggested that BU may be a zoonosis with humans acting as spill-over hosts connected directly or indirectly to possums via mosquitoes [40]. In 2011, just prior to the surge in new cases in Victoria on the Mornington Peninsula, Carson *et al.* validated this model in a new endemic area by showing that possum excreta at Sorrento and Blairgowrie on the Mornington peninsula was strongly PCR-positive close to the location of new human cases of BU [41]. Interestingly, no analogous small animal reservoir has yet been identified in Africa [42] but there is recent evidence that bandicoots[2] in the far north Queensland endemic focus of BU also excrete *M. ulcerans* in their faeces [43].

In Victoria, but not elsewhere in Australia several other naturally acquired BU infections have been reported in animals including at least three species of possums [44]. On the Bellarine Peninsula four domestic dogs have been diagnosed with BU [45], there has been a confirmed case in a cat from eastern Victoria [46], a long footed potoroo (small terrestrial marsupial) from eastern Victoria [46], two horses from eastern Victoria and at least three alpacas (Bellarine Peninsula and eastern Victoria) [35, 44–47]. Notably these animal infections have occurred across the same geographical region that is endemic for humans in Victoria (Fig. 1). Given the relatively wide experimental host range of *M. ulcerans* (lizards, amphibians, chick embryos, possums, armadillos, rats, mice, rabbits, guinea pigs, pigs and cattle [35, 48, 49]) it is surprising that naturally acquired disease in animals appears to be relatively rare and restricted. The failure (so far) to identify natural infections in animals outside Victoria is also surprising and remains unexplained. Possible explanations include the particular susceptibility of humans (and possums), something specific about Victorian strains of *M. ulcerans* compared with Queensland and Northern

[2] Bandicoots are small to medium sized native terrestrial marsupials.

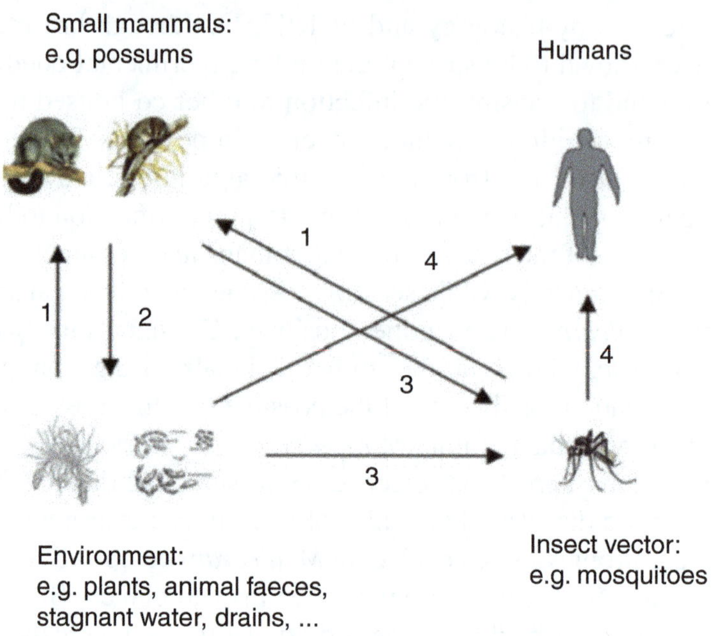

Fig. 4 Proposed transmission pathways of *M. ulcerans* between the environment, mosquitoes, possums and humans in Victoria Australia PLOS Neglected Tropical Diseases 2010 doi:https://doi.org/10.1371/journal.pntd.0000791.g005)

Territory strains [6], or that the concentration of *M. ulcerans* that accumulates in the environment in Victoria is higher than elsewhere. It is also possible that the disease does occasionally occur in animals outside Victoria but is not recognised or not brought to the attention of veterinarians familiar with the disease (Fig. 4).

5 Recent Epidemiology of Buruli Ulcer in Victoria

The rapid expansion of BU endemic areas and recent exponential increase in numbers of cases in humans is unprecedented and a significant cause of local concern. There is published and unpublished evidence that possums carry and excrete *M. ulcerans* DNA in high concentration in faeces at Point Lonsdale, Barwon Heads, Sorrento, Blairgowrie, Beaumaris and near Rosebud - all places where human BU has become endemic since 2002 (Fig. 2). Similar surveys outside endemic areas yield negative results [40]. The frequent acquisition of BU by visitors to endemic areas in Victoria as well as local people in almost equal proportion suggests that transmission of *M. ulcerans* is a chance event and that risk may not be present throughout the year. Possums are likely to carry *M. ulcerans* disease over prolonged periods yet from what we understand so far, the risk of acquiring infection appears to occur mainly in summer for both residents and visitors as cases in both groups present to doctors at similar times (winter and spring). This may be explained by an increase in vector activity in warmer weather [22] combined with greater exposure to the outdoor environment and/or less use of protective clothing. Interviews with visitors have identified very short exposure periods in some cases which have allowed us to estimate a median incubation period of 4.5 months but a range up to 9 months. A recent study of 649 BU lesions in 579 patients identified a highly non-random distribution with BU lesions in Victoria preferentially occurring

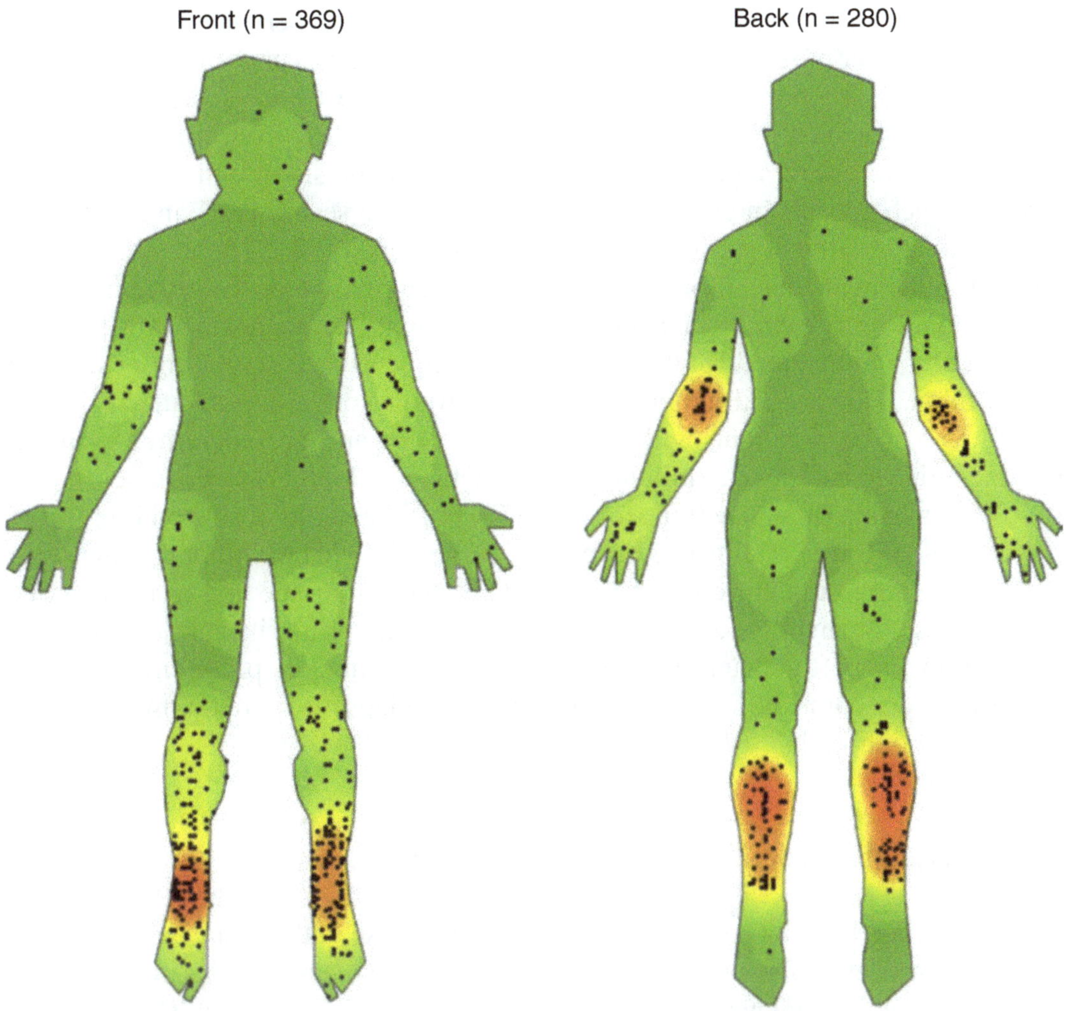

Fig. 5 Lesion distribution from 649 BU lesions occurring in 579 patients, Victoria, Australia. PLOS Neglected Tropical Diseases 2017: doi: https://doi.org/10.1371/journal.pntd.0005800.g002

on ankles, back of calves, forearms and elbows but rarely on the soles of feet, palms of hands or parts of the body that usually remain covered [19] (Fig. 5). One explanation for this distribution is targeting behaviour by biting insects although it is possible that there is more than one mode of transmission. A recent study of family clusters of BU in Victoria has suggested that exposure risk per household lasts only a short time and provides molecular typing evidence to support the long-standing epidemiological observation that *M. ulcerans* is not transmitted between individuals within a household [50].

6 Clinical Management of Buruli Ulcer in Australia

The large number of recent cases in Australia, particularly in Victoria and the relative complexity of treating BU has meant that a small group of Infectious Diseases Physicians and Surgeons have greatly increased their clinical experience and understanding of the management of BU. Much of this experience has been captured in a number of recent publications. We are fortunate to have universal healthcare coverage and rapid access to doctors and diagnostic tests. Many recently diagnosed cases are WHO category I (<5 cm), nevertheless from time to time we also see severe cases, sometimes due to delayed diagnosis or acute oedematous disease which appears in every respect to be similar to severe cases in Africa [19, 51]. National treatment guidelines have been developed [52, 53] and there is now general agreement that all-oral antibiotic regimens based on rifampicin with a companion drug (generally clarithromycin) are highly active against *M. ulcerans* [14, 54, 55], and that medical therapy alone is frequently curative without surgical intervention. However, drug therapy is not straightforward particularly in the elderly [51, 56] and patients with larger lesions often appear to deteriorate during treatment due to paradoxical inflammatory reactions [57, 58]. Short courses of oral steroids may help these reactions to settle [59, 60]. The optimal role and timing of surgery has still to be established but clearly surgery has an important role in assisting with healing through removing extensively necrotic tissue and to repair large skin defects. For small lesions the decision to treat with primarily surgical or medical therapy is partly determined by patient and clinician preference as both approaches are curative in the majority of cases. Relapse is more common if antibiotics are not used [17, 61, 62].

7 Buruli Ulcer: The Australian Paradox

BU is classified by WHO as a neglected tropical disease with most cases occurring in poor subsistence farming families in tropical river land regions in West and Central Africa. For unknown reasons rates of BU in Africa are now static or may even be in decline after significant epidemics during the past 20–30 years. In contrast, the south-eastern Australian state of Victoria is temperate and economically developed, yet case numbers are exponentially increasing. Most people affected live in or visit affluent coastal resort towns that have become newly endemic. Despite the contrast in epidemiology between Africa and Victoria the disease is quite similar in its clinical appearance, the suffering it causes and the complexity of treatment of severe forms of BU [63, 64].

References

1. Buckle G (1969) Notes on mycobacterium ulcerans. Aust N Z J Surg 38(4):320–323
2. Radford AJ (1975) Mycobacterium ulcerans in Australia. Aust NZ J Med 5(2):162–169
3. MacCallum P, Buckle G, Tolhurst JC, Sissons HA (1948) A new mycobacterial infection in man. J Pathol Bacteriol 60(1):93–122
4. Wallace JR, Mangas KM, Porter JL, Marcsisin R, Pidot SJ, Howden B et al (2017) Mycobacterium ulcerans low infectious dose and mechanical transmission support insect bites and puncturing injuries in the spread of Buruli ulcer. PLoS Negl Trop Dis 11(4):e0005553. https://doi.org/10.1371/journal.pntd.0005553

5. Lavender CJ, Senanayake SN, Fyfe JA, Buntine JA, Globan M, Stinear TP et al (2007) First case of Mycobacterium ulcerans disease (Bairnsdale or Buruli ulcer) acquired in New South Wales. Med J Aust 186(2):62–63
6. Lavender CJ, Stinear TP, Johnson PD, Azuolas J, Benbow ME, Wallace JR et al (2008) Evaluation of VNTR typing for the identification of Mycobacterium ulcerans in environmental samples from Victoria, Australia. FEMS Microbiol Lett 287(2):250–255. https://doi.org/10.1111/j.1574-6968.2008.01328.x
7. Vandelannoote K, Meehan CJ, Eddyani M, Affolabi D, Phanzu DM, Eyangoh S et al (2017) Multiple Introductions and Recent Spread of the Emerging Human Pathogen Mycobacterium ulcerans across Africa. Genome Biol Evol 9(3):414–426. https://doi.org/10.1093/gbe/evx003
8. Kaser M, Rondini S, Naegeli M, Stinear T, Portaels F, Certa U et al (2007) Evolution of two distinct phylogenetic lineages of the emerging human pathogen Mycobacterium ulcerans. BMC Evol Biol 7:177. https://doi.org/10.1186/1471-2148-7-177
9. Lavender CJ, Globan M, Johnson PD, Charles PG, Jenkin GA, Ghosh N et al (2012) Buruli ulcer disease in travelers and differentiation of Mycobacterium ulcerans strains from northern Australia. J Clin Microbiol 50(11):3717–3721. https://doi.org/10.1128/JCM.01324-12
10. Buultjens A, Vandelannoote K, Meehan CJ, Eddyani M, de Jong B, Fyfe JA et al (2018) Comparative genomics shows Mycobacterium ulcerans migration and expansion has preceded the rise of Buruli ulcer in south-eastern Australia. Appl Environ Microbiol 84(8):e02612–e02617
11. Francis G, Whitby M, Woods M (2006) Mycobacterium ulcerans infection: a rediscovered focus in the Capricorn Coast region of central Queensland. Med J Aust 185(3):179–180
12. Steffen CM, Smith M, McBride WJ (2010) Mycobacterium ulcerans infection in North Queensland: the 'Daintree ulcer'. ANZ J Surg 80(10):732–736. https://doi.org/10.1111/j.1445-2197.2010.05338.x
13. Smith M (1996) Studies on Mycobacterium ulcerans infection in the Douglas Shire of far North Queensland, Australia. James Cook University, Queensland
14. Jenkin GA, Smith M, Fairley M, Johnson PD (2002) Acute, oedematous Mycobacterium ulcerans infection in a farmer from far north Queensland. Med J Aust 176(4):180–181
15. Ramakrishnan A, Johnson PD, Hayman J, Coombs C (2004) Free rectus flap repair of cutaneous Mycobacterium ulcerans ulcer with joint involvement. ANZ J Surg 74(7):608–611. https://doi.org/10.1111/j.1445-2197.2004.03079.x
16. Russell FM, Starr M, Hayman J, Curtis N, Johnson PD (2002) Mycobacterium ulcerans infection diagnosed by polymerase chain reaction. J Paediatr Child Health 38(3):311–313
17. Steffen CM, Freeborn H (2016) Mycobacterium ulcerans in the Daintree 2009-2015 and the mini-epidemic of 2011. ANZ J Surg 88(4):E289–E293. https://doi.org/10.1111/ans.13817
18. Trubiano JA, Lavender CJ, Fyfe JA, Bittmann S, Johnson PD (2013) The incubation period of Buruli ulcer (Mycobacterium ulcerans infection). PLoS Negl Trop Dis 7(10):e2463. https://doi.org/10.1371/journal.pntd.0002463
19. Yerramilli A, Tay EL, Stewardson AJ, Kelley PG, Bishop E, Jenkin GA et al (2017) The location of Australian Buruli ulcer lesions-Implications for unravelling disease transmission. PLoS Negl Trop Dis 11(8):e0005800. https://doi.org/10.1371/journal.pntd.0005800
20. Loftus MJ, Tay EL, Globan M, Lavender CJ, Crouch SR, Johnson PDR, Fyfe JAM (2018) Epidemiology of Buruli ulcer infections, Victoria, Australia, 2011–2016. Emerg Infect Dis 24:1988–1997
21. Veitch MG, Johnson PD, Flood PE, Leslie DE, Street AC, Hayman JA (1997) A large localized outbreak of Mycobacterium ulcerans infection on a temperate southern Australian island. Epidemiol Infect 119(3):313–318
22. Johnson PD, Azuolas J, Lavender CJ, Wishart E, Stinear TP, Hayman JA et al (2007) Mycobacterium ulcerans in mosquitoes captured during outbreak of Buruli ulcer, southeastern Australia. Emerg Infect Dis 13(11):1653–1660. https://doi.org/10.3201/eid1311.061369
23. Boyd SC, Athan E, Friedman ND, Hughes A, Walton A, Callan P et al (2012) Epidemiology, clinical features and diagnosis of Mycobacterium ulcerans in an Australian population. Med J Aust 196(5):341–344

24. Hayman J (1991) Postulated epidemiology of Mycobacterium ulcerans infection. Int J Epidemiol 20(4):1093–1098
25. Johnson PD, Veitch MG, Leslie DE, Flood PE, Hayman JA (1996) The emergence of Mycobacterium ulcerans infection near Melbourne. Med J Aust 164(2):76–78
26. Flood P, Street A, O'Brien P, Hayman J (1994) Mycobacterium ulcerans infection on Phillip Island, Victoria. Med J Aust 160(3):160
27. Johnson PD, Veitch MG, Flood PE, Hayman JA (1995) Mycobacterium ulcerans infection on Phillip Island, Victoria. Med J Aust 162(4):221–222
28. Ross BC, Marino L, Oppedisano F, Edwards R, Robins-Browne RM, Johnson PD (1997) Development of a PCR assay for rapid diagnosis of Mycobacterium ulcerans infection. J Clin Microbiol 35(7):1696–1700
29. Ross BC, Johnson PD, Oppedisano F, Marino L, Sievers A, Stinear T et al (1997) Detection of Mycobacterium ulcerans in environmental samples during an outbreak of ulcerative disease. Appl Environ Microbiol 63(10):4135–4138
30. Fyfe JA, Lavender CJ, Johnson PD, Globan M, Sievers A, Azuolas J et al (2007) Development and application of two multiplex real-time PCR assays for the detection of Mycobacterium ulcerans in clinical and environmental samples. Appl Environ Microbiol 73(15):4733–4740. https://doi.org/10.1128/AEM.02971-06
31. Eddyani M, Lavender C, de Rijk WB, Bomans P, Fyfe J, de Jong B et al (2014) Multicenter external quality assessment program for PCR detection of Mycobacterium ulcerans in clinical and environmental specimens. PLoS One 9(2):e89407. https://doi.org/10.1371/journal.pone.0089407
32. Lavender CJ, Fyfe JA, Azuolas J, Brown K, Evans RN, Ray LR et al (2011) Risk of Buruli ulcer and detection of Mycobacterium ulcerans in mosquitoes in southeastern Australia. PLoS Negl Trop Dis 5(9):e1305. https://doi.org/10.1371/journal.pntd.0001305
33. Quek TY, Athan E, Henry MJ, Pasco JA, Redden-Hoare J, Hughes A et al (2007) Risk factors for Mycobacterium ulcerans infection, southeastern Australia. Emerg Infect Dis 13(11):1661–1666. https://doi.org/10.3201/eid1311.061206
34. Djouaka R, Zeukeng F, Daiga Bigoga J, N'Golo Coulibaly D, Tchigossou G, Akoton R et al (2017) Evidences of the low implication of mosquitoes in the transmission of Mycobacterium ulcerans, the causative agent of Buruli ulcer. Can J Infect Dis Med Microbiol 2017:1324310. https://doi.org/10.1155/2017/1324310
35. Portaels F, Chemlal K, Elsen P, Johnson PD, Hayman JA, Hibble J et al (2001) Mycobacterium ulcerans in wild animals. Rev Sci Tech 20(1):252–264
36. Mitchell PJ, McOrist S, Bilney R (1987) Epidemiology of Mycobacterium ulcerans infection in koalas (Phascolarctos cinereus) on Raymond Island, southeastern Australia. J Wildl Dis 23(3):386–390
37. McOrist S, Jerrett IV, Anderson M, Hayman J (1985) Cutaneous and respiratory tract infection with Mycobacterium ulcerans in two koalas (Phascolarctos cinereus). J Wildl Dis 21(2):171–173
38. Mitchell PJ, Jerrett IV, Slee KJ (1984) Skin ulcers caused by Mycobacterium ulcerans in koalas near Bairnsdale, Australia. Pathology 16(3):256–260
39. Bolliger A, Forbes BRV, Kirkland WB (1950) Transmission of a recently isolated mycobacterium to phalangers (Trichosurus vulpecula). Aust J Sci 12(4):146
40. Fyfe JA, Lavender CJ, Handasyde KA, Legione AR, O'Brien CR, Stinear TP et al (2010) A major role for mammals in the ecology of Mycobacterium ulcerans. PLoS Negl Trop Dis 4(8):e791. https://doi.org/10.1371/journal.pntd.0000791
41. Carson C, Lavender CJ, Handasyde KA, O'Brien CR, Hewitt N, Johnson PD et al (2014) Potential wildlife sentinels for monitoring the endemic spread of human buruli ulcer in South-East australia. PLoS Negl Trop Dis 8(1):e2668. https://doi.org/10.1371/journal.pntd.0002668
42. Durnez L, Suykerbuyk P, Nicolas V, Barriere P, Verheyen E, Johnson CR et al (2010) Terrestrial small mammals as reservoirs of Mycobacterium ulcerans in benin. Appl Environ Microbiol 76(13):4574–4577. https://doi.org/10.1128/AEM.00199-10

43. Roltgen K, Pluschke G, Johnson PDR, Fyfe J (2017) Mycobacterium ulcerans DNA in Bandicoot Excreta in Buruli Ulcer-Endemic Area, Northern Queensland, Australia. Emerg Infect Dis 23(12):2042–2045. https://doi.org/10.3201/eid2312.170780
44. O'Brien CR, Handasyde KA, Hibble J, Lavender CJ, Legione AR, McCowan C et al (2014) Clinical, microbiological and pathological findings of Mycobacterium ulcerans infection in three Australian Possum species. PLoS Negl Trop Dis 8(1):e2666. https://doi.org/10.1371/journal.pntd.0002666
45. O'Brien CR, McMillan E, Harris O, O'Brien DP, Lavender CJ, Globan M et al (2011) Localised Mycobacterium ulcerans infection in four dogs. Aust Vet J 89(12):506–510. https://doi.org/10.1111/j.1751-0813.2011.00850.x
46. Elsner L, Wayne J, O'Brien CR, McCowan C, Malik R, Hayman JA et al (2008) Localised Mycobacterium ulcerans infection in a cat in Australia. J Feline Med Surg 10(4):407–412. https://doi.org/10.1016/j.jfms.2008.03.003
47. O'Brien C, Kuseff G, McMillan E, McCowan C, Lavender C, Globan M et al (2013) Mycobacterium ulcerans infection in two alpacas. Aust Vet J 91(7):296–300. https://doi.org/10.1111/avj.12071
48. George KM, Chatterjee D, Gunawardana G, Welty D, Hayman J, Lee R et al (1999) Mycolactone: a polyketide toxin from Mycobacterium ulcerans required for virulence. Science 283(5403):854–857
49. Bolz M, Ruggli N, Ruf MT, Ricklin ME, Zimmer G, Pluschke G (2014) Experimental infection of the pig with Mycobacterium ulcerans: a novel model for studying the pathogenesis of Buruli ulcer disease. PLoS Negl Trop Dis 8(7):e2968. https://doi.org/10.1371/journal.pntd.0002968
50. O'Brien DP, Wynne JW, Buultjens AH, Michalski WP, Stinear TP, Friedman ND et al (2017) Exposure risk for infection and lack of human-to-human transmission of Mycobacterium ulcerans disease, Australia. Emerg Infect Dis 23(5):837–840. https://doi.org/10.3201/eid2305.160809
51. O'Brien DP, Friedman ND, Cowan R, Pollard J, McDonald A, Callan P et al (2015) Mycobacterium ulcerans in the elderly: more severe disease and suboptimal outcomes. PLoS Negl Trop Dis 9(12):e0004253. https://doi.org/10.1371/journal.pntd.0004253
52. Johnson PD, Hayman JA, Quek TY, Fyfe JA, Jenkin GA, Buntine JA et al (2007) Consensus recommendations for the diagnosis, treatment and control of Mycobacterium ulcerans infection (Bairnsdale or Buruli ulcer) in Victoria, Australia. Med J Aust 186(2):64–68
53. O'Brien DP, Jenkin G, Buntine J, Steffen CM, McDonald A, Horne S et al (2014) Treatment and prevention of Mycobacterium ulcerans infection (Buruli ulcer) in Australia: guideline update. Med J Aust 200(5):267–270
54. O'Brien DP, Hughes AJ, Cheng AC, Henry MJ, Callan P, McDonald A et al (2007) Outcomes for Mycobacterium ulcerans infection with combined surgery and antibiotic therapy: findings from a south-eastern Australian case series. Med J Aust 186(2):58–61
55. Gordon CL, Buntine JA, Hayman JA, Lavender CJ, Fyfe JA, Hosking P et al (2010) All-oral antibiotic treatment for buruli ulcer: a report of four patients. PLoS Negl Trop Dis 4(11):e770. https://doi.org/10.1371/journal.pntd.0000770
56. O'Brien DP, Friedman ND, Hughes A, Walton A, Athan E (2017) Antibiotic complications during the treatment of mycobacterium ulcerans disease in Australian patients. Intern Med J 47(9):1011–1019. https://doi.org/10.1111/imj.13511
57. O'Brien DP, Robson ME, Callan PP, McDonald AH (2009) "Paradoxical" immune-mediated reactions to Mycobacterium ulcerans during antibiotic treatment: a result of treatment success, not failure. Med J Aust 191(10):564–566
58. O'Brien DP, Robson M, Friedman ND, Walton A, McDonald A, Callan P et al (2013) Incidence, clinical spectrum, diagnostic features, treatment and predictors of paradoxical reactions during antibiotic treatment of Mycobacterium ulcerans infections. BMC Infect Dis 13:416. https://doi.org/10.1186/1471-2334-13-416
59. Trevillyan JM, Johnson PD (2013) Steroids control paradoxical worsening of Mycobacterium ulcerans infection following initiation of antibiotic therapy. Med J Aust 198(8):443–444

60. Friedman ND, McDonald AH, Robson ME, O'Brien DP (2012) Corticosteroid use for paradoxical reactions during antibiotic treatment for Mycobacterium ulcerans. PLoS Negl Trop Dis 6(9):e1767. https://doi.org/10.1371/journal.pntd.0001767
61. Steffen CM (2014) Risk factors for recurrent Mycobacterium ulcerans disease after exclusive surgical treatment in an Australian cohort. Med J Aust 200(2):85–86
62. Friedman ND, Athan E, Hughes AJ, Khajehnoori M, McDonald A, Callan P et al (2013) Mycobacterium ulcerans disease: experience with primary oral medical therapy in an Australian cohort. PLoS Negl Trop Dis 7(7):e2315. https://doi.org/10.1371/journal.pntd.0002315
63. Tai AYC, Athan E, Friedman ND, Hughes A, Walton A, O'Brien DP (2018) Increased severity and spread of Mycobacterium ulcerans, Southeastern Australia. Emerg Infect Dis 24(1):58. https://doi.org/10.3201/eid2401.171070
64. Loftus MJ, Kettleton-Butler N, Wade D, Whitby RM, Johnson PD. A severe case of Mycobacterium ulcerans (Buruli ulcer) osteomyelitis requiring a below-knee amputation. Med J Aust 2018;208:290–1

Mycobacterium ulcerans Infection in French Guiana; Current State of Knowledge

Pierre Couppié, Romain Blaizot, Camilla J. Velvin, Maylis Douine, Marine Combe, Mathieu Nacher, and Rodolphe E. Gozlan

1 Introduction

M. ulcerans infections (Buruli ulcer; BU) have been reported in French Guiana since the 1960s [1]. It is striking to see that French Guiana concentrates most of the cases of BU in America with only a few cases reported in Suriname, Peru, Mexico and Bolivia in the past [2, 3]. Either BU is underdiagnosed and underreported in most of the continent; or the specific environmental conditions in French Guiana represent a unique niche for the disease to appear; or a sublineage of *M. ulcerans* has established in French Guiana that has a higher virulence than other members of the ancestral lineage found in the Americas.

French Guiana is a tropical country located in South America on the Atlantic coast between Suriname and northern Brazil. It is mostly covered by Amazonian rainforest with the exception of the coastal strip, which is mainly composed of marshy savannah and mangroves. Most inhabitants reside along the coastal strip. There are several communities living in French Guiana, including Creoles, Maroons, Amerindians, mainland French people, Asians and many immigrants from South America and the Caribbean Islands. Since the early 1970s, French Guiana has seen

P. Couppié (✉) · R. Blaizot
Service de dermatologie, Cayenne Hospital, Cayenne Cedex, French Guiana

Université de Guyane, EA3593 Ecosystèmes Amazoniens et Pathologie Tropicale, Cayenne, French Guiana
e-mail: pierre.couppie@ch-cayenne.fr

C. J. Velvin · M. Combe · R. E. Gozlan
UMR ISEM, Université de Montpellier, CNRS, IRD, EPHE, Montpellier, France

M. Douine · M. Nacher
Université de Guyane, EA3593 Ecosystèmes Amazoniens et Pathologie Tropicale, Cayenne, French Guiana

Centre d'Investigation Clinique, CIC Inserm 1424, Cayenne Hospital, Cayenne Cedex, French Guiana

G. Pluschke, K. Röltgen (eds.), *Buruli Ulcer*,
https://doi.org/10.1007/978-3-030-11114-4_4

very rapid demographic and socioeconomic changes, with a population that increased from 45,000 in 1969 to 260,000 in 2017 and a gross domestic product (GDP) that grew by 7.2% per year between 1960 and 2002. Although the territory has developed rapidly, with improving living conditions such as better hygiene, gradual elimination of intestinal nematodes and easier access to health care, several migration waves from poor regions of Southern and Central America have led to pockets of poverty.

Here, we review current knowledge of BU in French Guiana with specific emphasis on links of *M. ulcerans* to the human host and the natural environment.

2 Links Between the Human Host and the Bacteria

2.1 Risk Factors of BU in French Guiana

A case-control study was conducted between 2002 and 2004 to determine relationships between the presence of *M. ulcerans* in the environment and human activities associated with the occurrence of BU in French Guiana [4]. The study involved 30 laboratory confirmed BU cases and 60 controls matched for age, gender and town of residence. Increased risk of contracting the disease was strongly associated with environmental factors and human behavioural factors. The environmental factors implicated in the transmission of *M. ulcerans* were proximity of the home or workplace of patients to natural freshwater habitats (e.g. marshes, rivers, and floodable areas). In addition, some behavioural factors such as hunting, fishing, recreational activity, and contact with river banks were associated with BU cases. These results are consistent with previous studies performed in Africa and Australia that have shown that people who are in close proximity to water sources or take part in outdoor activities are at higher risk of contracting BU. These findings suggest that informal professional or recreational activities like hunting or fishing and contact with those water sources are risk factors for contracting BU in French Guiana (Fig. 1).

2.2 Epidemiology and Clinical Aspects of BU in French Guiana

The epidemiology of BU in French Guiana since the first diagnosed case in 1969 was recently studied with the aim of characterising the incidence of the disease in the past 45 years, the temporal demographic trends of patients, and the associated clinical aspects of the disease [1]. Data were collected at Cayenne Hospital's dermatology department where 245 BU patients were reported between 1969 and 2013. Most cases occurred in coastal areas surrounded by marshy savannah. The highest number of cases was reported in Cayenne (47% of diagnosed cases), followed by Mana (20%), Sinnamary (16%), Kourou (11%), and forest areas (6%). The annual number of cases varied widely, with no cases in some years and up to 27 in others. Over the 45 years, the mean annual number of new cases was 5.4 per year and the

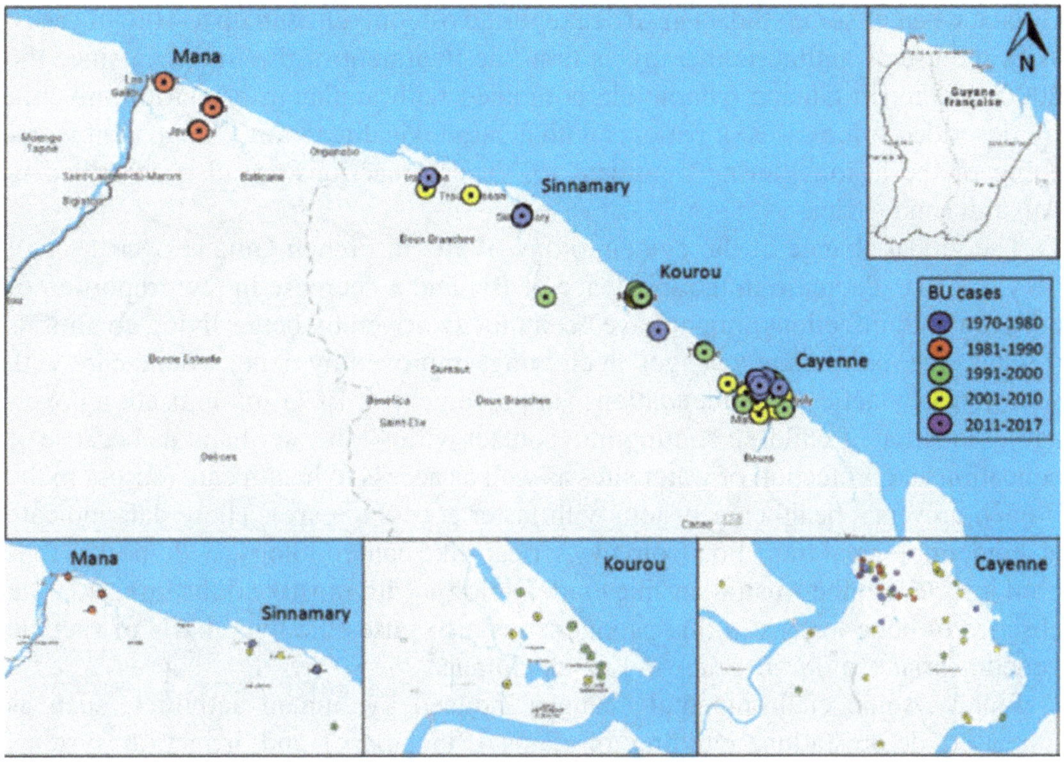

Fig. 1 Location of BU cases in French Guiana identified between 1969 and 2016. Precise geographic coordinates were available for 94 cases only. BU cases were sorted by periods of 10 years that are represented by different colors in the map. Top: location of the 94 BU cases along the coastline. Bottom: zoom on the Mana and Sinnamary region, Kourou and Cayenne city. The map was created online with BatchGeo

mean annual incidence rate was 4.3 per 100,000. However, the annual incidence of BU significantly decreased over time across all districts, from 6.1 infections per 100,000 persons in 1969–83 to 3.5 infections per 100,000 persons in 1999–2013. The highest mean annual incidence rate over the 45 years was found in the western coastal area around Sinnamary (21.1 per 100,000 persons) and Mana (21.2 per 100,000 persons). Annual incidence peaks were observed in the Sinnamary area between 1974 and 1978 (71.8 cases per 100,000 persons) and another in the Mana area between 1984 and 1988 (101.4 cases per 100,000 persons). Over time, the mean age of patients with BU increased significantly in French Guiana between 1969 and 2013. The proportion of children diagnosed with BU before 1984 was similar to that in Africa (more than 70%), but decreased with time, reaching around 21% in 1999–2013, which is closer to the proportion observed in Australia. Men and women were similarly affected in French Guiana, as observed in other countries and lesions were usually ulcerated (92%) mainly with single ulcers with undermined edges. Nodules and plaques were less frequent, sometimes ulcerated. Only 8% of the patients had multiple lesions. Lesions mainly affected limbs (lower limbs in 66% of patients; upper limbs in 24% of patients), which is similar to settings in Africa and Australia. Adults were significantly more often affected on the ankles than were children (14% vs 4%). No bone infections were diagnosed in French

Guiana whereas some studies in Africa reported osteomyelitis in up to 10% of cases. With the use of antibiotic therapy as first-line treatment of the infection since the 2000s in French Guiana (rifampicin combined with amikacin or clarithromycin), we described some cases of patients with a paradoxical reaction, an apparent worsening of the lesions during treatment [5]. These reactions are also described in Australia and Africa.

The global change of the epidemiology of BU in French Guiana over the past 45 years, (i.e. decrease in the incidence of BU and a decrease in the proportion of children with infections) might have been mostly driven by better living conditions (higher GDP per person, changes in clothing, improved hygiene, wound care with soap), prophylactic recommendations (improvement of BCG immunisation coverage), reduction of children coming into contact with freshwater habitats because of education and protection of water sites as well as access to health care (access to the French universal healthcare system with faster access to care). These data indicate an epidemiological transition from the African-like epidemiology, with mostly children infected, to the Australian-like epidemiology, with mostly adults infected. The absence of bone lesions in this patient series also raises the hypothesis of specific genetic variants of *M. ulcerans* in French Guiana.

Finally, some environmental changes induced by human activities, such as increased deforestation, building of dams (Sinnamary) and irrigation systems (Mana) that deeply modified the floodplain structure as well as modifications in rainfall patterns due to global climate change could also have influenced the change of BU incidence in French Guiana.

2.3 A Link Between Rainfall and BU Cases

Links between changes in rainfall and outbreaks of BU in French Guiana were studied from 1969–2012 [6]. During the study period, four inter-annual peaks in rainfall, each followed by three inter-annual periods of rainfall recessions were observed, with three corresponding peaks and recessions of BU cases. Intra-annual analysis showed bi-annual peaks in the number of new BU cases; after a peak in rainfall, the number of BU cases increased during a dry period. This was observed over several years but also annually, when BU cases are most likely to spike in the annual seasons. It has also been shown that climatic anomalies such as peaks in sea surface temperature (El Niño) create a decline in the oscillation of BU cases. This shows that outbreaks of BU in French Guiana can be triggered by combinations of rainfall patterns occurring on a long (i.e., several years) and short (i.e., seasonal) temporal scale, in addition to stochastic events driven by the El Niño-Southern Oscillation that may modulate these patterns.

Two hypotheses may explain the observed peaks and recessions of cases of BU. Firstly, the long periods of wet weather characterised by floods, during which the bacteria could be dispersed across the floodplain, are followed by a decrease in rainfall that leads the flooded areas to recede into a series of disconnected stagnant pools prone to bacterial development [7]. These rapid changes in aquatic habitat from flowing, well oxygenated habitats with rich animal communities to eutrophic, diversity poor communities have been shown to favour the development of *M. ulcerans* in the environment. Secondly, during the dry period these newly created freshwater pools become more easily accessible by foot for hunting or fishing or simply for recreational bathing. The combination of emergence of favourable habitats for *M. ulcerans* and increased contact of the local population with these habitats during dry periods could explain the observed link between rainfall and increased numbers of BU cases in French Guiana.

2.4 Genetic Diversity of *M. ulcerans* Among Local BU Cases

Only two clinical strains of *M. ulcerans* (ITM7922 and Mu_1G897) from French Guiana were genotyped prior to 2014 [8–10]. Comparative whole genome sequencing showed that the isolate Mu_1G897 belongs to the ancestral lineage of *M. ulcerans*. In the phylogenomic analysis it clustered with fish pathogens belonging to the sublineage 1 of the ancestral lineage, as defined by Doig et al. [8]. A recent study based on Multilocus Variable Number Tandem Repeat Analysis (MLVA) looked at the genetic diversity of *M. ulcerans* strains isolated from BU patients from French Guiana between 1994 and 2013 [11]. A total of 23 DNA samples obtained from 23 patients were purified from ulcer biopsies or derived from cultures. Three allelic combinations were characterised in this study: genotype I, which has been described previously (ITM7922 and Mu_1G897), genotype III, which is very similar to genotype I and genotype II which has distinctly different characteristics in comparison with the other two genotypes. In the MLVA based phylogenetic tree the classical lineage strains from Africa and Australia formed a separate cluster, indicating that all three genotypes from French Guiana belong to the ancestral lineage. This study revealed for the first time genetic variability between clinical isolates of *M. ulcerans* in French Guiana, a diversity that appears to be more pronounced than in Africa, where closely related local clonal complexes are observed. In addition, no correlations were observed between the MLVA-based phylogenetic clustering of the strains from French Guiana and the age of patients, date of infection or geographic location.

3 Environment and the Bacteria

3.1 First Detection of *Mycobacterium ulcerans* DNA in Environmental Samples from French Guiana

In French Guiana in 2014 a study identified for the first time the presence of *M. ulcerans* DNA in environmental samples from South America [12]. A total of 163 environmental samples, taken from 23 freshwater bodies were tested for the presence of the *M. ulcerans*-specific markers IS*2404* and KR by real-time (q) PCR. Sampling sites were selected by looking for (1) water-bodies near the homes of BU cases with habitats representing a high probability of human contact (including paths near or through the water site, presence of fishing or boating activities or other recreational uses and close proximity to human settlements) and (2) similar habitats in areas where there were no BU cases and with little human contact. Samples taken included water (from various sampling sites including bank-side, centre of the water body, water from within aquatic vegetation, shaded areas and sun exposed areas), soil (underwater sediment), detritus, dominant aquatic and semi-aquatic plant species, algae and biofilms. In addition, abiotic data such as dissolved oxygen, pH, conductivity and water temperature were measured.

Only five samples out of 163 were found positive for both IS*2404* and KR in three different water bodies from the Sinnamary and Tonate districts. All three positive sites were typologically similar, with areas that were highly stagnant, forming shallow water bodies. There was no statistical significance for differences in abiotic data between positive and negative sites. These three freshwater bodies with qPCR positive samples were on floodplains, suggesting that in French Guiana, this type of environment constitutes a source for environmentally-persistent mycobacteria or a "receptacle" concentrating these bacilli from further upstream (Fig. 2).

Fig. 2 Diversity of environmental sites positive for *M. ulcerans* DNA in French Guiana. For each site sediment, water and invertebrates samples were tested for *M. ulcerans* DNA by targeting the genetic markers IS*2404* and KR [6]

3.2 Biodiversity Drivers of *M. ulcerans* Distribution Across Freshwater Habitats

A study on a larger number of environmental samples was conducted to study the links between the drivers of freshwater foodwebs in French Guiana and the emergence of *M. ulcerans* in the environment. A collection of 3600 invertebrates and fish were sampled and catalogued from 17 freshwater sites across French Guiana [7, 13]. Landscape data as a measure of land use along a gradient from urban through agricultural to pristine rain forest were extracted from two maps including the CORINE 2006 European land cover map using satellite-derived data and the Hansen deforestation map. Each site was surveyed twice in the rainy season, during February and June 2013. Each insect taxon was positioned in the foodweb using a measure of stable isotopes (nitrogen $\delta^{15}N$ and carbon $\delta^{13}C$).

For each site, the *M. ulcerans* load was calculated for each taxon from the qPCR data for both IS*2404* and KR molecular markers. Each taxon's regional mean trophic niche width (metric of total available trophic niche space within a site), vulnerability (number of potential predator species for each prey taxon), connectance (a standard measure of foodweb complexity) and generality (number of potential prey species each predator taxon can consume) were calculated and weighted by the abundance at each site. The relationships between land cover data and trophic niche width were identified at the local site level using general additive models (GAMs) of the level of deforestation within each buffer zone in addition to urban and agricultural cover.

A total of 78 different freshwater taxa were identified, of which 383 specimens representing 44 taxa tested positive for both IS*2404* and KR. The highest concentrations of *M. ulcerans* were in species lower in the food chain, suggesting a diet high in aquatic algae, detritus, diatoms and similar food sources. These taxa were predominantly invertebrates in both adult and larval form.

Deforestation and either agricultural or urban intrusion were associated with a decline in local trophic niche width, resulting in decreased regional means of vulnerability and generality of taxa. This had an important effect on the potential local *M. ulcerans* load. Host taxa, which on average carried a high level of *M. ulcerans*, were most abundant at sites where there was a very low level of vulnerability and a mid-level of generality.

4 Conclusion

The epidemiology of BU in French Guiana has changed in the past 45 years, mainly with respect to a decrease in incidence, an increase in the average age of patients and peak incidences in geographical areas that have undergone environmental changes. Improved socioeconomic conditions may explain this transition from an African-type epidemiology to an Australian-type epidemiology. In addition, a link between the occurrence of human cases and climatic patterns has clearly been established. The influence of rainfall has been evidenced on an intra-annual

(6 months) and inter-annual (8 years) scale to which climatic anomalies such as El Niño are randomly added. Overall, episodes of high rainfall followed by relative drought are significantly associated with the increase in BU cases.

From a genetic perspective, the fairly large genetic diversity of *M. ulcerans* within French Guiana is rather unusual. The strains causing disease in French Guiana belong to the ancestral lineage and are markedly different from classical lineage strains from Africa and Australia. This seems to eliminate the possibility of an introduction of *M. ulcerans* into French Guiana with the arrival of populations from endemic areas of Africa. However, the diversity of strains present in the environment remains unknown and needs to be better understood in order to highlight potential selective processes shaping the genome of disease causing strains.

Current environmental data show a relatively high abundance of *M. ulcerans* DNA across freshwater habitats along the coastline of French Guiana. Human-induced environmental modifications (e.g. deforestation, agriculture, urbanization) have been found to be significant drivers of foodweb changes towards an increase in taxa with higher *M. ulcerans* loads.

Ongoing studies are looking to further our understanding of transmission pathways from the environment to the local human population. There may be multiple pathways, which may act as strong selective filters for *M. ulcerans* strains found in the environment.

References

1. Douine M, Gozlan G, Nacher M, Dufour J, Reynaud Y, Elguero E, Combe M, Velvin CJ, Chevillon C, Berlioz-Arthaud A, Labbé S, Sainte-Marie D, Guégan JF, Pradinaud R, Couppié P (2017) *Mycobacterium ulcerans* infection (Buruli ulcer) in French Guiana, South America, 1969–2013: an epidemiological study. Lancet Planet Health 1:e65–e73
2. Guerra H, Palomino JC, Falconi E, Bravo F, Donaires N, Van Marck E, Portaels F (2008) *Mycobacterium ulcerans* disease, Peru. Emerg Infect Dis 14:373–377
3. Röltgen K, Pluschke G (2015) Epidemiology and disease burden of Buruli ulcer: a review. Res Rep Trop Med 6:59–73
4. Elguero E, Broutin H, Nacher M, Chevillon C, Guegan JF, Couppié P (2009) Environment risk factors of Buruli ulcer. A case-control study in French Guiana. In: Second International Conference on Buruli Ulcer, Cotonou, Benin, March 30–April 3, 2009
5. Sambourg E, Dufour J, Edouard S, Morris A, Mosnier E, Reynaud Y, Sainte-Marie D, Nacher M, Guégan JF, Couppié P (2014) Réponses et réactions paradoxales au cours du traitement médicamenteux de l'infection à *Mycobacterium ulcerans* (ulcère de Buruli). Quatre observations en Guyane française. Ann Dermatol Venereol 141:413–418
6. Morris A, Gozlan RE, Hassani H, Andreou D, Couppié P, Guégan JF (2014) Complex temporal climate signals drive the emergence of human water-borne disease. Emerg Microbes Infect 3:e56
7. Combe M, Velvin CJ, Morris A, Garchitorena A, Carolan K, Sanhueza D, Roche B, Couppié P, Guégan JF, Gozlan RE (2017) Global and local environmental changes as drivers of Buruli ulcer emergence. Emerg Microbes Infect 6:e22
8. Doig K, Holt K, Fyfe J, Lavender C, Eddyani M, Portaels F, Yeboah-Manu D, Pluschke G, Seemann T, Stinear TP (2012) On the origin of *Mycobacterium ulcerans*, the causative agent of Buruli ulcer. BMC Genomics 13:258

9. De Gentile PL, Mahaza C, Rolland F, Carbonnelle B, Verret JL, Chabasse D (1992) Cutaneous ulcer from *Mycobacterium ulcerans*. A propos of 1 case in French Guiana. Bull Soc Pathol Exot 85:212–214
10. Ablordey A, Hilty M, Stragier P, Swings J, Portaels F (2005) Comparative nucleotide sequence analysis of polymorphic variable-number tandem-repeat loci in *Mycobacterium ulcerans*. J Clin Microbiol 43:5281–5284
11. Reynaud Y, Millet J, Couvin D, Rastogi N, Brown C, Couppié P, Legrand E (2015) Heterogeneity among *Mycobacterium ulcerans* from French Guiana revealed by multilocus variable number tandem repeat analysis (MLVA). PLoS One 10:e0118597
12. Morris A, Gozlan RE, Marion E, Marsollier L, Andreou D, Sanhueza D, Ruffine R, Couppié P, Guégan JF (2014) First detection of *Mycobacterium ulcerans* DNA in environmental samples from South America. PLoS Negl Trop Dis 8:e2660
13. Morris AL, Guégan JF, Andreou D, Marsollier L, Carolan K, Le Croller M, Sanhueza D, Gozlan RE (2016) Deforestation-driven food-web collapse linked to emerging tropical infectious disease, *Mycobacterium ulcerans*. Sci Adv 2:e1600387

Buruli Ulcer in Japan

Koichi Suzuki, Yuqian Luo, Yuji Miyamoto, Chiaki Murase, Mariko Mikami-Sugawara, Rie R. Yotsu, and Norihisa Ishii

1 Epidemiology and Bacteriological and Genomic Features of *M. ulcerans* subsp. *shinshuense*

1.1 Epidemiology

Japan is one of the few countries located in a temperate zone that reports cases of Buruli ulcer (BU). The first case, a 19-year-old female who presented with a chronic and necrotic ulcer on her left elbow, was reported by Mikoshiba et al. in 1982 [1]. This was considered to be an endemic infection due to the lack of a travel history outside the country. A taxonomic study using DNA hybridization assays was later performed on a mycobacterial strain (ATCC 33788) isolated from the skin ulcer lesion of this patient, revealing that this Japanese strain was highly similar to the classical *Mycobacterium ulcerans* strain ATCC 19423. However, the Japanese strain differed from the previously described *M. ulcerans* strain in mycolic acid

K. Suzuki · Y. Luo
Department of Clinical Laboratory Science, Teikyo University, Tokyo, Japan
e-mail: koichis0923@med.teikyo-u.ac.jp

Y. Miyamoto · N. Ishii (✉)
Leprosy Research Center, National Institute of Infectious Diseases, Tokyo, Japan
e-mail: yujim@nih.go.jp; norishii@niid.go.jp

C. Murase
Department of Dermatology, Nagoya University Graduate School of Medicine, Nagoya, Japan

M. Mikami-Sugawara
West Yokohama Sugawara Dermatology Clinic, Yokohama, Japan

Department of Environmental Immuno-Dermatology, Yokohama City University Graduate School of Medicine, Yokohama, Japan

R. R. Yotsu
National Sanatorium Suruga, Shizuoka, Japan

G. Pluschke, K. Röltgen (eds.), *Buruli Ulcer*,
https://doi.org/10.1007/978-3-030-11114-4_5

composition and some biological and biochemical characteristics; therefore, it was deemed to belong to a new subspecies of *M. ulcerans* named *M. ulcerans* subsp. *shinshuense* [2]. The disease was not noticed again in Japan until the second case of BU was identified and reported in 2003 [3]. Since then, the number of cases has increased gradually, summing to a total of 60 cases reported from 17 of the country's 47 prefectures till the end of 2016. *M. ulcerans* subsp. *shinshuense* has been isolated from 41 of the cases [4–7].

BU cases are rather sporadically distributed throughout the main and largest island of Japan, Honshu Island, with no particular focus [6, 8]. The Okayama Prefecture, which is localized in the central western regions of Japan facing the Seto Sea, has so far reported most of the cases (ten [17%] cases), followed by the Tottori Prefecture (six [10%] cases) and the Shiga Prefecture (seven [12%] cases). The Akita Prefecture is the most northern prefecture reporting BU cases, where the temperature can be below zero degrees Celsius during the coldest season of the year. All reported cases in Japan are considered to be domestic infections as none of the patients had a history of oversea travel to BU endemic countries before symptom onset. The acquisition of the pathogen from a suspicious aquatic environment remains undetermined in most cases. However, it is interesting to note that most BU cases in Japan were diagnosed in autumn and winter (48 cases), which may indicate a higher incidence of infection during summer considering the expected time lag between infection and symptom onset [9].

1.2 Genome

M. ulcerans and related mycobacterial strains (such as *M. liflandii*) share a characteristic genome structure consisting of a chromosome and a giant plasmid. This genome structure caused some difficulty when analyzing the whole genome sequences. Recently, with the help of next generation sequencing (NGS), the com plete genome sequences of two representative isolates, *M. ulcerans* Agy99 and *M. liflandii* 128FXT, were unraveled [10–12]. The complete genome sequence of *M. ulcerans* subsp. *shinshuense* (ATCC 33728) was also reported by a Japanese group [13]. Comparison of the whole genome of *M. ulcerans* subsp. *shinshuense* with those of the related strains revealed that it bears more similarity with *M. liflandii* 128FXT than with *M. ulcerans* Agy99 (Table 1). *M. ulcerans* subsp. *shinshuense* has a 5,899,681 bp chromosome with 65.64% GC content and a 166,617 bp giant plasmid with 62.76% GC content. The average nucleotide identities were 98.36% to *M. ulcerans* Agy99 and 99.10% to *M. liflandii* 128FXT [13]. The total number of coding DNA sequences (CDSs) in both *M. ulcerans* subsp. *shinshuense* and *M. liflandii* 128FXT were found to be roughly around 5000, approximately 800 CDSs more than in the *M. ulcerans* Agy99 chromosome (Table 1) [10–13]. With respect to insertion sequences (IS), IS*2606* was much more abundant in *M. ulcerans* Agy99 than in *M. ulcerans* subsp. *shinshuense* and *M. liflandii* 128FXT (Table 1) [10–13]. In contrast, more than 200 copies of IS*2404* were observed in *M. ulcerans* subsp. *shinshuense*, which share approximately 99% identity with the classical *M.*

Table 1 Genomic characteristics of *M. ulcerans* subsp. *shinshuense* and related mycobacteria

Characteristics	*M. ulcerans* subsp. *shinshuense* ATCC 33728	*M. ulcerans* Agy99	*M. liflandii* 128FXT
Chromosome			
Size (bp)	5,899,681	5,631,606	6,208,955
G + C (%)	65.64	65.47	65.62
No. of CDSs	5015	4160	4994
No. of pseudogenes	451	771	436
No. of IS*2404*	206	209	224
No. of IS*2606*	1	83	1
Plasmid			
Size (bp)	166,617	174,155	190,588
G + C (%)	62.76	62.5	62.9
No. of CDSs	72	72	95
No. of pseudogenes	7	7	22
No. of IS*2404*	4	4	15
No. of IS*2606*	1	8	3
References	[7], Yoshida et al. (unpublished)	[8]	[9, 10]
GenBank accession no.	AP017624 AP017625	CP000325 BX649209	NC_020133 NC_011355

CDS coding DNA sequence, *IS* insertion sequence

Table 2 Nucleotide substitutions observed in 16S rRNA sequences from *M. ulcerans* subsp. *shinshuense* and related mycobacteria

Species	Differentiating sequence position (underline)		GenBank accession no.
	492	1288	
M. ulcerans subsp. *shinshuense* ATCC 33728	TGG**G**GAA	TAA**G**GCC	AB548733
M. ulcerans Agy99	TGG**A**GAA	TAA**C**GCC	AB548729
M. liflandii 128FXT	TGG**A**GAA	TAA**A**GCC	CP003899
M. marinum ATCC 927	TGG**A**GAA	TAA**A**GCC	AB548717
M. pseudoshottsii JCM 15466	TGG**A**GAA	TAA**A**GCC	AB548713

ulcerans strains and *M. liflandii* [4, 7, 13]. Due to the abundance of IS*2404*, PCR amplification of a partial IS*2404* sequence is currently used for sensitive detection of *M. ulcerans* subsp. *shinshuense*, although differentiation from other closely-related IS*2404*-harboring species requires further investigation. The 16S rRNA sequences in *M. ulcerans* subsp. *shinshuense* were almost identical to those in *M. ulcerans* Agy99 (Table 2), except for two characteristic nucleotide substitutions at base positions 492 and 1288 found exclusively in *M. ulcerans* subsp. *shinshuense* (Table 2) [4, 7]. Including the listed representative isolate ATCC 33728 (Table 2), the same nucleotide substitutions have been found in ten clinical isolates of *M. ulcerans* subsp. *shinshuense* [7].

Comparison of eight pMUM001 gene sequences encoding lipid toxin mycolactone-producing enzymes (including *repA*, *parA*, serine/threonine protein kinase [STPK] gene, loading domain of *mls*, acyltransferase domain of *mls*, *rep* type II thioesterase gene, *rep* type III ketosynthase gene, and *rep* P450 hydroxylase gene) in the giant plasmid of *M. ulcerans* subsp. *shinshuense* and the African *M. ulcerans* strain by PCR amplification revealed the presence of seven pMUM001 gene sequences and the loss of the STPK gene in Japanese isolates [4, 7]. The elucidation of the complete genome sequence of *M. ulcerans* subsp. *shinshuense* further verified the above finding by showing that the STPK gene in *M. ulcerans* subsp. *shinshuense* is truncated, with low similarity to STPK in *M. ulcerans* Agy99 [13]. Despite the deficiency of a functional STPK gene in Japanese clinical isolates, no apparent bacteriological differences were observed, indicating the possibility that some signaling pathways compensate for STPK function in *M. ulcerans* subsp. *shinshuense* [14]. Intriguingly, a Chinese *M. ulcerans* strain has a 16S rRNA gene identical in sequence to that in *M. ulcerans* subsp. *shinshuense*, and also lacks the STPK gene in its giant plasmid like *M. ulcerans* subsp. *shinshuense* [4, 7]. This finding supports the suggestion that the Chinese strain is phylogenetically closest to *M. ulcerans* subsp. *shinshuense* among other *M. ulcerans* strains [15]. However, further characterization of the whole genome sequence of the Chinese strain is required for its classification.

The close resemblance, in addition to distinctive genomic features, between the classical *M. ulcerans* strains and the Japanese strain *M. ulcerans* subsp. *shinshuense* has evoked great interest in the intra-species evolutionary scenario for *M. ulcerans*. Phylogenetic analysis of irreversible genomic changes focusing on insertion-deletion polymorphisms in 12 regions of difference among *M. ulcerans* strains has unambiguously resolved a phylogenetic tree showing that *M. ulcerans* has evolved into at least two distinct lineages since divergence from the *M. marinum* progenitor [15]. In addition to *M. ulcerans* subsp. *shinshuense*, strains from China, South America, and Mexico were shown to belong to a lineage that is more closely related to *M. marinum*, namely the 'ancestral' lineage; while strains from Africa and Australia belong to the 'classical' lineage that has undergone major genomic rearrangement [15]. In agreement with this evolutionary scenario, 26,564 single nucleotide polymorphisms (SNPs) were found in a Japanese strain by a comparison with the reference genome of the classical lineage isolate Agy99 using a high-resolution phylogeny analysis based on genome-wide SNPs [16]. Moreover, calculation of time scales of the evolutionary process by the minimum evolution tree-based approach estimated that the divergence of Ghanaian subtypes (including Agy99) from the *M. marinum* progenitor occurred about 1000–3000 years ago, whereas *M. ulcerans* subsp. *shinshuense* diverged much earlier (about 394,000–529,000 years ago) [16].

1.3 Biochemical Properties

Mycolactone, secreted by *M. ulcerans* in the process of chronic infection, is an essential molecule for the virulence of BU [17]. Mycolactone is also secreted by strains originally designated *M. pseudoshottsii*, *M. liflandii*, and a subset of *M. marinum* strains [18–20]. *M. ulcerans* subsp. *shinshuense* produces mycolactone A/B, among

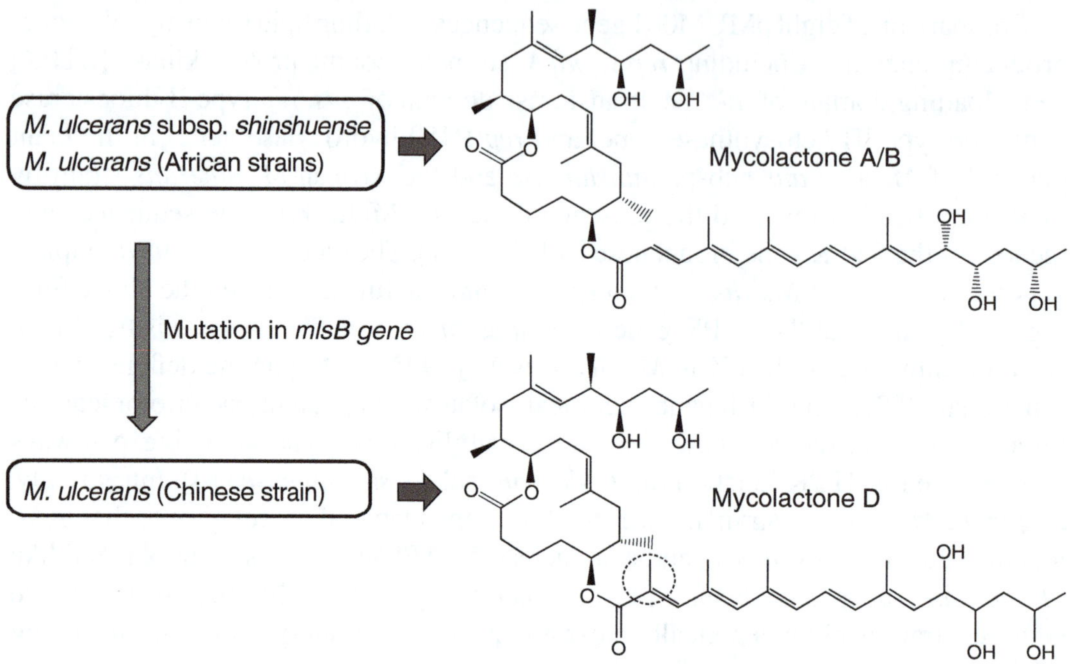

Fig. 1 Structural variation in mycolactones caused by a *mlsB* mutation uniquely found in a Chinese *M. ulcerans* strain. The dashed circle in mycolactone D represents the methyl group that distinctively differentiates it from mycolactone A/B produced by *M. ulcerans* subsp. *shinshuense* and African strains of *M. ulcerans*

five analogs of mycolactone, which was originally found in African *M. ulcerans* strains [12, 17, 18] (Fig. 1). In contrast, the Chinese strain of *M. ulcerans* produces a unique mycolactone D that is not observed in other *M. ulcerans* strains [21]. Mycolactone D possesses an additional methyl group in its acyl side chain, whereas the mycolactone A/B conserves an intact form (Fig. 1) [12, 17, 18, 21]. In the Chinese strain, nonsynonymous substitutions that could cause drastic functional changes have been found in the *mlsB* gene involved in synthesis of acyl side chains, while corresponding mutations were not observed in Japanese *M. ulcerans* subsp. *shinshuense* or the African strains, suggesting that the Chinese strain might have uniquely acquired the ability to modify mycolactone A/B to D through genetic alternations [12].

A variety of glycolipids are present in the mycobacterial cell wall that likely functions as a protective barrier from host immune attack [22, 23]. Mycolate is a type of long chain fatty acid covalently linked to various glycolipids as a major component in the mycobacterial cell wall [22, 23]. Mycobacteria produce three main types of mycolate: alpha-, keto- (which contains additional ketone groups), and methoxy-mycolate (which contains additional methoxy groups) [22, 23]. Tsukamura et al. reported that both *M. ulcerans* subsp. *shinshuense* and other *M. ulcerans* strains share the three types of mycolate, while the average number of carbons in mycolate in *M. ulcerans* subsp. *shinshuense* is higher than that in an Australian *M. ulcerans* strain (ATCC 19423) [2]. However, comprehensive comparison is needed to determine whether the distribution of the three mycolate types differs between *M. ulcerans* subsp. *shinshuense* and other *M. ulcerans* strains, and to unravel their implications in mycobacterial virulence.

As one major glycolipid component in the mycobacterial cell wall, phenolic glycolipids (PGLs) were found in *M. leprae*, *M. tuberculosis*, *M. marinum*, and *M. ulcerans* [24–26]. PGLs vary in their compositions of sugar moieties (including rhamnose, fucose, and glucose), which are highly species-specific [24–26]. Daffe et al. found that 10% of *M. ulcerans* strains produced PGL consisting of monosaccharide (3-*O*-methyl-rhamnose), supporting a close phylogenetic link between *M. ulcerans* and *M. marinum* [27]. However, the rest of the tested strains, including *M. ulcerans* subsp. *shinshuense*, did not produce PGL [27]. In *M. leprae,* PGL is known to induce a strong antibody response and to play a role in invasion into Schwann cells [28, 29]. In *M. tuberculosis*, clinical isolates have been divided into PGL-producing and non-producing groups, and the former is regarded as more virulent, suggesting that PGL plays a role in the pathogenicity of mycobacteria [30]. This suggests that PGL-producing *M. ulcerans* subsp. *shinshuense* strains may be found, albeit less frequently. Analysis of PGL in *M. ulcerans* subsp. *shinshuense* as well as in other *M. ulcerans* strains with reference to the clinical symptoms in each case might offer a new approach to understand the causes of invasive BU cases.

Drug susceptibility tests showed that *M. ulcerans* subsp. *shinshuense* strains were more sensitive to streptomycin (SM), kanamycin (KM), and clarithromycin (CAM) than reference *M. ulcerans* strains (ATCC 19423 and Agy99) [7]. Differing from other antimicrobial drugs, such as rifampicin (RFP) and levofloxacin (LVFX), in the mechanism of action, binding to ribosomal RNA is exclusively required for SM, KM, and CAM to exert their actions to inhibit mycobacterial protein synthesis. The specific nucleotide substitutions found in the 16S rRNA of *M. ulcerans* subsp. *shinshuense* might have affected the conformation of its transcribed ribosomal structure, rendering it more accessible to the drugs [4, 7].

M. ulcerans subsp. *shinshuense* strains and the Chinese strain were all positive for urease activity, whereas the *M. ulcerans* strains from Africa and Australia were negative [7]. This feature may allow scientists to distinguish *M. ulcerans* subsp. *shinshuense* from *M. ulcerans* strains by a simple urease activity test. It has been shown in some mycobacterial species that urease plays a key role in metabolizing urea as a sole nitrogen source under nutrition-limited conditions and in neutralizing the surrounding milieu to inhibit phagolysosomal maturation [31, 32], indicating that urease-positive strains may be more adapted to harsh environments and more capable of achieving intracellular survival. Further comparisons between urease-positive strains with urease-deficient mutants might clarify whether urease activity contributes to the persistence of *M. ulcerans* subsp. *shinshuense* in the host.

2 Clinical Features and Treatment of BU in Japan

2.1 Antimicrobial Treatment

In the BU cases diagnosed in Japan, skin lesions mostly developed on exposed body parts including upper and lower limbs, face, ear, jaw, and clavicle. Although multiple lesions were present in 14 cases, no single lesion larger than 5 cm in diameter was reported. Therefore, all cases fall into World Health Organization (WHO) category I or II. Advanced BU infection has not been seen in Japan. In contrast to the general impression that BU lesions are typically painless, approximately half of the cases confirmed to be infected by *M. ulcerans* subsp. *shinshuense* have reported pain [5, 8].

In Japan, a variety of antimicrobial drugs have been used to treat BU; e.g., RFP, minocycline (MINO), CAM, LVFX, gatifloxacin (GFXN), norfloxacin (NFX), ethambutol (EB), isoniazid (INH), ethionamide (ETH), and tosufloxacin (TFLX). However, surgical intervention was often unavoidable. Nakanaga et al. tested the drug susceptibilities of *M. ulcerans* subsp. *shinshuense* to major antimicrobial agents and found that it is most susceptible to RFP and CAM, while resistant to EB, ETH, and INH (Table 3) [7]. LVFX, a broad-spectrum fluoroquinolone antibiotic, which also showed satisfactory activity against *M. ulcerans* and *M. ulcerans* subsp. *shinshuense* (Table 3) [7], has been preferentially used for treating skin infections in Japan due to its better transferability to soft tissue and skin. Based on these reasons and in reference to WHO BU guidelines, a triple antimicrobial therapy with RFP (450 mg/day), LVFX (500 mg/day), and CAM (800 mg/day), called RLC therapy, has been recommended and has increased cure rates of BU without surgery (Table 4).

In our latest investigation on the efficacy of BU treatment in Japan, 50 cases (83% of the total reported cases) consisting of 19 male and 31 female patients, aged from 2 to 88 years (average: 45.8 ± 27.63 years) were included. Ten cases were excluded because of the absence of detailed information for treatment. These 50 BU cases were divided in two groups based on their antimicrobial regimens. One group (21 cases) received high-dose RLC (CAM at 800 mg/day). The other 29 cases received other treatments including low-dose RLC (CAM at 400 mg/day) or monotherapy of RFP, CAM or MINO; this group also included cases with interrupted RLC due to side effects and subsequent alternative chemotherapies. Five children received TFLX (12 mg/kg/day) substituted for LVFX, as TFLX (but not LVFX) is covered by medical insurance for children in Japan.

Table 3 Susceptibility testing results of *M. ulcerans* subsp. *shinshuense*

	MIC (μg/mL)	
	M. ulcerans subsp. *shinshuense*	
Antimicrobial agent	ATCC33728	501
SM	0.125	0.25
EB	16	8
KM	0.25	0.25
INH	8	8
RFP	0.06	0.06
LVFX	0.25	0.5
CAM	0.03	0.06
ETH	16	8
AMK	0.5	0.5

AMK amoxicillin, *CAM* clarithromycin, *EB* ethambutol, *ETH* ethionamide, *INH* isoniazid, *LVFX* levofloxacin, *KM* kanamycin, *MIC* minimal inhibitory concentration, *RFP* rifampicin, *SM* streptomycin

Table 4 Efficacy analysis of treatment for BU in Japan

Treatment	No.	Age	Sex (M/F)	Category (I/II)	Ulcer size (mm)	Cured completely without surgery
RLC (high dose)	21	42.1 ± 30.2	11/10	18/3	25.6 ± 23.8	15
Others	29	48.4 ± 26.3	8/21	24/5	27.3 ± 15.7	4
P value		0.43	0.087	1.0	0.79	6.95×10^{-5}

BU Buruli ulcer, *RLC* triple therapy with rifampicin (450 mg/day), levofloxacin (500 mg/day), and clarithromycin (800 mg/day)

Fifteen/twenty-one (71.4%) high-dose RLC-treated patients compared to 4/29 (13.7%) of the other patients had avoided surgery ($P < 0.0001$, Fisher's exact test). The distribution of age, sex, size of skin ulcer, or WHO category did not significantly differ for patients treated with high-dose RLC therapy and the others. Together, these results suggest that high-dose RLC therapy has been most effective among all chemotherapies (Table 4), which is in agreement with a previous investigation that included 40 BU cases in Japan by June 2013 [8]. The duration of high-dose RLC therapy ranged from 8 to 48 weeks (24 weeks in most cases), which was compared to the standard 8-week antibiotic treatment. The failure to determine a significant correlation between the duration of high-dose RLC therapy and the size of skin ulcers ($r = -0.2$, with Pearson correlation coefficient) indicates a tendency of overtreatment for BU in Japan.

The ineffectiveness of low-dose RLC may be attributed to a complex pharmacokinetic interaction between CAM and RFP. It has been reported that RFP induces production of cytochrome P450 3A4, an enzyme involved in the metabolism of CAM [33, 34], which would hasten the elimination of CAM from serum when used together [35, 36]. These reports indicate the need for a higher dose of CAM when used in combination with RFP. With reference to the WHO BU meeting in 2015, we currently recommend a high-dose RLC triple therapy consisting of RFP (10 mg/kg/day), LVFX (500 mg/day), and CAM (800 mg/day) in Japan [37].

2.2 Alternative Treatments of BU

Surgical excision and skin grafting were the only options for BU treatment until 2004 [38]. Since then, apart from chemotherapy also other therapeutic options, such as negative-pressure wound therapy (NPWT), ozone therapy, and hyperbaric oxygen therapy (HBOT) have been used or considered in the treatment of BU. Particularly for patients who have received surgical excision and still present large non-granulating wound areas, NPWT or ozone therapy could be a good choice for the next stage.

Since larger lesions or multiple ulcerative lesions often took a longer time to heal or failed to be cured by chemotherapy alone, surgical excision of the necrotic, ulcerated tissues had been performed in Japan. Among 31 cases that received surgical excision, 21 (67.7%) also received skin grafts (Table 5). Among the 21 patients, NPWT was tentatively performed in one patient for the purpose of wound bed preparation before skin grafting [39]. This patient had an ulcer on the right ankle with a necrotic

Table 5 Treatment for BU in Japan

Treatment	RLC (high dose) (n = 21)	Others (n = 29)	Total (n = 50)
Antimicrobial treatments only	15	4	19
Surgical excision	2	8	10
Surgical excision + skin graft	3	17	20
Surgical excision + NPWT + skin graft	1	0	1

BU Buruli ulcer, *NPWT* negative-pressure wound therapy, *RLC* triple therapy with rifampicin (450 mg/day), levofloxacin (500 mg/day), and clarithromycin (800 mg/day)

bed, and the Achilles tendon and the calcaneal bone were exposed. NPWT with the Vacuum-Assisted Closure (V.A.C.) Therapy System (Kinetics Concepts Inc.; KCI) was started as a pretreatment for a skin graft to be performed 22 days after surgical excision of the necrotic tissue. The vacuum suction was maintained at 125 mmHg. The wound dressings were changed every third day for 24 days. At the end of NPWT, acceptable granulation tissue had covered the ulcer bed and the skin graft was consequently successfully engrafted. The patient was cured without disability.

NPWT has been widely used in Japan for many kinds of ulcerative lesions or for wound treatment after skin grafting, as it can reduce exudate, improve local edema, and activate granulation. It is not to be used for wounds with active infection. Therefore, for BU treatment, NPWT is preferable for use after surgical debridement combined with appropriate antimicrobial chemotherapy. Dressings must be changed every 48–72 h. The absence of active wound infection should be checked each time at dressing. The exact frequency of dressing changes depends on the individual patient's circumstance, but it should not be less than three times a week. There are rental and purchasable NPWT (portable and disposable) models. Rental models include RENASYS by Smith & Nephew and V.A.C. by KCI, and purchasable models include PICO by Smith & Nephew. Otherwise healthy patients often must remain hospitalized for the sole purpose of receiving NPWT because the rental devices are only for hospital use. Thus, an affordable and purchasable NPWT device like PICO may be more cost-effective for an otherwise healthy patient. Nevertheless, its high cost is a large problem for patients in many countries (Table 6).

2.3 Other Suggested Therapies

Ozone therapy has been reported to shorten the course of wound healing and to promote lipid peroxidation and antioxidant protection indices [40]. Briefly, O_3 is topically administered by positioning a bag around the lesion and insufflating an O_2-O_3 mixture at a concentration of 25–30 μg/mL. To avoid gas leakage, the inflated bag is sealed just above the lesion. The bag is positioned and closed with an elastic band to give the gas mixture contact with the wound for about 20 min. When the bag is removed, the wound is covered with a sterile gauze. A case with a skin ulcer in Benin was reported to be healed after the ozone applications in 2 weeks [41].

Table 6 Approximate costs for NPWT per week

	Company and device		
	Smith and Nephew		KCI
Country	RENASYS (US$/week)	PICO (US$/week)	V.A.C. (US$/week)
Japan	620	420	700
South Africa (private)	N/A	N/A	600
South Africa (public)	450	175–300	450
Egypt (private)	290–340	220	N/A
Egypt (public)	145–155	N/A	N/A

Costs shown are approximate costs per week as of January 2017. The weekly costs include the costs for three sets of dressings and canisters and rental fee for the NPWT device (RENASYS and V.A.C.). The costs vary due to the size of lesions. *KCI* Kinetics Concepts Inc., *N/A* not available, *NPWT* negative-pressure wound therapy, *V.A.C.* vacuum-assisted closure therapy system

Compared to surgical treatment, NPWT and ozone therapy are easy to use procedures. NPWT requires only the NPWT device, dressings, and canisters, though the disposable dressings and canisters are expensive. In contrast, ozone therapy does not require such expensive disposable materials. It requires an oxygen tank, an ozone generator, sterile water, and plastic bags, together with a system to drive the gas into contact with the lesions. The oxygen generator is expensive (4000–8000 Euros), but the cost of the disposable items for ozone therapy is much lower than that for NPWT. This simpler and more affordable therapy may offer an effective option for managing BU as an alternative treatment in endemic areas.

HBOT has long been established to treat skin ulcers and wounds such as diabetic foot ulcers, burns, and serious soft tissue infections including gas gangrene and necrotizing fasciitis [42]. HBOT is the application of 100% oxygen at a pressure two or three times of the atmospheric pressure at sea level. This pressure increases arterial and tissue oxygen tension to exert multiple physiological and therapeutic effects [43]. HBOT has been experimentally used to treat mice with *M. ulcerans*-infected footpads [44, 45]. In human patients, Pszolla et al. reported that supportive therapy with HBOT promoted healing after surgical excision of the deep ulcers with osteomyelitis caused by *M. ulcerans* [45]. HBOT requires hyperbaric chambers, which may limit its application in endemic areas. However, if used in conjunction with other therapeutic procedures, HBOT may be an effective therapeutic adjunct in treating BU, and it may contribute to reduce large surgical interventions like amputations and to prevent permanent disabilities [46].

3 *M. ulcerans* subsp. *shinshuense* in the Environment of Japan

3.1 Detection of Environmental *M. ulcerans* subsp. *shinshuense* in Japan

So far, *M. ulcerans* has been detected in the backyard of a family house in a rare case of concurrent familial clustering in Japan [47, 48]. In November, three members of a family living in a rural town in Japan, developed BU symptoms almost simultaneously. At first, the 2-year-old daughter had a gradually enlarging eruption on her right cheek. Soon after her symptom onset, an asymptomatic indurated eruption appeared on the right forearm of her 5-year-old brother, and a gradually enlarging eruption developed on the right wrist of her 37-year-old mother, both of which showed acid-fast bacilli-positive histopathological features similar to the lesions found in the little girl. The patients were tentatively given oral anti-mycobacterial multi-drug therapy combined with surgical treatment and cured without recurrence. PCR amplification and gene sequencing afterwards determined the strains isolated from the patients' skin biopsy cultures to be *M. ulcerans* subsp. *shinshuense*, thus giving an indubitable diagnosis of BU. The concurrent familial cases strongly suggested the presence of the pathogen in their common living environment, although person-to-person transmission of *M. ulcerans* cannot be excluded. A field study involving environmental sampling was therefore carried out in summer (July and August) and autumn (October) of the following year in an attempt to determine the presence of *M. ulcerans* in the environment around the family's house.

The family was living in an old house surrounded by rich farmland and extensive irrigation channels. Immediate suspicions centered on a one-meter-wide agricultural water channel of stagnant water slowly flowing through the family's backyard, to which the family had routine access and where the children entertained themselves by catching small water creatures. Environmental samples, including water, mud, crayfish, earthworms, a freshwater snail, a hoverfly, a moth, and several kinds of aquatic or semi-aquatic insects, were collected from or near the water channel in the patients' backyard in summer and autumn. Genomic DNA was purified from each sample and subjected to whole genome amplification (WGA) to achieve higher sensitivity in PCR to detect trace amounts of mycobacterial DNA. PCR amplifications and DNA sequencing results showed that the *M. ulcerans* IS*2404* was detected in WGA-amplified DNA from a crayfish that was collected in summer (but not in the crayfish collected in autumn or in any other sample), presenting so far the only piece of evidence that links a contaminated aquatic environment to BU occurrence in Japan.

Intriguingly, *M. ulcerans* IS*2404* was detected in a crayfish but not in the water samples collected from the same channel at the same time [47], indicating that despite the presence of the pathogen in the contaminated aquatic environment, its potential presence in the water was nevertheless below the detection sensitivity of PCR amplification. In an aquatic community, due to biotic interactions between various habitants such as filter feeders, herbivorous, scavengers, and predators, *M. ulcerans* may be widespread across the whole community. In agreement with this point of view, in endemic areas *M. ulcerans* DNA has been detected in a wide variety of environment samples, such as aquatic insects, biofilms, soil, mosquitoes, crustaceans, detritus, fish, frogs, snails, worms, and various small mammals [49]. The transmission of pathogens within an aquatic environment, depending on biodiversity and the number of habitants, can drive the environmental load of the pathogens to be enriched or diluted in certain hosts, carriers, or biofilms. Additionally, certain keystone organisms could play an overwhelming role in the transmission and overall prevalence of *M. ulcerans* in a specific environment [50]. The crayfish (*Procambarus clarkii*) is known to have a wide food range, including plants, small fish, shrimp, plankton, benthos, and algae, and especially prefers mud and deceased aquatic organisms. Such an omnivorous eating habit may explain an enrichment of *M. ulcerans* in crayfish.

3.2 Seasonal Variation of BU in Japan

In addition to establishing the presence of *M. ulcerans* subsp. *shinshuense* in an aquatic environment in Japan for the first time, the findings of the field work were in line with a noticeable feature of seasonal variation of BU occurrence in Japan. Eighty percent of BU cases in Japan were diagnosed in autumn and winter (from September to February), but only a few in summer (July and August) [6, 8]. Under the assumption that the diagnosis is preceded for several months by *M. ulcerans* infection due to a long incubation period [9] and due to the delay in seeking medical care, it was speculated that the exposure to the pathogen occurred during summer. Such dynamics in *M. ulcerans* infection are putatively driven by a complex interplay of human activities and the prevalence of the pathogen in the environment. Japan has four distinct seasons with a rainy season throughout early summer (typically from early June to mid-July), and there are naturally more chances for bare skin to be exposed to an aquatic environment during the hot and humid summer time, especially in the water-rich countryside.

On the other hand, there was no direct information regarding the prevalence of *M. ulcerans* in the environment throughout the year in Japan before the field work. The results that *M. ulcerans* subsp. *shinshuense* was detected only in July, but not in any of the samples collected from the same water channel in October, hints at potential seasonally changing prevalence of environmental *M. ulcerans* subsp. *shinshuense* with a high peak in summer, which may have contributed to the observed seasonal variation of BU incidence in Japan. However, the scale of this fieldwork alone was far too small to make a solid prediction. Seasonal dynamics in environmental *M. ulcerans* prevalence can be driven by an extremely complex interplay between multiple biotic and abiotic factors, such as topographic factors (distribution and latitude of land cover of watersheds), climatic factors (temperature, precipitation), physical-chemical conditions (pH, dissolved oxygen, salinity) in stagnant and slow flowing ecosystems, and networks of aquatic taxa, as suggested by large-scale field studies conducted in BU endemic countries [51, 52]. In this respect, the seasonal dynamics of the prevalence of *M. ulcerans* in Japan (especially in a high latitude area as the site in the field work) may differ substantially from that reported in tropical/subtropical areas.

3.3 PCR Detection of *M. ulcerans* subsp. *shinshuense* DNA in the Environment

Field studies to detect environmental *M. ulcerans* subsp. *shinshuense* in Japan are still in their infancy compared to studies in BU endemic countries. Presently, the detection of *M. ulcerans* in the environment is mostly based on demonstrating by

PCR the presence of IS*2404*, an insertion sequence with over 200 copies in the whole genome of *M. ulcerans* [10]. IS*2404*-PCR is highly specific and sensitive for testing clinical specimens from human BU lesions; however, its application in the detection of environmental *M. ulcerans* is less straightforward, and the IS*2404*-PCR-based results should be interpreted with caution. IS*2404*-PCR detects the presence of IS*2404* regardless of the pathogenicity of the detected mycobacteria, which alone has little power in predicting endemicity of BU. *M. ulcerans* DNA was detected by IS*2404*-PCR extensively across southern Louisiana, U.S., leading to the conclusion that the distribution of environmental *M. ulcerans* is not restricted to areas where BU is endemic [53].

In a large-scale field study conducted in Benin, IS*2404*-positive samples were detected with similar prevalence in environmental samples in both non-endemic (9/10) and endemic (12/12) villages [49]. However, when IS*2404*-positive samples were further analyzed for a second PCR target, the enoyl reductase (ER) that is required for the synthesis of mycolactone, only 2/10 non-endemic villages had ER-positive samples, whereas 9/12 endemic villages had ER-positive samples. Thus, the PCR for IS*2404* alone falsely predicted 9/10 non-endemic villages to be endemic, and the additional use of the PCR for ER accurately predicted 8/10 non-endemic villages [49]. In a field study, the use of a single PCR target for detection of environmental *M. ulcerans* may easily signal a false alarm to public health authorities. Secondly, the results of conventional IS*2404* PCR analyses are only qualitative and do not reveal the abundance of the pathogen. They are thus inadequate to describe potential seasonal dynamics in the overall environmental load of *M. ulcerans*. Lastly, analysis of many IS*2404*-positive samples in Ghana have revealed the presence of mycolactone-producing mycobacterial species other than the ones commonly associated with BU disease in humans [54]. *Mycobacterial* strains closely related to the human pathogenic *M. ulcerans*, including ecovars originally designated *M. liflandii*, *M. pseudoshottsii*, and mycolactone-producing *M. marinum*, have been found to harbor IS*2404* [55, 56].

Thus, the analyses of *M. ulcerans* that rely solely upon the conventional PCR assay for IS*2404* are flawed. TaqMan Multiplex real-time PCR assays that target two insertion sequences of *M. ulcerans* (IS*2404*, IS*2606*) and a multicopy sequence encoding the ketoreductase B domain [55, 57], in which samples are considered positive only if the detected cycle threshold values are strictly lower than a default value based on external standard curves with serial dilutions of *Mycobacterial* DNA, can quantify the copy number of the targets to allow the differentiation of classical *M. ulcerans* from other IS*2404*-harboring strains in the environment [55, 57].

Moreover, a novel category of variable tandem repeats (VNTRs) called mycobacterial interspersed repetitive units (MIRUs) have been identified in the genomes of *M. ulcerans* and *M. marinum*. Analysis of strain-specific polymorphisms of MIRU loci has been used to tentatively genotype *M. ulcerans*, *M. marinum*, and an *M. marinum*-like organism that is considered a possible missing link between *M. ulcerans* and *M. marinum* [56, 58]. Ideas inspired by many studies conducted in the BU endemic countries, such as the use of multiple targets in quantitative PCR assays and MIRU-VNTR typing, will better equip us to evaluate the environmental distribution of *M. ulcerans* in future field studies in Japan. Distribution maps of Asian strains will play a key role in understanding the transmission mode of *M. ulcerans* and support control of BU.

4 BU in Asia and Future Perspectives

4.1 BU in China

In addition to BU cases reported from Japan, *M. ulcerans* subsp. *shinshuense* has also been identified as the causative agent of BU in a patient from China [59]. In 1997, a 40-year-old women, who had grown up in China and had been living in Europe, traveled to Shandong Province, China in the summer (July and August). According to the patient's narrative, she received many mosquito bites after walking barelegged in the field. Three months later, the patient noticed a swelling at the front of her left lower leg that afterwards developed a slightly depressed center with no other systemic symptoms. Four months after that, a surgeon excised her lesion. Reportedly, there was no noticeable bleeding during the procedure, and the excised tissue was pale and firm. The lesion did not heal after excision but further developed ulceration. After the patient went back to Europe, a doctor specializing in infectious diseases examined the lesion and suggested that it might be skin tuberculosis. Consequently, a smear of the ulcer base was made by the patient's general practitioner, and Ziehl-Neelsen staining revealed acid-fast bacteria scattered and in clumps. The patient was therefore referred to the Department of Dermatology, Section of Tropical Dermatology of the Academic Medical Center, Amsterdam, where diagnostic procedures showed BU-characteristic clinical, histopathological, and molecular findings [59].

At first examination, it was a single painless lesion (3 × 3.5 cm in diameter) on the lower leg with necrotic ulcer bed and undermined border. Histopathological examination of the biopsy specimens showed thickening and broadening of the epidermis, extensive eosinophilic necrosis in the deeper dermis and the subcutaneous tissue without granulomatous inflammatory reactions. Cultivation tests using a biopsy sample yielded pale-cream to yellow colonies after 40-day incubation on Coletsos medium only at 30°C. PCR analyses using DNA extracts both from ulcer biopsies and culture isolates confirmed the presence of the *M. ulcerans*-specific IS*2404* sequence. Using classical identification schemes the isolate was identified as *M. ulcerans* subsp. *shinshuense*. Treatment began with RFP + CAM and was later switched to ciprofloxacin (CPFX) + rifabutin (RFB) based on the results from drug susceptibility test of the original culture isolate. The patient was cured with an 8-months antibiotic treatment without reported recurrence [59].

The putative infection site of Shandong Province, China is above latitude 30°N. BU occurrence at such a high latitude is rare, and had only been reported in Japan, suggesting that *M. ulcerans* subsp. *shinshuense* is more adapted to environments at higher latitudes than the classical *M. ulcerans* that is observed to be confined to tropical/subtropical regions and temperate climatic areas. Moreover, the infection by *M. ulcerans* subsp. *shinshuense* in the patient from China was highly suspected to have occurred during her travel in summer, 3–4 months prior to symptom onset, which is in line with the proposed seasonal variation of BU occurrence in Japan. Thus, in contrast to the majority of BU cases that are caused by classical *M. ulcerans*, BU infections by *M. ulcerans* subsp. *shinshuense* have so far been only infrequently found in Japan and China, as apparent from the clinical reports [6, 8].

As stated before, it is possible that extensive reductive genomic changes have enabled the classical lineage to habituate in a more confined environment, resulting in clonal populations within Africa and Australia with genomic rearrangements rendering them more pathogenic and strengthening their ability to infect mammals [15]. In contrast, the ancestral lineage, closer to the *M. marinum* progenitor, may have remained as a largely environmental mycobacterium that only occasionally affects humans. From this point of view, it is reasonable to predict that large-scale field studies in Japan or China may reveal an unexpectedly extensive presence of *M. ulcerans* subsp. *shinshuense* in the environment that does not necessarily lead to an outbreak of BU.

4.2 Toward Detection of BU in Other Asian Countries

It appears that the stably increased number of reported BU cases from Japan is due to increased awareness of the disease among health workers rather than to an actual rise in the number of infections. On the other hand, the extremely low disease reporting rate in China and the absence of BU reporting from the neighboring Asian countries may be indicative of hidden cases. In non-endemic and less resourced Asian countries, awareness and knowledge of BU disease among health workers and the community are lacking, so there is the possibility of misdiagnosis. Moreover, insufficient health infrastructure and geographical challenges can also contribute to the underreporting of cases. Despite little evidence supporting a predictable future outbreak of BU in Asia, the importance of an early diagnosis cannot be overemphasized because of the potential life-long disfigurement and disabilities this disease can cause if left undiagnosed and untreated. Residents of Asian countries should be aware that BU is a mycobacterial infection that is not limited to tropical or subtropical regions.

Education of health workers, in particular dermatologists, in Asian countries is essential for early diagnosis of BU. Also, it is important to provide public health education to familiarize people, especially those living near rural aquatic environments, with BU symptoms. Since international travel has become more popular, the so-called imported tropical diseases may be seen more frequently than before. In addition to early diagnosis, the establishment of a standard treatment guideline optimized for BU infections in Asian countries is also needed to ensure efficacy without unnecessary surgery and overtreatment. Furthermore, a comprehensive comparison of the proteomes of *M. ulcerans* subsp. *shinshuense* with classical *M. ulcerans* may shed new light on the differences in their adaptive biology and response to antibiotic treatment.

For investigating the phylogenetics of *M. ulcerans* subsp. *shinshuense*, various approaches including genomic and biochemical analyses have been carried out [2, 4, 7, 12, 13, 15]. Only one strain of *M. ulcerans* subsp. *shinshuense* has undergone whole genome sequencing so far [13]. Genomic differences among the sequenced strain of *M. ulcerans* subsp. *shinshuense* and other ecovars of *M. ulcerans* have been found. Further comparative genomic analyses with multiple *M. ulcerans* subsp. *shinshuense* strains, including the Chinese strain, would enable a more precise phylogenetic characterization of *M. ulcerans* subsp. *shinshuense*, which could give information on the spread of these bacteria in Japan as well as in China. However, at the moment, only one Chinese strain has been recognized, so discovery of multiple Chinese strains might further contribute to the above-mentioned phylogenetic resolution. Additionally, it is still unclear whether the clinical features caused by *M. ulcerans* subsp. *shinshuense* differ from those of other *M. ulcerans* strains. Future attempts involving generation of recombinant and mutant strains of *M. ulcerans* subsp. *shinshuense* might clarify the unique characteristics of this *M. ulcerans* subspecies and could provide important clues into the pathogenic mechanisms of BU.

Acknowledgements The authors would like to thank Ms. Kayo Shinozaki for her assistance to prepare the manuscript.

References

1. Mikoshiba H, Shindo Y, Matsumoto H et al (1982) A case of typical mycobacteriosis due to *Mycobacterium ulcerans*-like organism (in Japanese). Jpn J Dermatol 92(5):557–565
2. Tsukamura M, Kaneda K, Imaeda T et al (1989) A taxonomic study on a mycobacterium which caused a skin ulcer in a Japanese girl and resembled *Mycobacterium ulcerans*. Kekkaku 64(11):691–697
3. Kazumi Y, Ohtomo K, Takahashi M et al (2004) *Mycobacterium shinshuense* isolated from cutaneous ulcer lesion of right lower extremity in a 37-year-old woman (in Japanese). Kekkaku 79(7):437–441
4. Nakanaga K, Ishii N, Suzuki K et al (2007) "*Mycobacterium ulcerans* subsp. *shinshuense*" isolated from a skin ulcer lesion: identification based on 16S rRNA gene sequencing. J Clin Microbiol 45(11):3840–3843. https://doi.org/10.1128/JCM.01041-07
5. Yotsu RR, Nakanaga K, Hoshino Y et al (2012) Buruli ulcer and current situation in Japan: a new emerging cutaneous Mycobacterium infection. J Dermatol 39(7):587–593. https://doi.org/10.1111/j.1346-8138.2012.01543.x
6. Yotsu RR, Murase C, Sugawara M et al (2015) Revisiting Buruli ulcer. J Dermatol 42(11):1033–1041. https://doi.org/10.1111/1346-8138.13049
7. Nakanaga K, Hoshino Y, Yotsu RR et al (2011) Nineteen cases of Buruli ulcer diagnosed in Japan from 1980 to 2010. J Clin Microbiol 49(11):3829–3836. https://doi.org/10.1128/JCM.00783-11
8. Sugawara M, Ishii N, Nakanaga K et al (2015) Exploration of a standard treatment for Buruli ulcer through a comprehensive analysis of all cases diagnosed in Japan. J Dermatol 42(6):588–595. https://doi.org/10.1111/1346-8138.12851
9. Trubiano JA, Lavender CJ, Fyfe JA et al (2013) The incubation period of Buruli ulcer (*Mycobacterium ulcerans* infection). PLoS Negl Trop Dis 7(10):e2463. https://doi.org/10.1371/journal.pntd.0002463
10. Stinear TP, Seemann T, Pidot S et al (2007) Reductive evolution and niche adaptation inferred from the genome of *Mycobacterium ulcerans*, the causative agent of Buruli ulcer. Genome Res 17(2):192–200. https://doi.org/10.1101/gr.5942807

11. Tobias NJ, Doig KD, Medema MH et al (2013) Complete genome sequence of the frog pathogen *Mycobacterium ulcerans* ecovar *Liflandii*. J Bacteriol 195(3):556–564. https://doi.org/10.1128/JB.02132-12
12. Pidot SJ, Hong H, Seemann T et al (2008) Deciphering the genetic basis for polyketide variation among mycobacteria producing mycolactones. BMC Genomics 9:462. https://doi.org/10.1186/1471-2164-9-462
13. Yoshida M, Nakanaga K, Ogura Y et al (2016) Complete genome sequence of *Mycobacterium ulcerans* subsp. *shinshuense*. Genome Announc 4(5):e01050-16. https://doi.org/10.1128/genomeA.01050-16
14. Arora G, Sajid A, Singhal A et al (2014) Identification of Ser/Thr kinase and forkhead associated domains in *Mycobacterium ulcerans*: characterization of novel association between protein kinase Q and MupFHA. PLoS Negl Trop Dis 8(11):e3315. https://doi.org/10.1371/journal.pntd.0003315
15. Kaser M, Rondini S, Naegeli M et al (2007) Evolution of two distinct phylogenetic lineages of the emerging human pathogen *Mycobacterium ulcerans*. BMC Evol Biol 7:177. https://doi.org/10.1186/1471-2148-7-177
16. Qi W, Kaser M, Roltgen K et al (2009) Genomic diversity and evolution of *Mycobacterium ulcerans* revealed by next-generation sequencing. PLoS Pathog 5(9):e1000580. https://doi.org/10.1371/journal.ppat.1000580
17. George KM, Chatterjee D, Gunawardana G et al (1999) Mycolactone: a polyketide toxin from *Mycobacterium ulcerans* required for virulence. Science 283(5403):854–857
18. Fidanze S, Song F, Szlosek-Pinaud M et al (2001) Complete structure of the mycolactones. J Am Chem Soc 123(41):10117–10118
19. Ranger BS, Mahrous EA, Mosi L et al (2006) Globally distributed mycobacterial fish pathogens produce a novel plasmid-encoded toxic macrolide, mycolactone F. Infect Immun 74(11):6037–6045. https://doi.org/10.1128/IAI.00970-06
20. Mve-Obiang A, Lee RE, Umstot ES et al (2005) A newly discovered mycobacterial pathogen isolated from laboratory colonies of *Xenopus* species with lethal infections produces a novel form of mycolactone, the *Mycobacterium ulcerans* macrolide toxin. Infect Immun 73(6):3307–3312. https://doi.org/10.1128/IAI.73.6.3307-3312.2005
21. Hong H, Spencer JB, Porter JL et al (2005) A novel mycolactone from a clinical isolate of *Mycobacterium ulcerans* provides evidence for additional toxin heterogeneity as a result of specific changes in the modular polyketide synthase. Chembiochem 6(4):643–648. https://doi.org/10.1002/cbic.200400339
22. Brennan PJ, Nikaido H (1995) The envelope of mycobacteria. Annu Rev Biochem 64:29–63. https://doi.org/10.1146/annurev.bi.64.070195.000333
23. Daffe M, Draper P (1998) The envelope layers of mycobacteria with reference to their pathogenicity. Adv Microb Physiol 39:131–203
24. Hunter SW, Brennan PJ (1981) A novel phenolic glycolipid from *Mycobacterium leprae* possibly involved in immunogenicity and pathogenicity. J Bacteriol 147(3):728–735
25. Daffe M, Lacave C, Laneelle MA et al (1987) Structure of the major triglycosyl phenol-phthiocerol of *Mycobacterium tuberculosis* (strain Canetti). Eur J Biochem 167(1):155–160
26. Dobson G, Minnikin DE, Besra GS et al (1990) Characterisation of phenolic glycolipids from *Mycobacterium marinum*. Biochim Biophys Acta 1042(2):176–181
27. Daffe M, Varnerot A, Levy-Frebault VV (1992) The phenolic mycoside of *Mycobacterium ulcerans*: structure and taxonomic implications. J Gen Microbiol 138(1):131–137. https://doi.org/10.1099/00221287-138-1-131
28. Cho SN, Yanagihara DL, Hunter SW et al (1983) Serological specificity of phenolic glycolipid I from *Mycobacterium leprae* and use in serodiagnosis of leprosy. Infect Immun 41(3):1077–1083
29. Ng V, Zanazzi G, Timpl R et al (2000) Role of the cell wall phenolic glycolipid-1 in the peripheral nerve predilection of *Mycobacterium leprae*. Cell 103(3):511–524

30. Reed MB, Domenech P, Manca C et al (2004) A glycolipid of hypervirulent tuberculosis strains that inhibits the innate immune response. Nature 431(7004):84–87. https://doi.org/10.1038/nature02837
31. Grode L, Seiler P, Baumann S et al (2005) Increased vaccine efficacy against tuberculosis of recombinant *Mycobacterium bovis* bacille Calmette-Guerin mutants that secrete listeriolysin. J Clin Invest 115(9):2472–2479. https://doi.org/10.1172/JCI24617
32. Lin W, Mathys V, Ang EL et al (2012) Urease activity represents an alternative pathway for *Mycobacterium tuberculosis* nitrogen metabolism. Infect Immun 80(8):2771–2779. https://doi.org/10.1128/IAI.06195-11
33. Benedetti MS (1995) Inducing properties of rifabutin, and effects on the pharmacokinetics and metabolism of concomitant drugs. Pharmacol Res 32(4):177–187
34. Chen J, Raymond K (2006) Roles of rifampicin in drug-drug interactions: underlying molecular mechanisms involving the nuclear pregnane X receptor. Ann Clin Microbiol Antimicrob 5:3. https://doi.org/10.1186/1476-0711-5-3
35. Alffenaar JW, Nienhuis WA, de Velde F et al (2010) Pharmacokinetics of rifampin and clarithromycin in patients treated for *Mycobacterium ulcerans* infection. Antimicrob Agents Chemother 54(9):3878–3883. https://doi.org/10.1128/AAC.00099-10
36. Wallace RJ Jr, Brown BA, Griffith DE et al (1995) Reduced serum levels of clarithromycin in patients treated with multidrug regimens including rifampin or rifabutin for *Mycobacterium avium-M. intracellulare* infection. J Infect Dis 171(3):747–750
37. World Health Organization (2015) WHO meeting on Buruli ulcer 2015 abstract. WHO, Geneva
38. Converse PJ, Nuermberger EL, Almeida DV et al (2011) Treating *Mycobacterium ulcerans* disease (Buruli ulcer): from surgery to antibiotics, is the pill mightier than the knife? Future Microbiol 6(10):1185–1198. https://doi.org/10.2217/fmb.11.101
39. Murase C, Kono M, Nakanaga K et al (2015) Buruli ulcer successfully treated with negative-pressure wound therapy. JAMA Dermatol 151(10):1137–1139. https://doi.org/10.1001/jamadermatol.2015.1567
40. Rosul MV, Patskan BM (2016) Ozone therapy effectiveness in patients with ulcerous lesions due to diabetes mellitus. Wiad Lek 69(1):7–9
41. Bertolottim A, Izzo A, Grigolato P et al (2013) The use of ozone therapy in Buruli ulcer had an excellent outcome. BMJ Case Rep 2013:bcr2012008249. https://doi.org/10.1136/bcr-2012-008249
42. Perdrizet GA (2014) Principles and practice of hyperbaric medicine: a medical practitioner's primer, part II. Conn Med 78(7):389–402
43. Dauwe P, Pulikkottil B, Lavery L et al (2014) Does hyperbaric oxygen therapy work in facilitating acute wound healing: a systematic review. Plast Reconstr Surg 133(2):208e–215e. https://doi.org/10.1097/01.prs.0000436849.79161.a4
44. Krieg RE, Wolcott JH, Confer A (1975) Treatment of *Mycobacterium ulcerans* infection by hyperbaric oxygenation. Aviat Space Environ Med 46(10):1241–1245
45. Pszolla N, Sarkar MR, Strecker W et al (2003) *Buruli ulcer*: a systemic disease. Clin Infect Dis 37(6):e78–e82. https://doi.org/10.1086/377170
46. Krieg RE, Wolcott JH, Meyers WM (1979) *Mycobacterium ulcerans* infection: treatment with rifampin, hyperbaric oxygenation, and heat. Aviat Space Environ Med 50(9):888–892
47. Luo Y, Degang Y, Ohtsuka M et al (2015) Detection of *Mycobacterium ulcerans* subsp. *shinshuense* DNA from a water channel in familial Buruli ulcer cases in Japan. Future Microbiol 10(4):461–469. https://doi.org/10.2217/fmb.14.152
48. Ohtsuka M, Kikuchi N, Yamamoto T et al (2014) Buruli ulcer caused by *Mycobacterium ulcerans* subsp. *shinshuense*: a rare case of familial concurrent occurrence and detection of insertion sequence *2404* in Japan. JAMA Dermatol 150(1):64–67. https://doi.org/10.1001/jamadermatol.2013.6816
49. Williamson HR, Benbow ME, Campbell LP et al (2012) Detection of *Mycobacterium ulcerans* in the environment predicts prevalence of Buruli ulcer in Benin. PLoS Negl Trop Dis 6(1):e1506. https://doi.org/10.1371/journal.pntd.0001506

50. Roche B, Benbow ME, Merritt R et al (2013) Identifying the Achilles' heel of multi-host pathogens: the concept of keystone "host" species illustrated by *Mycobacterium ulcerans* transmission. Environ Res Lett 8(4):045009. https://doi.org/10.1088/1748-9326/8/4/045009
51. Garchitorena A, Guegan JF, Leger L et al (2015) *Mycobacterium ulcerans* dynamics in aquatic ecosystems are driven by a complex interplay of abiotic and biotic factors. Elife 4:e07616. https://doi.org/10.7554/eLife.07616
52. Garchitorena A, Ngonghala CN, Texier G et al (2015) Environmental transmission of *Mycobacterium ulcerans* drives dynamics of Buruli ulcer in endemic regions of Cameroon. Sci Rep 5:18055. https://doi.org/10.1038/srep18055
53. Hennigan CE, Myers L, Ferris MJ (2013) Environmental distribution and seasonal prevalence of *Mycobacterium ulcerans* in Southern Louisiana. Appl Environ Microbiol 79(8):2648–2656. https://doi.org/10.1128/AEM.03543-12
54. Williamson HR, Benbow ME, Nguyen KD et al (2008) Distribution of *Mycobacterium ulcerans* in Buruli ulcer endemic and non-endemic aquatic sites in Ghana. PLoS Negl Trop Dis 2(3):e205. https://doi.org/10.1371/journal.pntd.0000205
55. Fyfe JA, Lavender CJ, Johnson PD et al (2007) Development and application of two multiplex real-time PCR assays for the detection of *Mycobacterium ulcerans* in clinical and environmental samples. Appl Environ Microbiol 73(15):4733–4740. https://doi.org/10.1128/AEM.02971-06
56. Stragier P, Ablordey A, Durnez L et al (2007) VNTR analysis differentiates *Mycobacterium ulcerans* and *IS*2404 positive mycobacteria. Syst Appl Microbiol 30(7):525–530. https://doi.org/10.1016/j.syapm.2007.06.001
57. Garchitorena A, Roche B, Kamgang R et al (2014) *Mycobacterium ulcerans* ecological dynamics and its association with freshwater ecosystems and aquatic communities: results from a 12-month environmental survey in Cameroon. PLoS Negl Trop Dis 8(5):e2879. https://doi.org/10.1371/journal.pntd.0002879
58. Stragier P, Ablordey A, Meyers WM et al (2005) Genotyping *Mycobacterium ulcerans* and *Mycobacterium marinum* by using mycobacterial interspersed repetitive units. J Bacteriol 187(5):1639–1647. https://doi.org/10.1128/JB.187.5.1639-1647.2005
59. Faber WR, Arias-Bouda LM, Zeegelaar JE et al (2000) First reported case of *Mycobacterium ulcerans* infection in a patient from China. Trans R Soc Trop Med Hyg 94(3):277–279

Population Genomics and Molecular Epidemiology of *Mycobacterium ulcerans*

Koen Vandelannoote, Miriam Eddyani, Andrew Buultjens, and Timothy P. Stinear

1 First Insights from the Complete Genome of *Mycobacterium ulcerans*

The first genome sequence of a *M. ulcerans* isolate was published in 2007 [1]. This finished and completely annotated genome represented an African clinical isolate (strain name Agy99), which was obtained in 1999 from a BU patient living in the Amansie West District of Ghana. An unexpected feature of the genome was the presence of a circular 174 kpb megaplasmid (named pMUM001) [1, 2]. The plasmid harboured three unusually large genes encoding the polyketide synthases (PKS) required for the biosynthesis of the major virulence factor, mycolactone [2, 3]. The 5.6 Mbp circular chromosome of the Agy99 genome also held some surprises, with the architecture of a bacterium undergoing reductive evolution. There was an abundance of pseudogenes (15% of the predicted ancestral protein-coding genes had been inactivated by accumulated mutations), evidence of large chromosome deletions and rearrangements, and extensive proliferation of two insertion sequences (IS*2404* and IS*2606*) [1]. Collectively, these features pointed to a bacterial population that had 'recently' passed through an evolutionary bottleneck and was adapting to a changed environment. Scrutiny of the types of genes lost by mutation or DNA deletion in the Agy99 genome, suggested a mycobacterium adapting to an environment where the proteins and pathways required to both survive in diverse aquatic environments and persist intracellularly are no longer required. This assessment also showed that *M. ulcerans* has lost many of the proteins and cell-wall associated molecules known to be potent antigens in other notable mycobacterial pathogens

K. Vandelannoote · M. Eddyani
Department of Biomedical Sciences, Institute of Tropical Medicine, Antwerp, Belgium

A. Buultjens · T. P. Stinear (✉)
Department of Microbiology and Immunology, Doherty Institute for Infection and Immunity, University of Melbourne, Melbourne, VIC, Australia
e-mail: tstinear@unimelb.edu.au

G. Pluschke, K. Röltgen (eds.), *Buruli Ulcer*,
https://doi.org/10.1007/978-3-030-11114-4_6

such as *Mycobacterium tuberculosis*, *Mycobacterium kansasii* and *Mycobacterium marinum*. These observations, combined with the presence of pMUM001 and the specific ability of *M. ulcerans* to make the immunosuppressive small molecule mycolactone, suggested a mycobacterium adapting to a protected niche environment where extracellular persistence and an immune evasion phenotype provide a survival advantage [4, 5].

2 An Aquatic Origin and Two Bottlenecks for a Recently Emerged, and Globally Distributed Pathogen

The initial *M. ulcerans* genome assessment was based on a comparison with the complete genome sequence of a single strain of the fish and human pathogen, *M. marinum*. From the mid-2000s onwards, scientists began reporting the presence of mycolactone-producing mycobacteria (MPMs) infecting fish, frogs and other ectotherms worldwide [6–11]. Subsequent comparative genetic and then comparative genomic studies of these mycobacteria with strains of *M. marinum* and *M. ulcerans* confirmed that all MPMs likely emerged during a single evolutionary event when a population of *M. marinum*-like bacteria acquired a pMUM plasmid and the specific ability to make mycolactones and then spread worldwide (Fig. 1) [12, 13]. The key genetic signatures of extensive pseudogene accumulation, expansion of IS*2404* and pMUM plasmid acquisition were present in the ancestor of all MPMs before they radiated around the world, represented by three main lineages, called lineages 1–3 (Fig. 1) [12]. The MPMs have been given a variety of species names including *M. pseudoshottsii*, *M. liflandii*, *M. shinshuense* and (confusingly) *M. marinum*. Based on their extensive shared genomic features and recent common heritage, it has been proposed that all MPMs should be considered under the single species banner of *M. ulcerans* [14]. The collective term *M. ulcerans–M. marinum* complex (abbreviated as MuMC) has also been proposed for all *M. marinum* and *M. ulcerans* as they share >97% nucleotide identity across a substantial 4.3 Mbp conserved core genome [12].

This first major evolutionary bottleneck that saw the emergence of *M. ulcerans* has been followed by at least one further population constriction that gave rise to lineage 3. This lineage, also designated the 'classical' lineage in the literature [15], represents the *M. ulcerans* isolates causing Buruli ulcer in Australia, Papua New

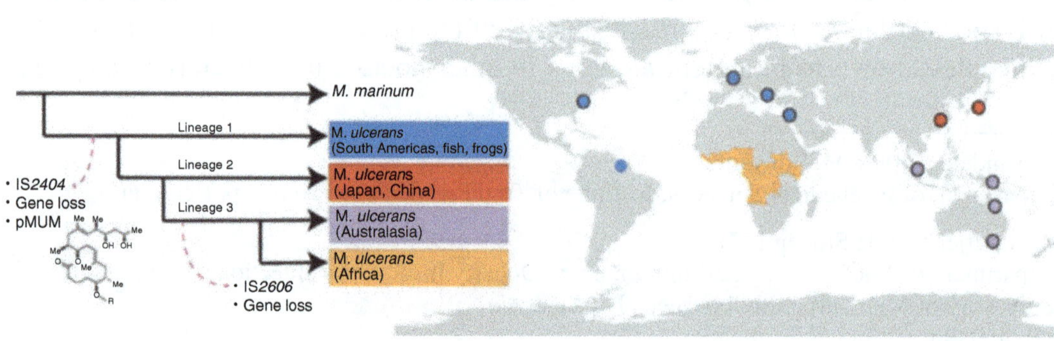

Fig. 1 A recently emerged and globally distributed pathogen

Guinea (PNG), and Africa, and is thus the lineage that accounts for most of the human cases of Buruli ulcer. This lineage is characterized by additional reductive evolution and expansion of IS *2606* to high copy numbers [12]. Interestingly, the 'success' of this lineage as the major human pathogen among all the MPMs is not due to the acquisition of additional genes. Lineage 3 contains no additional DNA sequences compared to isolates in lineages 1 and 2, thus more subtle genetic changes might underlie lineage 3 dominance. One potential genomic region for such changes is the mycolactone PKS gene locus on the pMUM plasmid. The different lineages of *M. ulcerans* produce different mycolactone sidechain structural variants with varying biological potencies (discussed in more detail in chapter "Mycolactone: More than Just a Cytotoxin" of this book). The genetic basis for some of these variants has been determined. Changes in acyltransferase domain substrate specificity within a particular PKS extension module or loss of an extension module within *mlsB*, the gene required for synthesis of the mycolactone sidechain, lead to biosynthesis of different mycolactones [16–18].

Given the ancestor of all *M. ulcerans* was an aquatic mycobacterium (*M. marinum*) and many of the MPMs are recovered from fish, frogs, and turtles, it seems likely that the origin of all MPM (including those lineages associated with Buruli ulcer) had an aquatic animal association. These ideas and inferences should be used to frame thinking around potential reservoirs of *M. ulcerans* in BU endemic areas. Of note too, the phylogeny of *M. ulcerans* from Australia and Papua New Guinea is ancestral to that of *M. ulcerans* from African countries, indicating that the spread of *M. ulcerans* throughout Australasia likely predates the spread of *M. ulcerans* across Africa [33].

3 New Understandings from Genomics on the Spread of *M. ulcerans* Across Africa

African countries carry the highest burden of BU, but until recently, knowledge on the spread of *M. ulcerans* across the continent has been sparse. Early studies of *M. ulcerans* populations using traditional genotyping techniques readily identified a highly clonal population structure, showing that *M. ulcerans* isolates had very conserved genomes that associated strongly with their geographic origin [19–24]. However, these techniques sampled only a small proportion of the mycobacterial genome, and thus lacked sufficient discriminatory power to crack the clonal population structure of this species at even the scale of a country, let alone at the scale of a BU endemic village. The advent of low cost, high throughput DNA sequencing has given researchers access to all the potential genetic variation arising within a *M. ulcerans* population and several studies have capitalised on this advance. Most recently, the genomes of an extraordinary collection of 165 *M. ulcerans* clinical isolates spanning 48 years and representing 11 endemic countries across Africa were sequenced and compared. Assessment of these genomes has produced the first detailed understanding of the introduction and continental spread of this pathogen. Key findings included the establishment of a molecular clock signal in the sequence

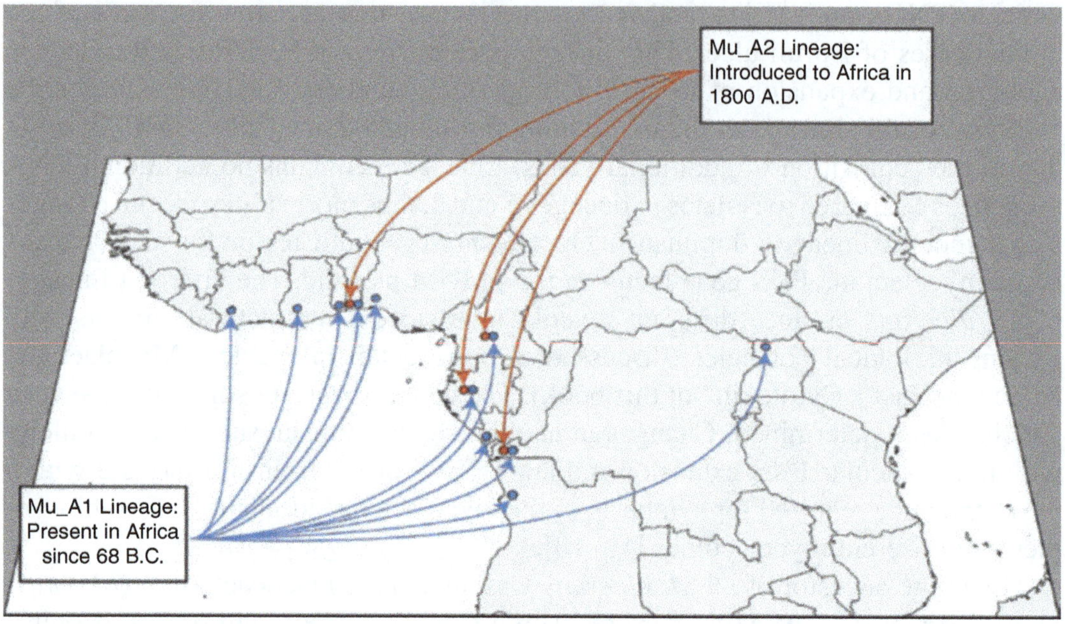

Fig. 2 Spread of two distinct *M. ulcerans* lineage 3 genotypes across Africa

data, suggesting that *M. ulcerans* is accumulating mutations at a rate of 0.33 SNPs per chromosome per year (approx. 6×10^{-8} substitutions/site/year). This rate of genetic change is comparable to *M. tuberculosis*, but 10–100 times slower than that of other pathogens such as *Staphylococcus aureus* [25]. Combining these temporal data with the high-resolution phylogeny inferred from genome comparisons and the geographic origins of the 165 isolates has permitted a reconstruction of the spread of *M. ulcerans* across Africa. The data suggest that there have been two distinct introductions of the bacteria to the African continent; the first occurring around 68 BC, with a likely origin in the area around current day Cameroon and Gabon, and then spreading outwards from these regions [26] (Fig. 2). This has been followed by a second, far more recent introduction during the 1800s. For both the early and recent sub-lineage (called Mu_A1 and Mu_A2 respectively; both representing sub-lineages of lineage 3), there were geographically localized but substantial clonal expansions of *M. ulcerans* populations in four particular hydrological basins (Congo, Kouffo, Oueme, and Nyong). These expansions occurred contemporaneously from the late 1800s onwards and in waves that mimicked interference in those specific regions by colonial powers during the 'scramble for Africa' [26].

Interestingly, the Mu_A2 genotype was also found in PNG, and phylogenetic inference suggests PNG may have been the origin of the Mu_A2 genotype. How the bacteria were transported across oceans and continents from south east Asia to Africa in the mid 1800s is not clear, although again, several European colonial powers were active in both these regions of the world at that time.

The presence of the Mu_A2 genotype in Africa indicates *M. ulcerans* can be mobilised and displaced across large distances, but the dominant characteristic of *M. ulcerans* populations is their strong, geographically restricted genotypes. In Africa, these constrained genotypes align with different hydrogeological basins

[26]. This observation is consistent with the well-described epidemiology of BU, where human disease is strongly associated with lentic and lacustrine environments, and also supports the notion of an aquatic reservoir for the bacterium (or at least a reservoir species restrained somewhat by these basins), with the hydrogeography likely providing a major barrier to reservoir (and therefore pathogen) movement [26, 27].

In South Eastern Australia, the native possum is a wildlife reservoir of *M. ulcerans* and there is a strong association between possums harbouring the bacteria and human Buruli ulcer [28, 29]. Genomics has shown that possum and patient *M. ulcerans* isolates have identical genotypes [29] indicating that humans and possums are part of the same transmission network.

4 Genomic Approaches to Micro-Molecular Epidemiological Investigations of BU

There were expectations at the dawn of the genomic revolution that comparisons of *M. ulcerans* genomes collected from BU patients in endemic areas would reveal striking patterns of bacterial spread that in turn would lead to a substantially deeper understanding of how this enigmatic disease is transmitted [15]. Several *M. ulcerans* population genomic studies have now been conducted in Africa and Australia at the descending scales of country, region and village. Important new insights have been made, whilst also raising new questions about how *M. ulcerans* is spreading.

Several teams have explored genomic variation of *M. ulcerans* populations in West Africa and they have all reported and confirmed the strong association between genotype and region as described above. This relationship has a fractal quality, where genotypes continue to associate with region across large spatial scales; from continental, to country, to regional levels. However, somewhat unexpectedly, micro epidemiological observations powered by the extreme resolution offered by whole genome sequence comparisons have shown that this association breaks down at very local scales (buffer size <50 km^2), where the distribution of *M. ulcerans* genotypes appears to become random. This finding is illustrated in several independent studies from Ghana and Cameroon that report clonal complexes aligned with specific river systems as predicted from earlier pre-genomic studies, but there is also mixing of genotypes at district and village levels [30–32]. Conclusions from these observations include the possibility that (1) the disease is being vectored by a somewhat mobile entity, perhaps an insect or (2) people are moving to different local areas and acquiring the infection. While the data are not conclusive for any specific hypothesis, there are examples of very young infants with no history of any travel developing BU and becoming infected with genotypes present in more distant villages, thus suggesting something(s) is vectoring *M. ulcerans* to susceptible human hosts [32].

In addition to the genotypic mixing at local scales, there are also several examples now of very distinctly different genotypes co-circulating with a 'local' endemic *M. ulcerans* clone. This was first described in the Asante Akim North district of

Ghana, where a genotype reminiscent of strains originating from Nigeria was detected alongside strains representing a local endemic clone [32]. The same phenomenon was recently described in south east Australia, with a distinct *M. ulcerans* clone from the far east of the country detected in a highly BU endemic area, co-circulating with a local clone [33]. These studies demonstrate the potential for *M. ulcerans* to be not recent at all mobilized and spread over large distances. The mechanisms of pathogen dispersal remain to be discovered, but perhaps it could be linked to the movement of human or other mammalian reservoirs [30, 31, 33].

M. ulcerans population genomic studies of highly active BU endemic areas in south east Australia have revealed some interesting features of the pathogen and the disease [33]. A study of 178 *M. ulcerans* genomes, collected over 70-years has provided a compelling reconstruction of the temporal and spatial spread of the pathogen in that region. The disease appears to have emerged in the early 1800s in the east of the country and then spread suddenly westward, moving into areas around the major population centre of Melbourne in the 1980s. Comparison of the temporal phylogeny with epidemiological data indicates that arrival of the bacteria in a specific region predates the appearance of human disease by 7–10 years. Similar to the African genomic studies, these observations and inferences suggest strongly that *M. ulcerans* is spreading, rather than awakening a reservoir of quiescent bacteria, perhaps following environmental disturbance. This information could help inform control strategies for BU, where environmental surveillance of the pathogen might help predict the occurrence of disease in humans.

5 Distinguishing Relapse from Reinfection and Familial Studies

An application of genomics has been to try and establish if a patient presenting twice with BU might have suffered a relapse of the previous infection or has been unlucky enough to be reinfected with a different population of bacteria. Distinguishing between these two scenarios is important for informing treatment options. Work in this area is just beginning, but in a study of four patients with recurrent BU episodes three were concluded to have relapsed, with 0–1 SNP differences between first and last obtained isolates from each patient, while the fourth patient was a possible example of re-infection with 20 SNPs different between first and last isolates [34]. This was the first study to deploy genomics in this way for informing BU treatment. The correct interpretation of these comparisons is obviously dependent on the availability of isolates from the patient over time, but equally important is a detailed understanding of the local population structure of *M. ulcerans* in that region. This understanding is necessary to give the proper context and therefore interpretation to any SNP differences discovered over time within a single patient. Another focused genomics study with similar aims was reported from Australia, where the research team sequenced and compared six *M. ulcerans* genomes recovered from three familial clusters [35]. The sequence data, combined with epidemiological information argued against person-to-person transmission and suggested that there was a relatively short time window of 1–2 months when family members were exposed to the pathogen [35].

6 Summary and Future Perspectives

M. ulcerans is likely a niche adapted mycobacterium. Its ability to make mycolactone is almost certainly critical to the persistence of the bacterium in that niche. The bacterium's primary host and reservoir is likely aquatic. This reservoir is probably restrained within lentic systems. BU spreads when the bacterium moves from one region to another. Pathogen spread can occur rapidly. These are some of the useful insights that genomic approaches have yielded. From glimpses into the unusual biology of the pathogen when the genome was first revealed, to the large, population-based studies that reconstruct pathogen spread, genomics has been the key tool to drive these new understandings. Future research should capitalize on the advent of cost-effective long-read sequencing to focus on establishing and comparing complete *M. ulcerans* genomes. The additional DNA sequence variation that potentially lies within complex regions of the *M. ulcerans* genome could prove very informative for efforts to track pathogen spread. More effort should be made too, to interrogate and predict metabolic pathways from genome sequence data that can then be exploited to improve efforts to detect and isolate *M. ulcerans* in pure culture from environmental sources. The pMUM plasmid in particular should be a focus for complete sequence assembly, as assessment of the mycolactone PKS genes will inform on the basis for the biosynthesis of mycolactone structural variants and differential pathogenesis.

References

1. Stinear TP, Seemann T, Pidot S, Frigui W, Reysset G, Garnier T et al (2007) Reductive evolution and niche adaptation inferred from the genome of *Mycobacterium ulcerans*, the causative agent of Buruli ulcer. Genome Res 17(2):192–200
2. Stinear TP, Mve-Obiang A, Small PL, Frigui W, Pryor MJ, Brosch R et al (2004) Giant plasmid-encoded polyketide synthases produce the macrolide toxin of *Mycobacterium ulcerans*. Proc Natl Acad Sci U S A 101(5):1345–1349
3. Stinear TP, Hong H, Frigui W, Pryor MJ, Brosch R, Garnier T et al (2005) Common evolutionary origin for the unstable virulence plasmid pMUM found in geographically diverse strains of *Mycobacterium ulcerans*. J Bacteriol 187(5):1668–1676
4. Stinear TP, Seemann T, Harrison PF, Jenkin GA, Davies JK, Johnson PD et al (2008) Insights from the complete genome sequence of *Mycobacterium marinum* on the evolution of *Mycobacterium tuberculosis*. Genome Res 18(5):729–741
5. Demangel C, Stinear TP, Cole ST (2009) Buruli ulcer: reductive evolution enhances pathogenicity of *Mycobacterium ulcerans*. Nat Rev Microbiol 7(1):50–60
6. Mve-Obiang A, Lee RE, Umstot ES, Trott KA, Grammer TC, Parker JM et al (2005) A newly discovered mycobacterial pathogen isolated from laboratory colonies of Xenopus species with lethal infections produces a novel form of mycolactone, the *Mycobacterium ulcerans* macrolide toxin. Infect Immun 73(6):3307–3312

7. Ranger BS, Mahrous EA, Mosi L, Adusumilli S, Lee RE, Colorni A et al (2006) Globally distributed mycobacterial fish pathogens produce a novel plasmid-encoded toxic macrolide, mycolactone F. Infect Immun 74(11):6037–6045
8. Rhodes MW, Kator H, McNabb A, Deshayes C, Reyrat JM, Brown-Elliott BA et al (2005) Mycobacterium pseudoshottsii sp. nov., a slowly growing chromogenic species isolated from Chesapeake Bay striped bass (*Morone saxatilis*). Int J Syst Evol Microbiol 55(Pt 3):1139–1147
9. Suykerbuyk P, Vleminckx K, Pasmans F, Stragier P, Ablordey A, Tran HT et al (2007) Mycobacterium liflandii infection in European colony of Silurana tropicalis. Emerg Infect Dis 13(5):743–746
10. Tobias NJ, Doig KD, Medema MH, Chen H, Haring V, Moore R et al (2013) Complete genome sequence of the frog pathogen *Mycobacterium ulcerans* ecovar Liflandii. J Bacteriol 195(3):556–564
11. Sakaguchi K, Iima H, Hirayama K, Okamoto M, Matsuda K, Miyasho T et al (2011) *Mycobacterium ulcerans* infection in an Indian flap-shelled turtle (*Lissemys punctata punctata*). J Vet Med Sci 73(9):1217–1220
12. Doig KD, Holt KE, Fyfe JA, Lavender CJ, Eddyani M, Portaels F et al (2012) On the origin of *Mycobacterium ulcerans*, the causative agent of Buruli ulcer. BMC Genomics 13:258
13. Yip MJ, Porter JL, Fyfe JA, Lavender CJ, Portaels F, Rhodes M et al (2007) Evolution of *Mycobacterium ulcerans* and other mycolactone-producing mycobacteria from a common *Mycobacterium marinum* progenitor. J Bacteriol 189(5):2021–2029
14. Pidot SJ, Asiedu K, Kaser M, Fyfe JA, Stinear TP (2010) *Mycobacterium ulcerans* and other mycolactone-producing mycobacteria should be considered a single species. PLoS Negl Trop Dis 4(7):e663
15. Roltgen K, Stinear TP, Pluschke G (2012) The genome, evolution and diversity of *Mycobacterium ulcerans*. Infection Genet Evol 12(3):522–529
16. Hong H, Spencer JB, Porter JL, Leadlay PF, Stinear T (2005) A novel mycolactone from a clinical isolate of *Mycobacterium ulcerans* provides evidence for additional toxin heterogeneity as a result of specific changes in the modular polyketide synthase. Chembiochem 6(4):643–648
17. Hong H, Stinear T, Skelton P, Spencer JB, Leadlay PF (2005) Structure elucidation of a novel family of mycolactone toxins from the frog pathogen *Mycobacterium sp.* MU128FXT by mass spectrometry. Chem Commun 34:4306–4308
18. Pidot SJ, Hong H, Seemann T, Porter JL, Yip MJ, Men A et al (2008) Deciphering the genetic basis for polyketide variation among mycobacteria producing mycolactones. BMC Genomics 9:462
19. Portaels F, Fonteyene PA, de Beenhouwer H, de Rijk P, Guedenon A, Hayman J et al (1996) Variability in 3' end of 16S rRNA sequence of *Mycobacterium ulcerans* is related to geographic origin of isolates. J Clin Microbiol 34(4):962–965
20. Stinear TP, Jenkin GA, Johnson PD, Davies JK (2000) Comparative genetic analysis of *Mycobacterium ulcerans* and *Mycobacterium marinum* reveals evidence of recent divergence. J Bacteriol 182(22):6322–6330
21. Hilty M, Yeboah-Manu D, Boakye D, Mensah-Quainoo E, Rondini S, Schelling E et al (2006) Genetic diversity in *Mycobacterium ulcerans* isolates from Ghana revealed by a newly identified locus containing a variable number of tandem repeats. J Bacteriol 188(4):1462–1465
22. Stragier P, Ablordey A, Bayonne LM, Lugor YL, Sindani IS, Suykerbuyk P et al (2006) Heterogeneity among *Mycobacterium ulcerans* isolates from Africa. Emerg Infect Dis 12(5):844–847
23. Kaser M, Hauser J, Small P, Pluschke G (2009) Large sequence polymorphisms unveil the phylogenetic relationship of environmental and pathogenic mycobacteria related to *Mycobacterium ulcerans*. Appl Environ Microbiol 75(17):5667–5675
24. Jackson K, Edwards R, Leslie DE, Hayman J (1995) Molecular method for typing *Mycobacterium ulcerans*. J Clin Microbiol 33(9):2250–2253
25. Duchene S, Holt KE, Weill FX, Le Hello S, Hawkey J, Edwards DJ et al (2016) Genome-scale rates of evolutionary change in bacteria. Microb Genom 2(11):e000094

26. Vandelannoote K, Meehan CJ, Eddyani M, Affolabi D, Phanzu DM, Eyangoh S et al (2017) Multiple introductions and recent spread of the emerging human pathogen *Mycobacterium ulcerans* across Africa. Genome Biol Evol 9(3):414–426
27. Vandelannoote K, Jordaens K, Bomans P, Leirs H, Durnez L, Affolabi D et al (2014) Insertion sequence element single nucleotide polymorphism typing provides insights into the population structure and evolution of *Mycobacterium ulcerans* across Africa. Appl Environ Microbiol 80(3):1197–1209
28. Carson C, Lavender CJ, Handasyde KA, O'Brien CR, Hewitt N, Johnson PD et al (2014) Potential wildlife sentinels for monitoring the endemic spread of human buruli ulcer in South-East Australia. PLoS Negl Trop Dis 8(1):e2668
29. Fyfe JA, Lavender CJ, Handasyde KA, Legione AR, O'Brien CR, Stinear TP et al (2010) A major role for mammals in the ecology of *Mycobacterium ulcerans*. PLoS Negl Trop Dis 4(8):e791
30. Lamelas A, Ampah KA, Aboagye S, Kerber S, Danso E, Asante-Poku A et al (2016) Spatiotemporal co-existence of two *Mycobacterium ulcerans* clonal complexes in the Offin River Valley of Ghana. PLoS Negl Trop Dis 10(7):e0004856
31. Bolz M, Bratschi MW, Kerber S, Minyem JC, Um Boock A, Vogel M et al (2015) Locally confined clonal complexes of *Mycobacterium ulcerans* in two Buruli ulcer endemic regions of Cameroon. PLoS Negl Trop Dis 9(6):e0003802
32. Ablordey AS, Vandelannoote K, Frimpong IA, Ahortor EK, Amissah NA, Eddyani M et al (2015) Whole genome comparisons suggest random distribution of *Mycobacterium ulcerans* genotypes in a Buruli ulcer endemic region of Ghana. PLoS Negl Trop Dis 9(3):e0003681
33. Buultjens AH, Vandelannoote K, Meehan CJ, Eddyani M, de Jong BC, Fyfe JAM et al (2018) Comparative genomics shows *Mycobacterium ulcerans* migration and expansion has preceded the rise of Buruli ulcer in south-eastern Australia. Appl Environ Microbiol 84(8):15
34. Eddyani M, Vandelannoote K, Meehan CJ, Bhuju S, Porter JL, Aguiar J et al (2015) A genomic approach to resolving relapse versus reinfection among four cases of Buruli ulcer. PLoS Negl Trop Dis 9(11):e0004158
35. O'Brien DP, Wynne JW, Buultjens AH, Michalski WP, Stinear TP, Friedman ND et al (2017) Exposure risk for infection and lack of human-to-human transmission of *Mycobacterium ulcerans* disease, Australia. Emerg Infect Dis 23(5):837–840

Mycolactone: More than Just a Cytotoxin

Laure Guenin-Macé, Marie-Thérèse Ruf, Gerd Pluschke, and Caroline Demangel

From their observation of necrotic areas around bacterial foci in Buruli ulcers (BUs), Connor and co-workers were the first, back in 1965, to suggest that *M. ulcerans* may produce a diffusible cytotoxin [1]. This hypothesis was later confirmed by injecting mycobacterial culture filtrates into the skin of guinea pigs, showing that this causes focal necrosis resembling that of naturally occurring human infections [1–3]. In 1999, George et al. succeeded in isolating a cytotoxic factor from *M. ulcerans* lipid extracts, and deciphered its chemical nature [4]. The *M. ulcerans* toxin was named mycolactone, based on its mycobacterial origin and macrolactone structure: a 12-membered lactone ring, to which a C5-O-linked polyunsaturated acyl side chain and a C-linked upper side chain comprising C12–C20 are appended (Fig. 1). Follow-up studies showed that *M. ulcerans*-derived mycolactone was in reality a mixture of two stereoisomers called A and B [6, 7] (Fig. 1). Since the initial discovery of mycolactone A/B, eight additional mycolactone congeners have been identified that are either produced by *M. ulcerans* strains of different geographical origins or genetically related fish and frog pathogens, which were initially given different species designations such as *M. pseudoshottsii* and *M. liflandii.* Comparative genome analysis later revealed that all mycolactone-producing mycobacteria evolved from a common *M. marinum*-like progenitor, and are therefore ecovars of a single *M. ulcerans* species [8]. While the macrolide core structure and upper side

L. Guenin-Macé
Immunobiology of Infection Unit, Institut Pasteur, INSERM, U1221, Paris, France

M.-T. Ruf · G. Pluschke
Swiss Tropical and Public Health Institute, Basel, Switzerland

University of Basel, Basel, Switzerland
e-mail: gerd.pluschke@swisstph.ch

C. Demangel (✉)
Immunobiology of Infection Unit, Institut Pasteur, Paris, France
e-mail: Demangel@pasteur.fr

G. Pluschke, K. Röltgen (eds.), *Buruli Ulcer*,
https://doi.org/10.1007/978-3-030-11114-4_7

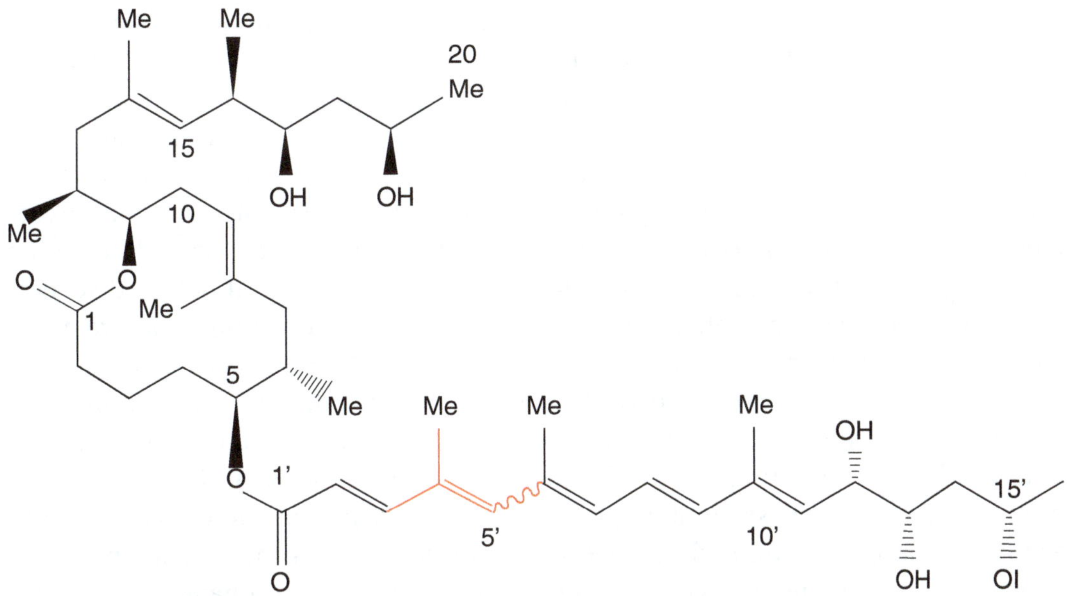

Fig. 1 Structure of mycolactone A/B. The red line indicates the region where stereoisomers differ (from [5])

chain are conserved, mycolactone populations from different *M. ulcerans* sub-lineages vary in the length, number and localization of hydroxyl groups and number of double bonds of the lower side chain. These modifications of the lower polyunsaturated acyl side chain cause pronounced changes in cytotoxicity [9–11]. The origin and chemistry of natural mycolactones, structure-activity relationships and the various approaches that were developed to generate synthetic mycolactones have been reviewed recently [12, 13]. These aspects will therefore not be further discussed here. Instead, this chapter provides an update on our understanding of mycolactone (A/B) biology, and discusses its proposed mechanisms of action in relation with BU pathogenesis.

1 Mycolactone and BU Disease

1.1 Pharmacodistribution

Mycolactone's diffusion in infected hosts was initially believed to coincide with its cytocidal action. However, in mice experimentally infected with *M. ulcerans* in footpads, the distinctive mass spectrometric signature of the toxin was detected in peripheral blood cells, spleen, liver and kidneys [14], showing that bacterially-produced mycolactone distributes far beyond infected tissues. The presence of mycolactone was assessed by extracting total lipids from homogenized organs with organic solvents, and analyzing the resulting acetone-soluble fractions by liquid chromatography tandem-mass spectrometry (LC-MS/MS). The body-wide distribution of mycolactone was further supported by animal studies using a radiolabeled form of the toxin, generated by feeding *M. ulcerans* cultures with [1–^{14}C] propionic acid and [1,2–^{14}C] acetic acid. In mice injected with

^{14}C-mycolactone via the subcutaneous, intravenous or intraperitoneal routes, radioactivity was indeed detected in peripheral blood, spleen, liver and kidneys after 24 h [14]. Using the LC-MS/MS approach, structurally-intact mycolactone was detected in human skin biopsies harvested from the center of pre-ulcerative BU lesions or the edge of viable skin around ulcers [15], in ulcer exudates and in serum samples from newly diagnosed patients and patients undergoing antibiotic treatment [16]. While toxin concentrations within lesions declined upon *M. ulcerans* killing by antibiotic treatment, mycolactone's presence was still detected several weeks after completion of therapy [15, 16], indicating a slow elimination rate. By spiking cell pellets or serum samples with purified toxin, we noted the poor yield of mycolactone extraction with organic solvents (around 10%). Assuming that 90% of mycolactone is lost during the process, serum concentrations of mycolactone would thus be estimated to fall within the 0–20 nM range at the start of antibiotic therapy [16]. These numbers should nevertheless be considered with caution, as they are based on a limited number of measurements. How mycolactone traffics from bacteria infecting the skin to peripheral blood and distant organs, and whether it accumulates preferentially in certain cell populations or tissues during the course of BU disease remains largely unknown.

At the cellular level, mycolactone is currently believed to diffuse passively through the plasma membrane. In support of this model, fluorescently labeled mycolactones were shown to penetrate fibroblasts in a non-saturable and non-competitive manner within minutes [17] and to accumulate in the cytoplasmic compartment [9, 18]. Two studies using molecular dynamics simulations on the one hand, and Langmuir monolayers as membrane models on the other hand, recently suggested that the passage of mycolactone across cellular membranes may nevertheless alter their dynamic properties, and cause mechanical and physical perturbations [19, 20].

1.2 Mycolactone Contribution to BU Disease Manifestations

1.2.1 Pathogenesis and Histopathology of BU Lesions

BU is a chronic necrotizing skin disease affecting primarily subcutaneous adipose tissue. Lesions usually start as single, painless, subcutaneous nodule or papule. The dermis and epidermis overlying the lesion eventually degenerates and sloughs off, leading to the development of ulcers with undermined edges and a necrotic slough at the base. Contiguous coagulation necrosis of the deep dermis and subcutaneous fat tissue with destruction of blood vessels and interstitial edema, epidermal hyperplasia and the presence of fat cell ghosts and extracellular clusters of acid fast bacilli (AFB) constitute hallmarks of BU disease. In advanced lesions the necrotic process may extend through deep fascia and expose deeper structures like muscle or bone. In early lesions, necrosis of dermal collagen is leading to a fibrillary appearance, but gradually all cellular and structural details are disappearing and an amorphous coagulum is developing in the center of the lesion [21]. Clusters of extracellular acid-fast bacilli are primarily found in the deep layers of the necrotic adipose tissue [22], but tissue destruction extends far beyond (Fig. 2), which is ascribed to the

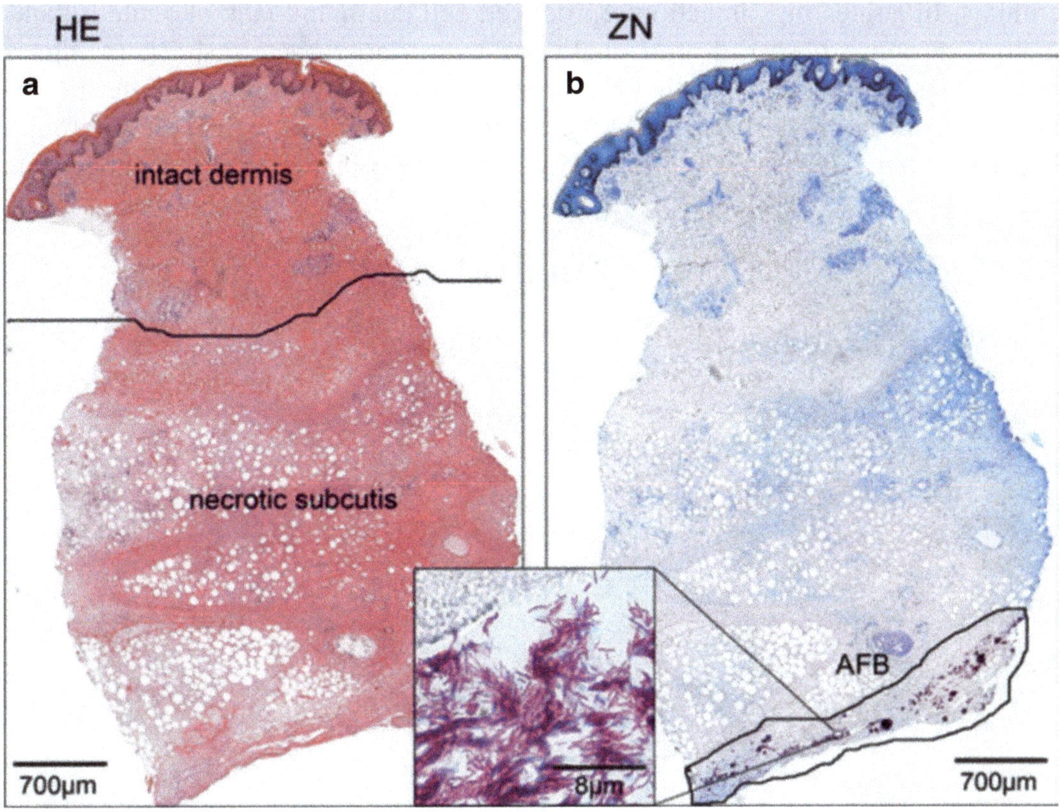

Fig. 2 Histopathological features of a tissue punch biopsy from an untreated human BU plaque lesion. Histological sections were stained with Haematoxylin-Eosin (HE) or Ziehl-Neelsen (ZN). (**a**) The subcutaneous tissue presents with the typical signs of BU histopathology: massive necrosis, fat cell ghosts and edema. The epidermis presents with epidermal hyperplasia. (**b**) A band of extracellular ZN positive AFB is present focally in the deep layer of the necrotic subcutis (adapted from [23])

diffusion of mycolactone. In the center of active lesions, inflammatory cells are rarely found (Fig. 2) and cellular responses are defective at both the local and the systemic levels. There is multiple evidence, summarized below, that mycolactone contributes directly to these disease manifestations.

The most commonly used experimental infection models for BU is the mouse model, in which inoculation of *M. ulcerans* into the footpads, ears or the tail causes lesions that develop over several weeks. Evaluation of antibiotic treatment modalities in this model has been shown to be predictive for treatment efficacy in humans. However the mouse model has some limitations for the investigation of the pathogenesis of BU in humans, since the composition of the tissue at the injection sites differs markedly from human skin. Therefore the pig has been recently evaluated as a model to study the early pathogenesis of BU [24]. Pig and human skin share many similarities like thickness, general structure of the epidermis, dermis and the subcutaneous tissue [25], making porcine skin a preferred model for burn and wound healing studies [26]. Also guinea pigs have been used to study early host-pathogen interaction and in particular the effect of mycolactone [27]. However both in pigs and guinea pigs no chronic infections are establishing [28].

The key role of mycolactone in the pathogenesis of *M. ulcerans* infection has been first demonstrated by injection of this molecule into the skin of experimental animals. In guinea pigs injection reproduced cell death and lack of acute inflammatory response observed after challenge with viable bacteria [27]. In pigs a necrotic lesion is developing that is surrounded by an infiltration belt after mycolactone injection (Fig. 3b), resembling closely the histopathological picture found in early human lesions (Fig. 3a). The key role of mycolactone in BU pathogenesis

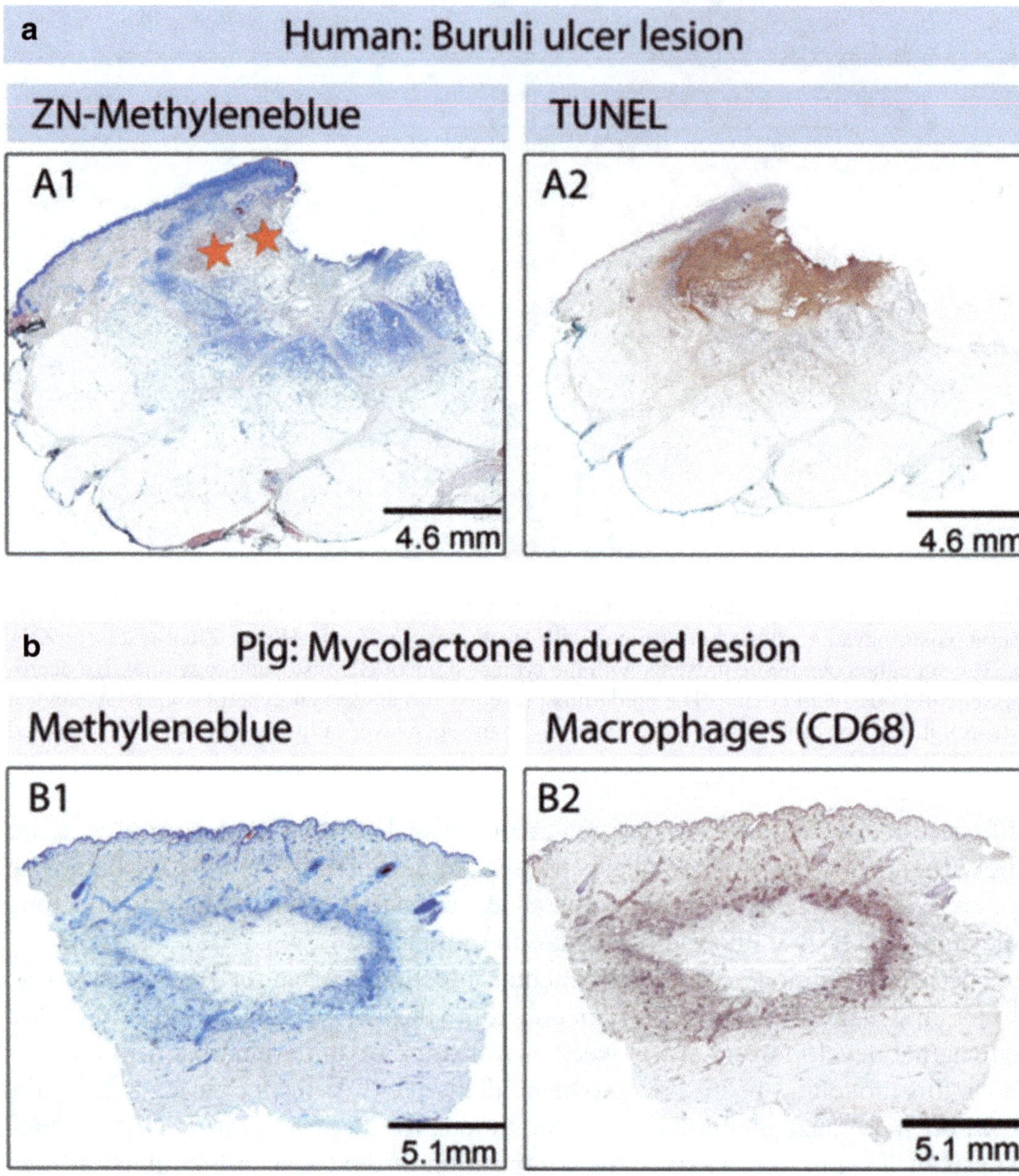

Fig. 3 Formation of a necrotic lesion core surrounded by an infiltration belt. Histological sections were stained with Ziehl-Neelsen (ZN)/Methyleneblue (*A1*) or Methyleneblue alone (*B1*), the "In situ cell death detection kit" (*A2*) or by immunohistochemistry for the macrophage marker CD68 (*B2*). (**a**) Early human BU lesion presenting with a necrotic core containing extracellular clusters of AFB (location indicated by red stars) surrounded by a belt of infiltrating leukocytes demonstrated by the blue staining of nuclei (*A1*). Strong TUNEL staining inside the necrotic core and in some of the infiltrating cells of the infiltration belt (*B1*). (**b**) Lesions in the skin of a pig (Sus scrofa) induced by a single subcutaneous injection of 50 µg of synthetic mycolactone A/B. The toxin caused formation of a lesion with a necrotic core (*B1* and *B2*) surrounded by a dense belt of infiltrating cells consisting primarily of CD68 positive macrophages (*B2*) (adapted from [29])

has been further demonstrated by infection of experimental animals with mycolactone negative *M. ulcerans* mutants. While wild-type bacteria cause an extracellular infection, the mycolactone negative mutants produce an intracellular inflammatory infection [30, 31] similar to that of *M. marinum*, the progenitor of *M. ulcerans* [8].

M. ulcerans is multiplying in the sanctuary of the central necrotic core of the lesion, which is usually free of living infiltrating leukocytes, although large numbers of macrophages and lymphocytes are collecting already in early lesions at the margin of the necrotic center of the lesions [29]. These infiltrates form a belt, which separates the highly confined necrotic core from the still intact tissue surrounding the lesion (Fig. 3a). Infiltration of leukocytes may be reduced by sublethal concentrations of mycolactone affecting chemokine production by skin dendritic cells (DCs) [32] (see Sect. 1.2.3), while cells entering the necrotic core may undergo apoptosis caused by mycolactone's diffusion from bacterial foci in the center of the lesions (see below). Some of the leukocytes in the infiltration belt are TUNEL (terminal deoxynucleotidyl transferase dUTP nick end labeling) staining positive [29], which is regarded as a sign for double-strand DNA breaks generated during apoptosis. The necrotic core of BU lesions shows a very strong, but diffuse TUNEL staining (Fig. 3a). At the periphery of BU lesions, interaction of phagocytes with AFBs and intracellular *M. ulcerans* can be observed and in particular in resolving BU lesions, granulomatous inflammation can be found. During antibiotic treatment of BU, intra-lesional concentrations of mycolactone decline [15]. This allows leukocytes to reach the extracellular mycobacteria located in the necrotic core of the lesion, leading to their phagocytosis and destruction [33]. In turn, chronic leukocyte infiltration cumulates in the development of ectopic lymphoid structures [34]. Gathering of DCs and development of defined granuloma structures indicate that antigen recognition and processing is leading to active *M. ulcerans* specific immune responses. Three main types of mixed infiltrates have been identified in antibiotic treated lesions, highly organized epithelioid granulomas, diffuse infiltrates and dense lymphocyte clusters in proximity to vessels. AFB already start to be internalized by phagocytes after 2 weeks of treatment, when *M. ulcerans* can still be cultured from tissue homogenates. Antibiotic treatment may act rapidly on the bacterial production of mycolactone, whereas complete killing of bacteria may take longer.

The cellular mechanism(s) of mycolactone-induced ulceration of the skin were investigated by exposing fibroblasts, epithelial cells and endothelial cells to purified mycolactone in vitro and monitoring dose-response effects on cell viability. Prolonged exposure (>48 h) to nanomolar concentrations (>10 nM) of mycolactone A/B induced the detachment and death of all skin cells studied. This included mouse (L929) and human (HDF) fibroblast cell lines, human epithelial (HeLa) and keratinocyte (HaCaT) cell lines [17, 18, 27, 35–37], and human dermal microvascular endothelial cells [38]. Primary human keratinocytes were relatively more susceptible to mycolactone-mediated cytotoxicity, with cell mortality being detected after only 24 h of treatment [39, 40]. Notably, cell detachment was preceded by alterations of the actin cytoskeleton [17, 18, 35]. In HeLa cells exposed to mycolactone, production of spike-like protrusions was activated after 30 min, while actin polymerization abnormally localized in the perinuclear area after 4 h, coinciding with a near-complete incapacity of the cells to adhere to plastic wells [18]. Longer treatments (16–24 h) induced defects in epithelial cell capacity to establish cell-cell contacts and close wounds in vitro [18, 41]. In the mouse ear model, intradermal injection of mycolactone caused a dramatic loss of E-cadherin–adhesive contacts in the stratum spinosum, coupled to thinning of the external stratum granulosum and lucidum, resulting in a significant decrease in epidermis width after 54 h [18]. In accordance with the coagulative necrosis in BU lesions, exposing human dermal microvascular endothelial cells to mycolactone depleted thrombomodulin from the cell surface [38]. Mycolactone also decreased the production of collagen biosynthesis enzymes by L929 fibroblasts [35]. Together, these studies suggest that mycolactone provokes BU formation by concomitantly reducing the viability and the healing potential of skin cells, altering the strength of cell-cell and cell-extracellular matrix adhesion, and affecting coagulation control.

1.2.2 Local Analgesia

Early reports favored the hypothesis that BU-associated analgesia was primarily due to nerve destruction. Indeed, the histopathological examination of BU biopsies showed local axonal damages, with loss of myelin in 24% of the patients [42]. Consistently, mouse footpad infection with *M. ulcerans* induced nerve fiber degeneration in advanced ulcers [43], and injection of purified mycolactone in mouse footpads triggered neurological damages associated with hyposensitivity [44]. The possibility that mycolactone causes analgesia by directly killing sensory neurons and Schwann cells, the peripheral nervous system (PNS) cells generating and transmitting pain signals, was investigated in vitro. Anand and co-workers observed significant neurite degeneration in rat and human primary dorsal root ganglion (DRG) sensory neurons after 24 h of exposure to a 100 nM concentration of toxin [45]. Consistently, we found that exposing mouse DRG to 25 nM mycolactone for 16 h downregulated the production of multiple proteins mediating interactions between neurons and the extracellular matrix [46]. Longer treatments (>48 h) induced massive mortality of primary DRGs in both studies [45, 47], although a third study reported a minimal loss of viability following exposure to mycolactone doses of up to 70 μM [48]. Schwann cells, the principal glia of the PNS, are important in many aspects of peripheral nerve biology including the conduction of nervous impulses along axons, nerve regeneration upon injury and production of the nerve extracellular matrix. Prolonged (>48 h) exposure to mycolactone also caused significant mortality in the mouse Schwann cell line SW10 [49], and primary rat Schwann cells (IC_{50} of 1 nM) [47]. Of note, microglia showed comparable susceptibility to mycolactone-induced toxicity, with decreased viability manifesting after 48 h of treatment with mycolactone concentrations varying between 3 and 10 nM [47]. Whether mycolactone diffuses into peripheral nerves and reaches the central nervous system (CNS) in vivo is unknown. However, these cellular studies fully support the view that BU-associated analgesia may be due, at least partially, to cytopathic effects at the lesion level.

Notably, Marsollier, Brodin and co-workers showed that infection with *M. ulcerans* can induce local hypoesthesia at early stages of disease, in the absence of apparent nerve destruction [50]. Here, pain was generated by applying a focused radiant heat source onto infected footpads, and pain perception was monitored by an adaptation of the tail-flick procedure, in mice that had been previously anesthetized. Injection of 5 μg mycolactone in mouse footpads produced the same effects after 2 h and during 2 days, with normal nociceptive abilities being recovered after 8 days. Parallel to this, we showed that systemic delivery of mycolactone (2 μg) limits the development of distantly generated inflammatory pain [10]. Mycolactone was administered intraperitoneally 1 h before formalin injection into the hind paws, and stereotypical licking and biting behaviors were monitored after 10 min, the time that it takes for inflammatory processes to develop. In addition to demonstrating the intrinsic analgesic properties of mycolactone, these animal studies indicated that bacterially-derived mycolactone reduces pain perception in infected hosts by other mechanisms than cytotoxicity. Section 2 reviews the molecular targets that have been proposed to explain these observations.

1.2.3 Local and Systemic Immunomodulation

Transient defects in the systemic production of IFN-γ have been repeatedly reported in BU patients with progressive ulcers [51–55]. The relative lack of IFN-γ in ex vivo-stimulated blood samples from BU patients reversed at late stages of disease, which may explain why it was not observed in two studies [56, 57]. Of note, IFN-γ production defects resolved after surgical excision of the lesions, indicating their association with *M. ulcerans* bacteria [54]. These defects were observed with antigen-specific [51, 52] and non-specific [53–55] activation stimuli (such as the T-cell mitogen PHA), suggesting that *M. ulcerans* infection induces T cell anergy. In line with this hypothesis, multi-analyte profiling of PHA-stimulated whole blood culture supernatants revealed that most T cell-derived cytokines were suppressed during BU progression [55]. Moreover, patients with BU displayed a distinctive immunosuppressive signature in their serum, marked by a downregulation of multiple mediators of inflammation [55, 58]. In all, these studies revealed that *M. ulcerans* impairs the normal development of inflammatory and cellular immune responses to infection.

Downregulation of cytokines and chemokines in patient sera was still observed 1 month after the end of antibiotic therapy [55, 58]. Since mycolactone was detected in lesions and serum of patients several weeks after they had completed antibiotic treatment [15, 16], the concept emerged that mycolactone may be the bacterial factor causing immunological defects in infected individuals. This hypothesis was supported by mouse studies showing that infection with wild-type *M. ulcerans*, but not with a mycolactone-deficient mutant, causes systemic defects in IL-2 production [14]. Pioneering work conducted by Foxwell and co-workers showed that mycolactone prevents the LPS-induced release of TNF and IL-10 by human monocytes, and production of IL-2 by activated T lymphocytes, under conditions not altering cell viability [59]. Subsequent studies by our group and others examined in detail the cell-type specific effects of mycolactone on immune functions, with the following conclusions. Comparably to skin cells and cells of the nervous system, human primary macrophages and DCs were susceptible to prolonged (>48 h) treatment with mycolactone (>10 nM) [10, 32]. In contrast, the viability of human primary T cells and polymorphonuclear neutrophils treated under the same conditions was minimally affected [10], showing that certain cell types are more resistant than others against mycolactone toxicity. Notably, non-cytotoxic treatment with mycolactone prevented the phenotypic and functional maturation of DCs, limiting their ability to activate T cells and produce inflammatory chemokines in response to TLR stimulation [32]. It decreased DC expression of MHC class I and II, leading to impaired direct and indirect antigen presentation [60]. In monocytes and macrophages, mycolactone blocked the activation-induced production of cytokines and chemokines post-transcriptionally and irrespective of the activation stimulus [61, 62]. Mycolactone also downregulated the basal expression of TCR and homing receptor L-selectin (CD62L) by resting T cells, leading to altered responsiveness to TCR stimulation and impaired capacity to reach peripheral lymph nodes in vivo [10, 63, 64]. It also limited the T cell capacity to produce cytokines in response to activation stimuli that bypass the TCR, at the post-transcriptional level. In mice, the systemic administration of mycolactone conferred protection against chronic skin inflammation and rheumatoid arthritis, demonstrating the anti-inflammatory potency of mycolactone in vivo [10]. Mycolactone also displayed potent anti-inflammatory activity on the nervous system, with nanomolar concentrations of mycolactone preventing the activation-induced production of pro-inflammatory mediators by activated DRGs, Schwann cells and microglia in conditions that did not affect cell viability [47]. In intrathecally-injected rats, mycolactone decreased significantly the production of inflammatory cytokines in the spinal cord without inducing detectable cytotoxicity [47]. The overall conclusion of these studies was that mycolactone represents a potent natural immunosuppressor operating by a novel molecular mechanism. They supported the view that *M. ulcerans* produces mycolactone to evade recognition and control by the host immune system.

2 Molecular Targets and Mechanisms of Action

In recent years, three distinct molecular targets have been proposed to explain the diverse biological effects of mycolactone. Proposed mechanisms of action are discussed below, starting from the most recently discovered, the Sec61 translocon, as we feel it is important to re-examine previous findings in the light of these new data.

2.1 Sec61 Blockade

Simmonds and co-workers made a breakthrough in 2014 by discovering that mycolactone blocks the production of TNF and other model secretory proteins by preventing protein translocation into the endoplasmic reticulum (ER), leading to their subsequent degradation in the cytosol by the ubiquitin:proteasome system

[62]. Using cell-free systems, High and co-workers then showed that mycolactone causes selective defects in the cotranslational insertion of Sec61 secretory substrates [65]. Sec61 is a heterotrimeric complex which in eukaryotes mediates the transport of almost all secretory proteins (secreted proteins and most ER-, Golgi-, endosome- and lysosome-resident proteins, as well as proteins containing glycosylphosphatidylinositol-anchors) and integral transmembrane proteins (TMPs) into the ER [66]. We found that mycolactone operates by targeting the pore-forming subunit of the Sec61 translocon, and identified a single amino acid mutation conferring resistance to mycolactone, localizing its interaction site near the lumenal plug of Sec61 [67]. Expression of this Sec61 mutant in mycolactone-treated T cells rescued their homing potential and effector functions, providing definitive genetic evidence that Sec61 is the host receptor mediating the immunomodulatory effects of mycolactone [67]. Compared to previously known Sec61 inhibitors (cotransin and derivatives, Apratoxin A), mycolactone was more effective at blocking the production of cytokines and homing receptors by immune cells [46, 67]. Mycolactone is therefore the first Sec61 inhibitor that is produced by a human pathogen, and the most potent inhibitor to date.

TMPs can be divided into Type I, II or III, according to the presence of a signal peptide (SP) and the orientation of the protein's N-terminus at the ER membrane [68]. Using in vitro assays of single-pass TMP translocation, High and co-workers observed that secretory proteins, Type I and Type II TMPs were generally susceptible to mycolactone-mediated Sec61 blockade, while in contrast, mycolactone had no effect on Type III TMP integration [65, 69]. This led the authors to hypothesize that protein resistance to mycolactone essentially depends on how the protein initially engages the translocon. Our profiling of mycolactone's signature in the proteome of $CD4^+$ T lymphocytes, DCs and DRG neurons broadly supported this mechanistic model for the effects of mycolactone on different classes of single-spanning TMPs, and extended its validity to multi-pass TMPs [46, 60, 67]. Importantly, these global analyses also highlighted proteomic alterations beyond Sec61 substrates, resulting from secondary effects of protein translocation blockade. For instance, by blocking the production of both IFN-γ and IFN-γ receptor in activated lymphocytes, mycolactone affected the autocrine activation of multiple IFN-responsive genes [67]. In macrophages, it resulted in altered IFN-γ-induced production of nitric oxide synthase (iNOS) [67], which is essential for the control of *M. ulcerans* infection. In addition to highlighting the critical importance of Sec61 activity for immune cell function, migration and communication, these data thus provided a molecular explanation for the immunological defects of BU patients.

Mycolactone was recently shown to promote Bim-dependent cell apoptosis through the mTORC2-Akt-FoxO3 axis [36]. Using L929 fibroblasts as a model, Bieri et al. found that mycolactone inhibits the mechanistic Target of Rapamycin (mTOR) signaling pathway, leading to inactivation of the serine/threonine kinase Akt, subsequent activation of the transcription factor FoxO3, and up-regulation of pro-apoptotic Bim. The importance of Bim-dependent apoptosis in the pathogenesis was demonstrated by *M. ulcerans* infection of Bim knock-out mice, which did not develop necrotic BU lesions and was able to contain the mycobacterial multiplication [36].

How mycolactone interferes with mTORC2 signaling nevertheless remained elusive. Subsequent studies by our group showed that expression of mycolactone-resistant mutants of Sec61 in HEK293 cells protected them against the cytopathic effects of the toxin [67], establishing the critical role of Sec61 blockade in mycolactone-induced cell death. Recently, Simmonds and co-workers reported that inhibition of Sec61-dependent translocation by mycolactone rapidly activates an integrated stress response (ISR) in RAW264.7 and HeLa cells without activating ER stress sensors, driving cytotoxicity via the ATF4/CHOP/Bcl-2/Bim route [70]. Our studies in DCs led to comparable conclusions with regard to mycolactone stimulating ATF4/CHOP signaling, however we detected ER stress-specific activation signals within hours of treatment [46]. Mycolactone-driven ER stress nevertheless differed from conventional unfolded protein response (UPR) by the downregulation of GRP78/BIP, a master regulator of the UPR that is normally induced by canonical ER stress. Whether mycolactone-driven ATF4 induction results primarily from ISR or UPR may thus depend on the cell type. Of note, there is convincing evidence that chronic UPR causes Akt inhibition through complex interactions with mTORC1/C2 signaling [71], providing a possible explanation for how Sec61 blockade activates the mTORC2-Akt-FoxO3 axis, leading to apoptosis by a parallel route downstream of the UPR [36]. Importantly, both studies support the view that mycolactone-mediated cytotoxicity is a late consequence of Sec61 blockade, resulting from the induction of chronic stress triggering apoptosis via the ATF4/CHOP/Bcl-2/Bim signaling pathway.

In summary, by blocking Sec61, mycolactone downregulates the production of a large array of secretory and membrane receptors. One major effect of this molecular blockade is that the host is unable to mount an effective immune response to the underlying mycobacterial infection. Prolonged cell exposure to mycolactone is however cytotoxic, since it triggers stress responses inducing apoptosis.

2.2 AT2R Stimulation

In 2014, Marsollier, Brodin and co-workers identified type 2 angiotensin II receptors (AT2R) as additional targets of mycolactone, mediating its analgesic properties [50]. In primary hippocampal neurons, mycolactone (>7 nM) provoked cell hyperpolarization within 20 min, through activation of KCN4 (TRAAK) potassium channels. The authors showed that mycolactone operates by activating AT2R, leading to phospholipase A2-mediated arachidonic acid (AA) liberation, and generation of prostaglandin E2 (PGE2) from AA by cyclooxygenase-1, which activates KCN4. Mycolactone also caused the hyperpolarization of primary DRGs, at concentrations superior to 1 μM [48]. The importance of the AT2R signaling cascade in mycolactone-driven hypoesthesia was demonstrated by use of an AT2R selective blocker in mice infected with *M. ulcerans*, or injected with mycolactone, and AT2R-KO animals as controls [50]. AT2R silencing in HeLa cells had no effect on mycolactone-driven defects in IFN-γ receptor expression nor MCP-1 chemokine production, indicating that AT2R is not involved in mycolactone-mediated immunomodulation [67]. Based on our current understanding of mycolactone-mediated Sec61 blockade, the biogenesis of the multi-spanning membrane protein AT2R may resist its inhibitory effects on protein translocation [69]. Together with the recent evidences of Sec61-dependent anti-inflammatory activity of mycolactone on the immune and nervous systems, and toxicity in neurons (see Sect. 2.1), these data suggest that mycolactone-mediated activation of AT2R and inhibition of Sec61 are two independent mechanisms contributing to BU-associated analgesia.

2.3 N-WASP Activation

In 2013, we reported that mycolactone activates the Wiskott-Aldrich syndrome protein (WASP) family of actin-nucleating factors, leading to uncontrolled assembly of actin [18]. WASP and N-WASP belong to a family of scaffold proteins mediating the dynamic remodeling of the actin cytoskeleton, via interaction of their C-terminal verprolin-cofilin-acidic (VCA) domain with the ARP2/3 actin-nucleating complex. While WASP expression is restricted to hematopoietic cells, N-WASP is widely expressed. In basal conditions, WASP and N-WASP are auto-inhibited by intramolecular interactions sequestering the VCA domain from ARP2/3. Binding of activated GTPases or phosphoinositide lipids to N-terminal target sequences triggers conformational changes resulting in release of the VCA, thereby enabling binding and activation of ARP2/3. Using in vitro assays of actin polymerization, we showed that mycolactone mimics endogenous GTPase CDC42 in disruption of WASP/N-WASP auto-inhibition. The possible contribution of WASP/N-WASP activation to mycolactone immunosuppressive effects was ruled out by silencing or inhibiting the proteins in cellular assays of cytokine production [18]. In contrast, N-WASP inhibition by wiskostatin in HeLa epithelial cells partly relieved mycolactone-induced alterations in actin polymerization and cell adhesion. A Sec61 blockade is unlikely to explain such early, wiskostatin-sensitive effects of mycolactone. We propose that a fraction of mycolactone may bind to cytosolic WASP/N-WASP following its diffusion through the plasma membrane, leading to the uncontrolled assembly of actin and defective cell-matrix adhesion. In support of this model, a limited, yet significant co-localization of mycolactone with active WASP was observed after 1 h of treatment in HeLa cells [18]. Mycolactone injection in mouse epidermis induced structural changes in the multilayered organization of this tissue that were suppressed by wiskostatin, suggesting that mycolactone-driven activation of N-WASP and subsequent defects in cell adhesion occur in vivo. While this mechanism may not be central to BU pathogenesis, we speculate that it may synergize with Sec61 blockade to impair skin integrity.

3 Conclusions

Recent years have witnessed tremendous progress in our understanding of the molecular mechanisms underpinning mycolactone biology (summarized in Fig. 4), and therefore BU pathogenesis. Notably, studies on mycolactone have revealed a

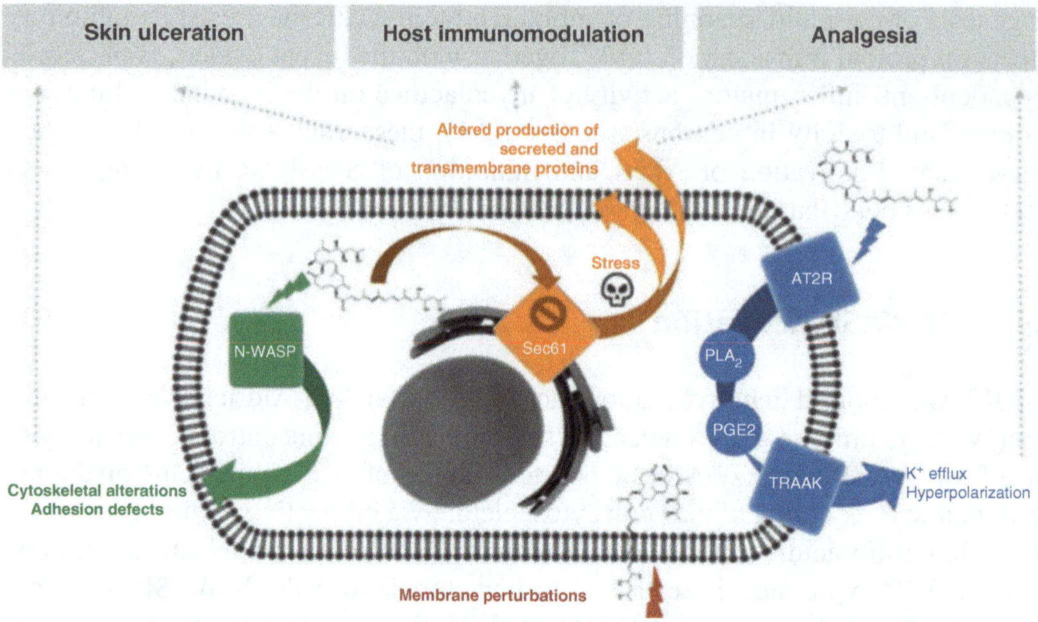

Fig. 4 Proposed molecular targets and mechanisms of action for mycolactone-mediated ulcerative, immunosuppressive and analgesic properties

novel mechanism of immunomodulation based on protein translocation blockade, which may be exploited therapeutically to limit inflammation and pain. Meanwhile, attempts have been made to develop a live vaccine for BU based on immunization with mycolactone-negative *M. ulcerans* strains [8]. However, vaccination only transiently delayed the onset of pathology induced by a highly virulent *M. ulcerans* strain [72]. While mycolactone is an obvious target antigen for a BU subunit vaccine, initial attempts by several groups to develop an immunogenic formulation of the toxoid failed [73]. Recent studies nevertheless indicated that toxin neutralizing antibodies can be elicited by protein-carrier conjugates of synthetic mycolactone [37], providing new prospects for the development of a BU vaccine.

References

1. Connor DH, Lunn HF (1965) *Mycobacterium ulcerans* infection (with comments on pathogenesis). Int J Lepr 33(Suppl):698–709
2. Krieg RE, Hockmeyer WT, Connor DH (1974) Toxin of *Mycobacterium ulcerans*. Production and effects in guinea pig skin. Arch Dermatol 110:783–788
3. Read JK, Heggie CM, Meyers WM, Connor DH (1974) Cytotoxic activity of *Mycobacterium ulcerans*. Infect Immun 9:1114–1122
4. George KM, Chatterjee D, Gunawardana G, Welty D, Hayman J, Lee R, Small PL (1999) Mycolactone: a polyketide toxin from *Mycobacterium ulcerans* required for virulence. Science 283:854–857
5. Kishi Y (2011) Chemistry of mycolactones, the causative toxins of Buruli ulcer. Proc Natl Acad Sci U S A 108:6703–6708
6. Benowitz AB, Fidanze S, Small PLC, Kishi Y (2001) Stereochemistry of the core structure of the mycolactones. J Am Chem Soc 123:5128–5129

7. Fidanze S, Song FB, Szlosek-Pinaud M, Small PLC, Kishi Y (2001) Complete structure of the mycolactones. J Am Chem Soc 123:10117–10118
8. Doig KD, Holt KE, Fyfe JA, Lavender CJ, Eddyani M, Portaels F, Yeboah-Manu D, Pluschke G, Seemann T, Stinear TP (2012) On the origin of *Mycobacterium ulcerans*, the causative agent of Buruli ulcer. BMC Genomics 13:258
9. Chany AC, Casarotto V, Schmitt M, Tarnus C, Guenin-Mace L, Demangel C, Mirguet O, Eustache J, Blanchard N (2011) A diverted total synthesis of mycolactone analogues: an insight into Buruli ulcer toxins. Chemistry 17:14413–14419
10. Guenin-Mace L, Baron L, Chany AC, Tresse C, Saint-Auret S, Jonsson F, Le Chevalier F, Bruhns P, Bismuth G, Hidalgo-Lucas S, Bisson JF, Blanchard N, Demangel C (2015) Shaping mycolactone for therapeutic use against inflammatory disorders. Sci Transl Med 7: 289ra285
11. Scherr N, Gersbach P, Dangy JP, Bomio C, Li J, Altmann KH, Pluschke G (2013) Structure-activity relationship studies on the macrolide exotoxin mycolactone of *Mycobacterium ulcerans*. PLoS Negl Trop Dis 7:e2143
12. Gehringer M, Altmann KH (2017) The chemistry and biology of mycolactones. Beilstein J Org Chem 13:1596–1660
13. Saint-Auret S, Chany AC, Casarotto V, Tresse C, Parmentier L, Abdelkafi H, Blanchard N (2017) Total syntheses of Mycolactone A/B and its analogues for the exploration of the biology of Buruli ulcer. Chimia (Aarau) 71:836–840
14. Hong H, Coutanceau E, Leclerc M, Caleechurn L, Leadlay PF, Demangel C (2008) Mycolactone diffuses from *Mycobacterium ulcerans*-infected tissues and targets mononuclear cells in peripheral blood and lymphoid organs. PLoS Negl Trop Dis 2:e325
15. Sarfo FS, Phillips RO, Zhang J, Abass MK, Abotsi J, Amoako YA, Adu-Sarkodie Y, Robinson C, Wansbrough-Jones MH (2014) Kinetics of mycolactone in human subcutaneous tissue during antibiotic therapy for *Mycobacterium ulcerans* disease. BMC Infect Dis 14:202
16. Sarfo FS, Le Chevalier F, Aka N, Phillips RO, Amoako Y, Boneca IG, Lenormand P, Dosso M, Wansbrough-Jones M, Veyron-Churlet R, Guenin-Mace L, Demangel C (2011) Mycolactone diffuses into the peripheral blood of buruli ulcer patients - implications for diagnosis and disease monitoring. PLoS Negl Trop Dis 5:e1237
17. Snyder DS, Small PL (2003) Uptake and cellular actions of mycolactone, a virulence determinant for Mycobacterium ulcerans. Microb Pathog 34:91–101
18. Guenin-Mace L, Veyron-Churlet R, Thoulouze MI, Romet-Lemonne G, Hong H, Leadlay PF, Danckaert A, Ruf MT, Mostowy S, Zurzolo C, Bousso P, Chretien F, Carlier MF, Demangel C (2013) Mycolactone activation of Wiskott-Aldrich syndrome proteins underpins Buruli ulcer formation. J Clin Invest 123(4):1501–1512
19. Lopez CA, Unkefer CJ, Swanson BI, Swanson JMJ, Gnanakaran S (2018) Membrane perturbing properties of toxin mycolactone from *Mycobacterium ulcerans*. PLoS Comput Biol 14:e1005972
20. Nitenberg M, Benarouche A, Maniti O, Marion E, Marsollier L, Gean J, Dufourc EJ, Cavalier JF, Canaan S, Girard-Egrot AP (2018) The potent effect of mycolactone on lipid membranes. PLoS Pathog 14:e1006814
21. Hayman J (1993) Out of Africa: observations on the histopathology of *Mycobacterium ulcerans* infection. J Clin Pathol 46:5–9
22. Ruf MT, Bolz M, Vogel M, Bayi PF, Bratschi MW, Sopho GE, Yeboah-Manu D, Boock AU, Junghanss T, Pluschke G (2016) Spatial distribution of *Mycobacterium ulcerans* in Buruli ulcer lesions: implications for laboratory diagnosis. Plos Negl Trop Dis 10:e0004767
23. Ruf MT, Sopoh GE, Brun LV, Dossou AD, Barogui YT, Johnson RC, Pluschke G (2011) Histopathological changes and clinical responses of Buruli ulcer plaque lesions during chemotherapy: a role for surgical removal of necrotic tissue? PLoS Negl Trop Dis 5:e1334
24. Bolz M, Ruggli N, Borel N, Pluschke G, Ruf MT (2016) Local cellular immune responses and pathogenesis of Buruli ulcer lesions in the experimental *Mycobacterium ulcerans* pig infection model. PLoS Negl Trop Dis 10:e0004678

25. Liu Y, Chen JY, Shang HT, Liu CE, Wang Y, Niu R, Wu J, Weiv H (2010) Light microscopic, Electron microscopic, and Immunohistochemical comparison of Bama Minipig (Sus scrofa domestica) and human skin. Comp Med 60:142–148
26. Sullivan TP, Eaglstein WH, Davis SC, Mertz P (2001) The pig as a model for human wound healing. Wound Repair Regen 9:66–76
27. George KM, Pascopella L, Welty DM, Small PL (2000) A *Mycobacterium ulcerans* toxin, mycolactone, causes apoptosis in guinea pig ulcers and tissue culture cells. Infect Immun 68:877–883
28. Silva-Gomes R, Marcq E, Trigo G, Goncalves CM, Longatto-Filho A, Castro AG, Pedrosa J, Fraga AG (2015) Spontaneous healing of *Mycobacterium ulcerans* lesions in the guinea pig model. PLoS Negl Trop Dis 9:e0004265
29. Ruf MT, Steffen C, Bolz M, Schmid P, Pluschke G (2017) Infiltrating leukocytes surround early Buruli ulcer lesions, but are unable to reach the mycolactone producing mycobacteria. Virulence 8:1918–1926
30. Coutanceau E, Marsollier L, Brosch R, Perret E, Goossens P, Tanguy M, Cole ST, Small PL, Demangel C (2005) Modulation of the host immune response by a transient intracellular stage of *Mycobacterium ulcerans*: the contribution of endogenous mycolactone toxin. Cell Microbiol 7:1187–1196
31. Adusumilli S, Mve-Obiang A, Sparer T, Meyers W, Hayman J, Small PLC (2005) *Mycobacterium ulcerans* toxic macrolide, mycolactone modulates the host immune response and cellular location of M-ulcerans in vitro and in vivo. Cell Microbiol 7:1295–1304
32. Coutanceau E, Decalf J, Martino A, Babon A, Winter N, Cole ST, Albert ML, Demangel C (2007) Selective suppression of dendritic cell functions by *Mycobacterium ulcerans* toxin mycolactone. J Exp Med 204:1395–1403
33. Schutte D, Pluschke G (2009) Immunosuppression and treatment-associated inflammatory response in patients with *Mycobacterium ulcerans* infection (Buruli ulcer). Expert Opin Biol Ther 9:187–200
34. Schutte D, Um-Boock A, Mensah-Quainoo E, Itin P, Schmid P, Pluschke G (2007) Development of highly organized lymphoid structures in Buruli ulcer lesions after treatment with rifampicin and streptomycin. PLoS Negl Trop Dis 1:e2
35. Gama JB, Ohlmeier S, Martins TG, Fraga AG, Sampaio-Marques B, Carvalho MA, Proenca F, Silva MT, Pedrosa J, Ludovico P (2014) Proteomic analysis of the action of the *Mycobacterium ulcerans* toxin mycolactone: targeting host cells cytoskeleton and collagen. PLoS Negl Trop Dis 8:e3066
36. Bieri R, Scherr N, Ruf MT, Dangy JP, Gersbach P, Gehringer M, Altmann KH, Pluschke G (2017) The macrolide toxin mycolactone promotes bim-dependent apoptosis in Buruli ulcer through inhibition of mTOR. ACS Chem Biol 12:1297–1307
37. Dangy JP, Scherr N, Gersbach P, Hug MN, Bieri R, Bomio C, Li J, Huber S, Altmann KH, Pluschke G (2016) Antibody-mediated neutralization of the exotoxin Mycolactone, the main virulence factor produced by *Mycobacterium ulcerans*. PLoS Negl Trop Dis 10:e0004808
38. Ogbechi J, Ruf MT, Hall BS, Bodman-Smith K, Vogel M, Wu HL, Stainer A, Esmon CT, Ahnstrom J, Pluschke G, Simmonds RE (2015) Mycolactone-dependent depletion of endothelial cell thrombomodulin is strongly associated with fibrin deposition in Buruli ulcer lesions. PLoS Pathog 11:e1005011
39. Bozzo C, Tiberio R, Graziola F, Pertusi G, Valente G, Colombo E, Small PL, Leigheb G (2010) A *Mycobacterium ulcerans* toxin, mycolactone, induces apoptosis in primary human keratinocytes and in HaCaT cells. Microbes Infect 12:1258–1263
40. Gronberg A, Zettergren L, Bergh K, Stahle M, Heilborn J, Angeby K, Small PL, Akuffo H, Britton S (2010) Antioxidants protect keratinocytes against *M. ulcerans* mycolactone cytotoxicity. PLoS One 5:e13839
41. Graziola F, Colombo E, Tiberio R, Leigheb G, Bozzo C (2017) *Mycobacterium ulcerans* mycolactone interferes with adhesion, migration and proliferation of primary human keratinocytes and HaCaT cell line. Arch Dermatol Res 309:179–189

42. Zavattaro E, Boccafoschi F, Borgogna C, Conca A, Johnson RC, Sopoh GE, Dossou AD, Colombo E, Clemente C, Leigheb G, Valente G (2012) Apoptosis in Buruli ulcer: a clinicopathological study of 45 cases. Histopathology 61:224–236
43. Goto M, Nakanaga K, Aung T, Hamada T, Yamada N, Nomoto M, Kitajima S, Ishii N, Yonezawa S, Saito H (2006) Nerve damage in *Mycobacterium ulcerans*-infected mice - probable cause of painlessness in Buruli ulcer. Am J Pathol 168:805–811
44. En J, Goto M, Nakanaga K, Higashi M, Ishii N, Saito H, Yonezawa S, Hamada H, Small PL (2008) Mycolactone is responsible for the painlessness of *Mycobacterium ulcerans* infection (buruli ulcer) in a murine study. Infect Immun 76:2002–2007
45. Anand U, Sinisi M, Fox M, MacQuillan A, Quick T, Korchev Y, Bountra C, McCarthy T, Anand P (2016) Mycolactone-mediated neurite degeneration and functional effects in cultured human and rat DRG neurons: mechanisms underlying hypoalgesia in Buruli ulcer. Mol Pain 12:1744806916654144
46. Morel JD, Paatero AO, Wei J, Yewdell JW, Guenin-Macé L, Van Haver D, Impens F, Pietrosemoli N, Paavilainen VO, Demangel C (2018) Proteomics reveals scope of mycolactone-mediated Sec61 blockade and distinctive stress signature. Mol Cell Proteomics 17(9):1750–1765. https://doi.org/10.1074/mcp.RA118.000824
47. Isaac C, Mauborgne A, Grimaldi A, Ade K, Pohl M, Limatola C, Boucher Y, Demangel C, Guenin-Mace L (2017) Mycolactone displays anti-inflammatory effects on the nervous system. PLoS Negl Trop Dis 11:e0006058
48. Song OR, Kim HB, Jouny S, Ricard I, Vandeputte A, Deboosere N, Marion E, Queval CJ, Lesport P, Bourinet E, Henrion D, Oh SB, Lebon G, Sandoz G, Yeramian E, Marsollier L, Brodin P (2017) A bacterial toxin with analgesic properties: hyperpolarization of DRG neurons by Mycolactone. Toxins (Basel) 9:E227
49. En J, Kitamoto S, Kawashima A, Yonezawa S, Kishi Y, Ishii N, Goto M (2017) Mycolactone cytotoxicity in Schwann cells could explain nerve damage in Buruli ulcer. PLoS Negl Trop Dis 11:e0005834
50. Marion E, Song OR, Christophe T, Babonneau J, Fenistein D, Eyer J, Letournel F, Henrion D, Clere N, Paille V, Guerineau NC, Saint Andre JP, Gersbach P, Altmann KH, Stinear TP, Comoglio Y, Sandoz G, Preisser L, Delneste Y, Yeramian E, Marsollier L, Brodin P (2014) Mycobacterial toxin induces analgesia in buruli ulcer by targeting the angiotensin pathways. Cell 157:1565–1576
51. Gooding TM, Johnson PD, Campbell DE, Hayman JA, Hartland EL, Kemp AS, Robins-Browne RM (2001) Immune response to infection with *Mycobacterium ulcerans*. Infect Immun 69:1704–1707
52. Prevot G, Bourreau E, Pascalis H, Pradinaud R, Tanghe A, Huygen K, Launois P (2004) Differential production of systemic and intralesional gamma interferon and interleukin-10 in nodular and ulcerative forms of Buruli disease. Infect Immun 72:958–965
53. Westenbrink BD, Stienstra Y, Huitema MG, Thompson WA, Klutse EO, Ampadu EO, Boezen HM, Limburg PC, van der Werf TS (2005) Cytokine responses to stimulation of whole blood from patients with buruli ulcer disease in Ghana. Clin Diagn Lab Immun 12:125–129
54. Yeboah-Manu D, Peduzzi E, Mensah-Quainoo E, Asante-Poku A, Ofori-Adjei D, Pluschke G, Daubenberger CA (2006) Systemic suppression of interferon-gamma responses in Buruli ulcer patients resolves after surgical excision of the lesions caused by the extracellular pathogen *Mycobacterium ulcerans*. J Leukoc Biol 79:1150–1156
55. Phillips R, Sarfo FS, Guenin-Mace L, Decalf J, Wansbrough-Jones M, Albert ML, Demangel C (2009) Immunosuppressive signature of cutaneous *Mycobacterium ulcerans* infection in the peripheral blood of patients with Buruli ulcer disease. J Infect Dis 200(11):1675–1684
56. Nausch N, Antwi-Berko D, Mubarik Y, Abass KM, Owusu W, Owusu-Dabo E, Debrah LB, Debrah AY, Jacobsen M, Phillips RO (2017) Analysis of *Mycobacterium ulcerans*-specific T-cell cytokines for diagnosis of Buruli ulcer disease and as potential indicator for disease progression. PLoS Negl Trop Dis 11:e0005415
57. Phillips R, Horsfield C, Kuijper S, Sarfo SF, Obeng-Baah J, Etuaful S, Nyamekye B, Awuah P, Nyarko KM, Osei-Sarpong F, Lucas S, Kolk AH, Wansbrough-Jones M (2006) Cytokine

response to antigen stimulation of whole blood from patients with *Mycobacterium ulcerans* disease compared to that from patients with tuberculosis. Clin Vaccine Immunol 13:253–257
58. Phillips RO, Sarfo FS, Landier J, Oldenburg R, Frimpong M, Wansbrough-Jones M, Abass K, Thompson W, Forson M, Fontanet A, Niang F, Demangel C (2014) Combined inflammatory and metabolic defects reflected by reduced serum protein levels in patients with Buruli ulcer disease. PLoS Negl Trop Dis 8:e2786
59. Pahlevan AA, Wright DJ, Andrews C, George KM, Small PL, Foxwell BM (1999) The inhibitory action of *Mycobacterium ulcerans* soluble factor on monocyte/T cell cytokine production and NF-kappa B function. J Immunol 163:3928–3935
60. Grotzke JE, Kozik P, Morel JD, Impens F, Pietrosemoli N, Cresswell P, Amigorena S, Demangel C (2017) Sec61 blockade by mycolactone inhibits antigen cross-presentation independently of endosome-to-cytosol export. Proc Natl Acad Sci U S A 114:E5910–E5919
61. Simmonds RE, Lali FV, Smallie T, Small PL, Foxwell BM (2009) Mycolactone inhibits monocyte cytokine production by a posttranscriptional mechanism. J Immunol 182:2194–2202
62. Hall BS, Hill K, McKenna M, Ogbechi J, High S, Willis AE, Simmonds RE (2014) The pathogenic mechanism of the *Mycobacterium ulcerans* virulence factor, mycolactone, depends on blockade of protein translocation into the ER. PLoS Pathog 10:e1004061
63. Boulkroun S, Guenin-Mace L, Thoulouze MI, Monot M, Merckx A, Langsley G, Bismuth G, Di Bartolo V, Demangel C (2010) Mycolactone suppresses T cell responsiveness by altering both early signaling and posttranslational events. J Immunol 184:1436–1444
64. Guenin-Mace L, Carrette F, Asperti-Boursin F, Le Bon A, Caleechurn L, Di Bartolo V, Fontanet A, Bismuth G, Demangel C (2011) Mycolactone impairs T cell homing by suppressing microRNA control of L-selectin expression. Proc Natl Acad Sci U S A 108:12833–12838
65. McKenna M, Simmonds RE, High S (2016) Mechanistic insights into the inhibition of Sec61-dependent co- and post-translational translocation by mycolactone. J Cell Sci 129(7):1404–1415
66. Park E, Rapoport TA (2012) Mechanisms of Sec61/SecY-mediated protein translocation across membranes. Annu Rev Biophys 41:21–40
67. Baron L, Paatero AO, Morel JD, Impens F, Guenin-Mace L, Saint-Auret S, Blanchard N, Dillmann R, Niang F, Pellegrini S, Taunton J, Paavilainen VO, Demangel C (2016) Mycolactone subverts immunity by selectively blocking the Sec61 translocon. J Exp Med 213: 2885–2896
68. Goder V, Spiess M (2001) Topogenesis of membrane proteins: determinants and dynamics. FEBS Lett 504:87–93
69. McKenna M, Simmonds RE, High S (2017) Mycolactone reveals the substrate-driven complexity of Sec61-dependent transmembrane protein biogenesis. J Cell Sci 130:1307–1320
70. Ogbechi J, Hall BS, Sbarrato T, Taunton J, Willis AE, Wek RC, Simmonds RE (2018) Inhibition of Sec61-dependent translocation by mycolactone uncouples the integrated stress response from ER stress, driving cytotoxicity via translational activation of ATF4. Cell Death Dis 9:397
71. Appenzeller-Herzog C, Hall MN (2012) Bidirectional crosstalk between endoplasmic reticulum stress and mTOR signaling. Trends Cell Biol 22:274–282
72. Fraga AG, Martins TG, Torrado E, Huygen K, Portaels F, Silva MT, Castro AG, Pedrosa J (2012) Cellular immunity confers transient protection in experimental Buruli ulcer following BCG or mycolactone-negative *Mycobacterium ulcerans* vaccination. PLoS One 7:e33406
73. Einarsdottir T, Huygen K (2011) Buruli ulcer. Hum Vaccin 7:1198–1203

The Immunology of Buruli Ulcer

João Fevereiro, Alexandra G. Fraga, and Jorge Pedrosa

Buruli ulcer (BU) represents a unique human mycobacteriosis. Caused by *Mycobacterium ulcerans*, the disease spectrum is dominated by the activity of mycolactone, a dermanecrotic toxin that has shown the ability to interfere with the immune response. This poses an additional difficulty to the understanding of the immunological determinants for the outcome of infection, a fundamental step to develop better preventive or curative strategies. In this chapter, the immune response against *M. ulcerans* is reviewed, both with a series of clinical observations and experimental infection models and going through several other lines of evidence, including epidemiological and genetic studies. This holistic approach is expected to shed further light on the intriguing pathophysiology of this disease and help guide future research efforts.

1 Buruli Ulcer: The First Histological Observations of a Necrotic Track

Mycobacterium ulcerans has always been a distinct pathogen among mycobacteria. The classical clinical description of *M. ulcerans* infection, Buruli ulcer (BU), reports a lesion that usually begins as an indurated subcutaneous papule which slowly ulcerates, presenting as an ulcer with undermined edges [1–3]. Typical skin biopsies reveal epidermal destruction concurring with areas of hyperplasia and, in the deeper layers of the skin, a widespread panniculitis with many thrombotic capillaries and in some cases calcification of fat cells [1, 3–6]. In the necrotic core, a large number of bacteria are usually present in the extracellular space amidst apoptotic cell debris,

J. Fevereiro · A. G. Fraga · J. Pedrosa (✉)
Life and Health Sciences Research Institute (ICVS), School of Medicine, University of Minho, Braga, Portugal

ICVS/3B's—PT Government Associate Laboratory, Braga, Portugal
e-mail: id5937@alunos.uminho.pt; afraga@med.uminho.pt; jpedrosa@med.uminho.pt

G. Pluschke, K. Röltgen (eds.), *Buruli Ulcer*,
https://doi.org/10.1007/978-3-030-11114-4_8

sometimes surrounded by a belt of scattered neutrophils and macrophages with intracellular bacilli, as well as T cells and foci of B cells [1, 4, 7–10]. As such, suspicions early arose of BU pathology being dominated by an exotoxin that would diffuse and destroy the surrounding tissues [4, 11].

Indeed, purified in 1999, a polyketide-derived macrolide named mycolactone was shown to reproduce many of the histopathological hallmarks of BU disease, inducing extensive tissue necrosis and microvascular thrombosis, in a dose-dependent way [12, 13]. Specifically, mycolactone provokes apoptotic cell death by driving expression of pro-apoptotic proteins BCL2L11 (Bim) and Fas and inhibiting cell cycle progression [14–16]. Furthermore, this toxin has the remarkable ability to bind to the alpha subunit of the Sec61 translocon in the endoplasmatic reticulum (ER), impeding post-translational modifications, such as glycosylation, and translocation of newly synthesized membrane and secretory proteins into and across the ER membrane [17, 18]. Eventually, cytokines interferon gamma (IFN-γ), tumor necrosis factor alpha (TNF-α), interleukin (IL) 6 and cyclooxygenase (Cox) 2, as well as a variety of other secreted proteins, end up in the cytosol, being marked for proteasome degradation [17, 18]. Lastly, mycolactone also disturbs several pathways related to stress response, collagen biosynthesis and cytoskeleton dynamics, the latter mainly achieved by hyperactivation of the Wiskott–Aldrich syndrome protein (WASP) [19, 20].

The pleiotropic effects of mycolactone make *M. ulcerans* infection a unique paradigm of successful manipulation of the host immune orchestra. By inducing cellular death and preventing the generation and processing of crucial signaling molecules, the ability of the host to cope with the pathogen becomes severely crippled. Thus, what the actual role of the immune system is in BU, is a question of utmost importance to understand the disease.

2 Host Attempts to Control *Mycobacterium ulcerans* Infection

2.1 Local Immune Response

Due to its indolent course and various clinical presentations, characterizing the kinetics of the immune response against *M. ulcerans* in humans has always been a challenge. It is therefore not surprising that much of what is known about BU pathogenesis today came from studies resorting to animal models, which seem to present several resemblances to the human findings [1, 21–23].

Although understanding the very early stages of the contact of *M. ulcerans* with the host is still of extreme difficulty, it is currently thought that the pathogen depends on skin surface injuries or transmission vectors to penetrate the epidermal layer of the skin, highly rich in keratinocytes [24–26]. The importance of these cells for BU pathogenesis became evident when it was discovered that upon recognition of *M. ulcerans* through dectin-1, toll-like receptor (TLR) 2 and to a lesser extent TLR4, keratinocytes produced bactericidal reactive oxygen species and cathelicidin, as well as chemokines IL-8 and the monocyte chemoattractant protein (MCP) 1 [27].

Once mycolactone-producing strains of *M. ulcerans* reach the subcutaneous tissue, there is an influx, within the first hours, of neutrophils to the center of the lesion and of major histocompatibility complex (MHC) class II^{+} cells to the periphery of the infection foci, followed by a sustained recruitment of macrophages for at least a week [22, 28]. During this stage, mycobacteria are phagocytosed and, with the contribution of the activity of its superoxide dismustase and catalase enzymes, transiently replicate within phagosomes [9, 29]. Upon replication, the pathogen is released to the extracellular space, accumulating in the newly formed necrotic areas where it perpetuates the cycle of mycolactone production, leading to the generation of acellular foci that continuously expand [9, 22]. Mycolactone further exacerbates this process by unbalancing the levels and activation state of thrombomodulin, protein C and other factors belonging to the coagulation cascade, favoring fibrin deposition and subsequent vessel thrombosis [30, 31]. As a consequence, the central necrotic area shows a high number of extracellular bacilli, surrounded by an infiltrate of neutrophils, macrophages and lymphocytes located in the interface between the necrotic area and healthy tissue [22]. With the inability of the immune system to reach the mycolactone producing-mycobacteria, the infection then progresses with the appearance of edema and ulceration of the epidermis, which can be potentiated by the coalescence of several foci of infection, if present [22].

Production of different congeners of mycolactone by different *M. ulcerans* lineages may alter the aggressiveness of the disease or even lead to distinct phenotypes, as it was initially observed in humans and later confirmed in mice [22, 23, 32–34]. In fact, in a proteomics study comparing a wild type (WT) mycolactone-producing strain with a mycolactone-negative strain isolated from a transposon library, several proteins related to virulence and stress response factors were found to be upregulated in the mycolactone-negative strain [35]. In practice, in the absence of mycolactone, as in the case of the natural mutant Mexican isolate 5114, infection with *M. ulcerans* resembles other mycobacterial infections [22, 23, 36]. Indeed, after an initial influx of neutrophils, well-formed granuloma-like structures, mostly composed of macrophages with epithelioid transformation and of lymphocytes, organize in the following days to weeks and surround the proliferating mycobacteria [12, 22, 23]. These inflammatory infiltrates eventually distribute homogeneously in the subcutaneous and muscle tissue, with no apparent necrosis or alterations occurring in the epidermal layer of the skin [12, 22, 23]. Surprisingly, the fact that mice are seemingly unable to eliminate the mycolactone-negative mutants for at least 12 months suggests that the bacteria use additional mycolactone-independent mechanisms to evade the immune system [22]. It should be noted that many of the classical mycobacterial virulence factors, such as the 6 kDa early secretory antigenic target (ESAT-6), the 10 kDa culture filtrate antigen (CFP-10) or alpha-crystallin-like protein (HspX) present in the Mexican ancestral lineage strain, are also highly immunogenic, thus indicating that this *M. ulcerans* lineage has evolved to survive regardless of their presence [37, 38].

These studies are unveiling how mycolactone shapes the cellular landscape of BU, but dose-response analyses reveal in more detail the high degree of pleiotropism presented by this toxin. Right from the inception of the immune response, mycolactone is able to affect the phagocytic capacity of murine macrophages at high multiplicities of infection and to inhibit the production of a series of human and murine monocytic/macrophagic molecules, including the proinflammatory cytokine TNF-α [9, 12, 15, 28, 39, 40]. Moreover, it hinders the production of nitrous oxide species, assumed to be important to control the infection, according to in vitro experiments and a clinical trial with BU patients [41–43].

Mycolactone also possesses dose-dependent immunomodulatory effects on T cells and dendritic cells (DC). Besides suppressing DCs' maturation, migration and production of the chemokines macrophage inflammatory protein (MIP) 1α and MIP-1β, regulated on activation normal T cell expressed and secreted (RANTES), IFN-γ–inducible protein 10 kD and MCP-1, Sec61 blockade in DCs results in the suppression of direct and cross-presentation of synthetic peptides to $CD8^+$ T cells by downregulation of MHC class I and β2 microglobulin [44, 45]. As for T cells, mycolactone represses TNF-α activity over factor nuclear kappa B (NF-κB) and transcription and post translation of the T cell receptor, ultimately resulting in the blockage of IL-2 production, an essential cytokine for T cell proliferation [15, 46].

Although all of these mechanisms hinder cell-mediated immunity (CMI), none has been as extensively tested in in vivo models as IFN-γ, the hallmark cytokine of type 1 T helper (T_H1) responses. Indeed, mice infected with *M. ulcerans* strains of low and intermediate virulence produce increased levels of IFN-γ, which induce phagosome maturation and acidification, as well as nitrite production [43]. Likewise, IFN-$\gamma^{-/-}$ mice display a severe impairment in the growth control of lower virulence strains when compared with WT mice [43, 47]. Importantly, none of these phenomena are observed upon inoculation with highly virulent strains or high doses of mycolactone, which demonstrates the extent to which mycolactone can damage an otherwise protective immune response [43, 47].

Responses against *M. ulcerans* infection are therefore dependent on the amount/variant of mycolactone produced by the pathogen. While many authors recognize the importance of CMI, the accumulation of mycolactone disturbing a wide array of CMI processes and the occupation by the mycobacteria of the extracellular space for most of the infection period places several question marks on what is or would be the ideal protective immune response. As such, other degrees of evidence are needed to develop a more accurate picture.

2.2 Regional and Systemic Responses

Viable *M. ulcerans* bacilli can be seen in the draining lymph node (DLN) of mice very early after subcutaneous infection, where cells initiate an immune response, as measured by their production of IFN-γ upon ex-vivo stimulation [28]. However histological analyses of the DLN also reveal the typical pattern of cell depletion and/or necrosis, affecting in this case both the T and the B cell compartments [48]. Remarkably, albeit T cells show an impairment in their proliferative capacity to expand upon antigenic stimuli, they are deemed to be relatively resistant to cell death by mycolactone [46, 49]. In contrast, mycolactone interferes with murine T cell expression of L-selectin (CD62-L), a necessary molecule to home these cells to the peripheral lymph nodes [49]. Whilst these findings in animal models of infection were so far not validated in humans, the indication that mycolactone can target cell homing mechanisms warrants further investigation [31, 49].

Both in humans and mice, *M. ulcerans* is occasionally found in internal organs, but probably due to its restrictive growth temperature, its viability there decreases over time and infection does not effectively establish [50–54]. Relevantly, mycolactone also disseminates systemically and, in spite of not causing major cell death or compromising immunity against other intracellular pathogens, it does alter the cytokine production of circulating cells [52, 54, 55]. As a result, some authors scrutinized the cytokine production pattern of peripheral blood cells from BU-infected patients, which did not lead to consistent results. Understandably, this can be attributed to differences in the genetic background of the study populations, the virulence of endemic *M. ulcerans* strains, the cells or stimuli applied in the assays used or even the power of the studies. IFN-γ, the most thoroughly analyzed cytokine until now, is the paradigm of this aspect, with some studies reporting opposite outcomes [56–63]. Nonetheless, the most consensual view until now is that BU is associated with an impaired T_H1 response, resulting in a lower proliferation of peripheral blood mononuclear cells (PBMCs) and production of IFN-γ upon stimulation [57, 60, 63, 64]. A deeper analysis of published studies further reveals that altough this response is not correlated with the stage of the lesion, it systematically returns in patients with healed lesions to the profile found in controls from endemic regions [59, 63]. Moreover, reports indicate that there is a correlation between the histological characteristics of a lesion and cytokine production, as patients who present well-formed granulomas stain more positively for IFN-γ, whereas patients without these structures tended to have more IL-10^+ cells [61, 65]. Interestingly, ex vivo IL-10 production by circulating leukocytes was found to increase during active disease in many of the studies, even if this was not statistically significant in some of them [56, 58–60, 62].

Consistent with some data in rodents, it was found that the production of chemokines such as IL-8, MIP-1β and MCP-1 was suppressed in BU patients throughout the different stages of the disease, strongly supporting the notion that *M. ulcerans* infection leads to a defect in immune cell recruitment in advanced stages of the infectious process [27, 44, 59]. A more recent analysis in humans with active BU confirmed the downregulation of MIP-1β and MCP-1 in the serum of patients, opening up a possibility to use certain chemokines as a molecular signature of the disease [31].

More extensive multi-analyte profiling of serum proteins in BU patients and endemic controls revealed that, although the disease does not influence significantly the leukocyte composition of the peripheral blood, it impacts an even wider array of circulating molecules [31, 66]. Indeed, several of these analytes contribute to acute phase reaction, metabolism, coagulation and tissue remodeling, with some of them already having been implicated in healing speed [31, 66]. Specifically regarding metabolic factors, *M. ulcerans* not only interferes with energy-generation, but also with peptide, lipid and nucleotide pathways [66]. Even though many of these alterations seem to be in line with the expected response of the host to infectious processes, others, as the proposed impairment of the tricarboxylic acid (TCA) cycle, could be the result of a direct effect of mycolactone [66, 67]. In reality, among TCA cycle intermediates, patients with BU have decreased citrate levels, whose relevance in other infectious diseases is associated with generating prostaglandins, nitric oxide and other antimicrobial peptides [66, 67].

All of these studies provide indications that the repercussions of *M. ulcerans* are more widespread than initially thought. While the disease focus is in the subcutaneous tissue, the ability of mycolactone to exert its actions beyond this local site most certainly contributes to the establishment of a permissive environment for the growth of the pathogen. This further hints on many other aspects of infection, such as immune cell metabolism, which are likely determinants for the way host cells interact with the pathogen.

3 Diagnostics Research Make it Evident: The Triggering of Cellular and Humoral Arms

Early attempts to find a better BU diagnostic method culminated in the development of burulin, a whole cell lysate of *M. ulcerans* that induced a delayed-type hypersensitive response in individuals from BU-endemic regions, more noticeable in people with already healed lesions [68–70]. Unfortunately, patients presented an equally strong reaction to tuberculin, consequently making the test of little usefulness in regions where tuberculosis is also prevalent or where the majority of the population is vaccinated with Bacillus Calmette-Guérin (BCG) [69, 70]. Furthermore, PBMCs from BU patients were later shown to proliferate very little and produce only small amounts of IFN-γ upon stimulation with live or dead *M. ulcerans* when compared to healthy individuals with a positive tuberculin test [64]. In contrast, a significant proportion of the evaluated patients possessed *M. ulcerans*-specific antibodies, which contrary to the cellular responses, can be detected in significant amounts throughout the different stages of the disease [64, 69]. Interestingly, incubation of serum from healthy endemic family controls with an *M. ulcerans* culture filtrate also revealed a notable amount of specific immunoglobulin (Ig) G, but no IgM, implying that this latter class of Igs has more potential for the identification of new BU cases [71]. In accordance, most epidemiological studies managed to screen populations in contact with *M. ulcerans* with moderate success, by focusing on detecting IgG against the 18 kDa small heat shock protein (shsp), a constitutive *M. ulcerans* cell wall protein involved in biofilm production [72–75]. Screenings of *M. ulcerans* whole protein lysates also identified the 65 kDa heat shock protein (Hsp65) as a potential candidate for a serological approach to determine active BU disease [73]. However, this protein has homologs in other mycobacterial species, thus prompting further analysis on the specificity of such antibodies [77].

All in all, in addition to reinforcing a role for CMI in BU, the search for novel diagnostic methods has put some visibility on the presence of antibodies in the context of *M. ulcerans* exposure and infection. Regarding each of these aspects, it remains to be established why not all exposed individuals develop the disease and how antibodies contribute to the control of *M. ulcerans* infections.

4 Antibiotic Treatment in the Aid of the Host Immune Response

Antibiotic regimens based on the conjoint use of rifampicin with another drug have been used with success in the treatment of BU, rapidly killing *M. ulcerans* and restoring a vast array of processes needed for tissue homeostasis [78, 79]. Probably the most critical achievement of these drugs is to break the chain of mycolactone production, which allows mycolactone to wane from the system [80]. Indeed, direct correlations between mycolactone concentration and the time needed to heal BU lesions were established, suggesting this is a crucial factor for the immune response to re-establish [80]. Following this idea, both in humans and experimentally infected mice, treatment is associated with restoration of systemic levels of several inflammatory molecules, including IL-4, IL-7, granulocyte-macrophage colony-stimulating factor (GM-CSF), as well as the chemokines IL-8 and MCP-1 [59, 81]. Simultaneously, in contrast to the typical absence of local inflammatory infiltrates seen in the central area of active untreated BU lesions, patients undergoing chemotherapy display very diverse organizations of cellular immune infiltrates, including classical granulomas, diffuse infiltrates or dense lymphocyte aggregations [79, 82, 83]. While this likely corresponds to different stages of the healing process, it is not clear if these might also represent inter-individual differences among BU patients [79, 83, 84]. Nevertheless, it is perceivable amidst the different histological observations in both humans and experimentally infected mice under treatment that phagocytes are needed throughout the process for uptaking bacteria and cellular debris, paving the way for wound healing [79, 83–85]. In mice, the regional preservation of DLN structures is noticeable, including germinal centers, a pattern not common to untreated groups, possibly due to the persistence of viable mycolactone-producing bacteria in the organ [85].

The immune break caused by mycolactone, in several reports described as paradoxical reactions in BU patients, has frequently been mistaken for treatment failure [86–89]. In fact, patients under standard regimens of antibiotherapy may display temporary clinical deterioration with local signs of increasing inflammation, ulceration and even development of new lesions [86, 87, 90]. These reactions appear to be due to the presence of large amounts of antigens, that once mycolactone starts to wane from the system, elicit a potent inflammatory response [86, 87]. In accordance, many of these paradoxical reactions appear to be related with age, duration of treatment, rapidly progressing forms of the disease such as edema, or even incomplete surgical lesion excisions that leave mycobacteria in surrounding tissues [86–88, 91, 92]. However, attempts to find biomarkers that could help to predict paradoxical reactions, such as neopterin, a catabolic product released by macrophages upon IFN-γ stimulation, have so far failed [93]. While this makes it undeniably more difficult to prevent such phenomena, patients' symptoms have been successfully managed with corticosteroid adjunctive therapy [94, 95]. In a similar way, BU patients co-infected with human immunodeficiency virus (HIV) have benefited from such an approach, since they develop a comparable immune reconstitution syndrome once treatment with antiretroviral therapy starts [96, 97]. Although evidence is still relatively limited, HIV infected patients tend to generally develop BU more frequently and with worse clinical outcomes when compared to the general population, in spite of no direct correlation with the degree of immunosuppression having been established [96, 98–100]. On the other hand, immunomodulation with corticosteroids has been suggested to favor activation of latent *M. ulcerans* infection [101]. Moreover, these immunosuppressive drugs prolonged the time needed to completely resolve infection during antibiotherapy in mice, an effect attributed primarily to the disturbance of immune cell recruitment to the site of infection [101, 102].

As a whole, this reinforces the notion that CMI is critical for the control of BU. In this line, innate effector cells emerge as one of the most, if not the main responsible determinant for containing bacterial proliferation at the site of lesion. Still, to what degree their interaction with mycolactone-compromised T and B cells influences their activity and consequently the outcome of BU treatment is yet to be clarified.

5 Epidemiological Clues and the Search for Novel Resistance and Susceptibility Markers

More than ever, there has been an effort in integrating epidemiological and pathophysiological observations in a more comprehensive understanding of diseases [103]. The demography of BU has been however a source of debate, as efforts to identify risk populations have not always been conclusive [104]. Interestingly though, large observational studies appear to converge in some aspects: children under 15 years and the elderly are at a higher risk of contracting the disease, with the latter group also presenting more frequently severe phenotypes [92, 105–111]; and gender distribution of cases varies according to age group, with males tending to predominate in the younger ages and females in the adulthood [105, 107, 108, 110–113]. While these differences could be a product of behavioral exposure, they could too indicate an underlying physiological basis, which would be in agreement with the ascertained roles of age and sex in numerous infectious diseases, including other mycobacterioses [114, 115].

The incidence of tuberculosis in humans is also higher in children and older adults, albeit the clinical features tend to be distinct, probably as a consequence of the likewise distinct underlying immune response [116]. Nevertheless, in both these populations, *Mycobacterium tuberculosis* growth and active disease is predicted to be favored by an unbalance in the percentage of IFN-γ^+ $CD4^+$ T lymphocytes recruited to the infection site [117–120]. These findings were further replicated in aged murine models of tuberculosis infection and ultimately match the postulated importance of IFN-γ in BU, which if not totally ablated by mycolactone, confers some degree of protection against *M. ulcerans* [43, 47, 115, 121]. Therefore, taken all of this into account, stratification by age of the cellular and molecular profile of BU patients could provide a critical avenue to obtain new information on the determinants of susceptibility *vs* protective immune responses, both collectively as well as in different age groups.

Parallel to age, the influence of sex over biological processes such as the immune response can be observed in a broad range of conditions, from infectious diseases to vaccination [122]. Even though some differences in the immune system of both genders are already established at birth, the effects become much more pronounced after puberty [122, 123]. This suggests that differences in exposure to environmental reservoirs of *M. ulcerans* explain better the higher incidence of BU among male pre-pubertal children than sex-related biological differences. After puberty however, sex hormones start to exert a powerful influence over several pathways of both the innate and acquired immune system arms resulting, for instance, in a bias towards T_H2 cells in females and T_H1 cells in males [122]. Strikingly, both male humans and mice appear to be more vulnerable to mycobacterial infections than their female counterparts, an observation that has been attributed to the effects of testosterone on, among others, macrophage motility [124, 125]. However, considering that BU pathogenesis and clinical features are mainly attributed to the activities of mycolactone, any effect of sex hormones in the immune response should pass unnoticeable unless it is in relationship with the mechanism of action of the toxin. In this line, estrogen has been shown to be able to activate neural WASP through a phosphorylation cascade induced by focal adhesion kinase, which could then render the cells more prone to the actions of mycolactone [20, 126]. In the end, this might be just one of several explanations for the seemingly increased susceptibility of adult women to the disease than men [127, 128].

Another debatable topic in BU is disease transmission. Although not the main focus of this chapter, some epidemiological records and experimental laboratory results point to environmental exposure as the main source for BU acquisition, namely through contact of injured skin with water bodies and through insect bites [25, 26, 113, 129–133]. However, some work in Africa, Asia and Australia additionally raised the possibility of host genetics playing a major part in BU development [92, 134–137]. Even though no evidence apart from anecdotal reports supports human-to-human transmission, the odds ratio (OR) of acquiring BU was determined as five times higher in people with family history of BU [25, 137–139]. Curiously, when this relationship was explored more thoroughly, strong associations with grandparent BU history but not contemporaneous family members were found [137]. In parallel, BU WHO category III lesions were recently suggested to constitute distinct clinical entities from categories I and II. When compared to both milder categories, they showed no dependency on time to develop and take considerably longer to heal [89, 109, 140]. Altogether, this strongly implies that additional host-intrinsic factors influence BU outcome, a lead followed by some studies evaluating the role of single nucleotide polymorphisms (SNPs) in genes related to the immune function (Table 1). For instance, confirmation of the critical importance of nitric oxide (NO) and IFN-γ to BU pathogenesis was achieved, as SNPs leading to lower promoter activity of *iNOS* and *IFNG* were associated with higher susceptibility to the disease [141].

However, genetic analyses are also helping to unravel the key function of other undisclosed molecular players of the immune system. Actually, the first ever human gene to be associated with BU was the solute carrier family 11, member 1 (*SLC11A1* or *NRAMP1*), whose main function is to transport divalent cations, including Zn^{2+}, Fe^{2+} and Mn^{2+}, to the late endosomal/lysosomal compartment [142, 143]. Since lysosomes eventually fuse with the phagosomes, where *M. ulcerans* resides transiently, this provides an important clue for the importance of the intramacrophage phase for the containment of *M. ulcerans* infection [9, 44, 143]. Accordingly, in a Beninese cohort of 208 patients, SNPs in autophagy-related genes E3 ubiquitin-protein ligase (*PARK2*) and autophagy-related protein 16-1 (*ATG16L1*) were associated with susceptibility to disease acquisition and phenotype aggressiveness [144]. In the same line, two different polymorphisms in the gene encoding the nucleotide-binding oligomerization domain-containing 2 (NOD2) protein, which is involved in bacterial recognition, were also associated with disease phenotype [144]. In fact, in other infections with intracellular pathogens including some species of mycobacteria, NOD2, PARK2 and the autophagy process have been described to help to control bacterial load and to improve host survival [145, 146]. Overall, this underlines a prominent role for xenophagy, the recognition and targeting of bacteria to autophagosomes, in the control of *M. ulcerans* infection, hence requiring further research (Table 1).

Table 1 Summary of genes significantly associated with BU disease susceptibility and or phenotype severity

First author and year	Country (Continent)	Population (n cases)	Gene	SNP rs# number	Association	Reported adjusted OR (95% CI)
Stienstra et al. (2006)	Ghana (Africa)	182	*SLC11A1* (*NRAMP1*)	rs17235409	Disease susceptibility	2.89 (1.41–5.91)
Capela et al. (2016)	Benin (Africa)	208	*PARK2*	rs1333955	Disease susceptibility	1.43 (1.00–2.06)
			NOD2	rs9302752	WHO lesion category	2.23 (1.14–4.37)
				rs2066842	WHO lesion category	12.7 (0.60–269)
			ATG16L1	rs2241880	Lesion ulceration	0.35 (0.13–0.90)
Bibert et al. (2017)	Ghana (Africa)	96	*iNOS*	rs9282799	Disease susceptibility	1.99 (1.22–3.26)
			IFNG	rs2069705	Disease susceptibility	1.56 (1.14–1.99)

Given the contribution of many genes to a clinical phenotype, gene association analyses can also present some interpretation challenges: single studies often lack the necessary power to control for all interactions, and genetic polymorphisms can not only vary in frequency, but additionally have different functional consequences among different populations [147]. A good example is the lack of association of *PARK2* with susceptibility to BU, as recently described in Ghana [141]. Although seemingly conflicting with previous results, the study populations were different and the tested polymorphisms were not the same, thus making direct data comparison difficult. Taken as a whole nonetheless, these population studies hold potential for a better understanding of host immune responses against BU, due to their ability to suggest, identify, or confirm potentially relevant targets for patient stratification or even disease treatment that could provide indispensable for future public health interventions.

6 (Un)successful Preventive Approaches

As of now, the only available vaccine targeting mycobacterial diseases is the BCG vaccine, a live attenuated bovine tuberculosis bacillus (*Mycobacterium bovis*) widely deployed against tuberculosis, but under an intense scrutiny for its unsatisfactory results [148]. In the case of BU, even though BCG appears to decrease the incidence of osteomyelitis forms, the incidence of new BU cases remains mostly unaltered, reinforcing the urge for better alternatives [149–155].

Efforts to develop a specific vaccine against *M. ulcerans* started several decades ago. Initial immunization studies in mice with different live mycobacterial species, including BCG and *M. ulcerans*, looked very promising in preventing the development of footpad swelling upon later challenge with *M. ulcerans* [156]. However, more recent experimental data clarified that even if a booster dose of the vaccine is administered or if the dose of *M. ulcerans* challenge is lowered, mice are only able to contain the infection for a short period of time, which is in accordance with observations in humans [48, 157, 158]. Nevertheless, these data provide important hints on the determinants of a protective immune response, some of which in agreement with what is seen during *M. ulcerans* clearance upon antibiotic treatment. Indeed, BCG-immunized mice display visible local cellular infiltrates together with a more rapid and sustained increase of IFN-γ and TNF-α levels and, in later time points, of the chemotactic MIP-2 and of inducible nitric oxide synthase (iNOS)$^+$ cells [48]. Concomitantly, a prolonged presence of CD4$^+$ T cells producing IFN-γ is evident in murine DLN, indicating that the trafficking of the bacteria to the DLN and subsequent presentation to the T cell compartment is temporarily being achieved [48]. Interestingly, immunization with a mycolactone-negative *M. ulcerans* strain appeared to be less effective than BCG in attracting cells to the site of the lesion and in inducing iNOS$^+$ cell accumulation [48]. On the other hand, in a study employing dewaxed *M. ulcerans* one month before infection, a strong generation of IgG against the bacilli was observed, with very little bacilli present in footpad sections and no phenotype development [159]. However, this study only evaluated protection for 28 days, and thus it remains to be investigated whether protection is long-lasting. Moreover, it is unclear if protective immunogens could have been removed during the dewaxing process, in spite of a general lack of protection detected in immunization interventions employing surface proteins, such as the 18 kDa shsp and the 27 kDa laminar binding protein [32, 160, 161]. Still, these results add to others confirming the extreme hydrophobicity of the *M. ulcerans* cell wall, which cumulatively with mycolactone represents a virulence factor and aids in shielding the pathogen from immune surveillance mechanisms, consequently adding another layer of complexity to the development of preventive strategies [162].

Other immunization alternatives have thus been tested, such as vaccination with the mycolyl-transferase antigen 85A (Ag85A), a major secreted member of a family of proteins with important cell wall synthetic activity in several mycobacterial species, including *M. ulcerans* [163, 164]. Whether in the form of a plasmid DNA vaccine or as a recombinant protein, vaccination strategies with Ag85A conferred some degree of protection by eliciting a cellular response with high levels of IFN-γ, IL-2 and Ag85-specific IgGs, but, similarly to BCG, eventually failed to deter disease development [163–166]. Likewise, a vaccine employing DNA plasmids of polyketide synthase domains of *Mycobacterium leprae* Hsp65, previously tested in a murine model of tuberculosis, conferred protection against BU only for a few months, despite the high homology of the amino acid sequence with the *M. ulcerans* orthologue [167, 168]. Therefore, better targets for immunization are still to be found and, to this end, a comparative genomics screening study identified several essential components of *M. ulcerans* that could provide important hints to direct future vaccination strategies [169].

So, why did not any of these immunization attempts succeed? Even if a transient protective response against *M. ulcerans* is generated, bacteria are consistently able to overcome it at some point. This likely implies that, unless elimination of the replication of the bacilli is achieved at the very early stages post-infection, accumulation of mycolactone will eventually begin to significantly damage the surrounding tissue including immune cells, thus defending the pathogen and downplaying any protective immune mechanism. Indeed, the administration of either antibiotics or bacteriophages in *M. ulcerans* infected mice corroborate this interpretation, since the killing of *M. ulcerans* is accompanied by an increase in the cellular immune response associated with cytokines such as IFN-γ [85, 170, 171].

In this sense, neutralization of mycolactone itself seems to be an appealing target both for passive or active immunization strategies, as it would open the necessary opportunity for the immune response against *M. ulcerans* to develop and maintain effectively. This is not a novel concept, but has been for long hampered by the lipid-like nature of mycolactone, which makes it poorly immunogenic [168, 172]. However, recently, anti-mycolactone IgG antibodies were generated in mice through the administration of a protein conjugate of a truncated non-cytotoxic form of the toxin [173]. Although these antibodies were able to bind to mycolactone and to neutralize its effect in in vitro assays with L929 fibroblast cell lines, it remains to be clarified if the antibodies are protective in vivo and if immunization with the truncated form of mycolactone induces a long-lasting protective humoral response [173]. Nevertheless, these are promising steps for future strategies targeting both active immunization efforts of endemic populations and passive immunization of BU patients.

7 Novel Models for Investigation and Future Perspectives

Research on BU includes a wide range of animal models, ranging from mammals and reptiles to birds, in an attempt to find one mimicking human disease best and allowing a detailed study of the pertaining cellular and molecular events [3, 174–177]. Nonetheless, most of the evidence on BU pathophysiology and the host immune response has originated from experimentally infected mice, which, in spite of mimicking most of the clinical features of human BU, such as erythema, edema and ulcers, tend to progress very rapidly to necrosis [50]. In an attempt to directly tackle this issue, in vitro sub-culturing of a *M. ulcerans* Ghanaian isolate (NM20/02) resulted in an attenuated strain that allowed a more prolonged evaluation of the cellular infiltrates [32]. Additionally, the extended kinetics facilitated the confirmation of other studies disclosing murine strain-specificities in the response against *M. ulcerans* infection, with overall evidence indicating that BALB/c mice are able to delay bacterial growth and respond more effectively to BCG immunization than their C57BL/6 counterparts [21, 32, 157]. This critical point to be considered when extrapolating data to the human pathophysiology, can eventually be overcome by the addition of information from other animal models with more similarities with the human skin, such as pigs and guinea pigs [174, 175, 178, 179].

Intriguingly, whereas many of the above models have been very useful to learn about the disease pathogenesis, very little progress was made concerning the mechanisms behind natural resistance to the disease. In fact, for long it has been known that natural resistance to *M. ulcerans* infection occurs in some animal species and some studies also indicated the potential for spontaneous healing of BU lesions in humans [3, 86, 176, 180–183]. On the other hand, it is estimated that even with proper antibiotic and surgical treatment, more than 20% of patients develop permanent sequelae [111]. What causes lie between these two extremes? Two recent studiesaimed to shed some light on this question, by exploring the ability of the Hartley guinea pigs and the FVB/N mice to spontaneously heal BU [21, 184]. Strikingly, whereas the guinea pigs were able to achieve sterilizing immunity, FVB/N mice appear to halt the production of mycolactone by *M. ulcerans* without being able to completely eliminate the pathogen [21, 184]. Considering for instance, that changes in the sugar content of the growth medium of *M. ulcerans* can affect its production of mycolactone and antigens, this raises the hypothesis whether FVB/N mice deal with the infection by manipulating the local environment [185]. It is intriguing to observe that both models control the infection regardless of the initial inoculum, which most likely suggests that there are key host factors that can effectively hamper the progression of the infection and that have never been so far considered as preventive or therapeutic approaches [21, 184]. Hence, it is highly likely that answers to this conundrum lie within the genetic makeup of BU patients that spontaneously recover from the disease, but with the current widespread use of antibiotics, the task of identifying such cases is extremely difficult. Nevertheless, the existence of animal models of resistance already constitutes an exciting opportunity that should be seized to unravel what factors could be exploited in order to overturn a progressive disease into a state of full resolution.

8 Conclusions

For a long time, research on *M. ulcerans* infection has been highly focused on experimental models. While these led to many important advances, integration of recent progress in epidemiological, clinical, cellular and molecular research constitutes an opportunity to further our understanding of BU pathogenesis. In this context, several lines of evidence point towards the importance of CMI in the local interaction with invading *M. ulcerans* bacilli, but the scope of this contact is still widely unknown. Likewise, bearing in mind that mycolactone can act beyond the site of infection, the consequences of its influence on the systemic production of cytokines or metabolic networks of the immune cells are yet to be understood. As such, it would be important to determine how immune recognition of *M. ulcerans* occurs at the molecular level, what host factors influence cytokine production upon this recognition and what the effect of this interaction is on CMI effector mechanisms. As for the humoral immunity, it would also be paramount to understand what its importance is and to what extent it prevents or delays the progression of the disease. It is only if answers for these fundamental questions are found, that the way to tailor more effective vaccination and prophylactic measures will become clear.

Acknowledgements This work was developed under the scope of the project NORTE-01-0145-FEDER-000013, supported by the Northern Portugal Regional Operational Programme (NORTE 2020), under the Portugal 2020 Partnership Agreement, through the European Regional Development Fund (FEDER); FEDER, through the Competiveness Factors Operational Programme (COMPETE) and by National funds, through the Foundation for Science and Technology (FCT), under the scope of projects POCI-01-0145-FEDER-007038 and POCI-01-0145-FEDER-031312; Infect-ERA grant Infect-ERA/0002/2015; and FCT through the PhD grant PD/BD/113699/2015 to JF and the postdoctoral grant SFRH/BPD/112903/2015 to AGF.

References

1. Dodge OG (1964) Mycobacterial skin ulcers in Uganda: histopathological and experimental aspects. J Pathol Bacteriol 88:169–174
2. Hayman J (1993) Out of Africa: observations on the histopathology of Mycobacterium ulcerans infection. J Clin Pathol 46:5–9

3. MacCallum P, Tolhurst JC (1948) A new mycobacterial infection in man. J Pathol Bacteriol 60:93–122
4. Connor DH, Lunn HF (1965) Mycobacterium ulcerans infection (with comments on pathogenesis). Int J Lepr 33(Suppl):698–709
5. Guarner J, Bartlett J, Whitney EAS et al (2003) Histopathologic features of Mycobacterium ulcerans infection. Emerg Infect Dis 9:651–656
6. Hayman J, McQueen A (1985) The pathology of Mycobacterium ulcerans infection. Pathology (Phila) 17:594–600
7. Rondini S, Horsfield C, Mensah-Quainoo E et al (2006) Contiguous spread of Mycobacterium ulcerans in Buruli ulcer lesions analysed by histopathology and real-time PCR quantification of mycobacterial DNA. J Pathol 208:119–128. https://doi.org/10.1002/path.1864
8. Ruf M-T, Steffen C, Bolz M et al (2017) Infiltrating leukocytes surround early Buruli ulcer lesions, but are unable to reach the mycolactone producing mycobacteria. Virulence 8:1918–1926. https://doi.org/10.1080/21505594.2017.1370530
9. Torrado E, Fraga AG, Castro AG et al (2007) Evidence for an intramacrophage growth phase of Mycobacterium ulcerans. Infect Immun 75:977–987. https://doi.org/10.1128/IAI.00889-06
10. Walsh DS, Meyers WM, Portaels F et al (2005) High rates of apoptosis in human Mycobacterium ulcerans culture-positive Buruli ulcer skin lesions. Am J Trop Med Hyg 73:410–415
11. Krieg RE, Hockmeyer WT, Connor DH (1974) Toxin of Mycobacterium ulcerans. Production and effects in Guinea pig skin. Arch Dermatol 110:783–788
12. Adusumilli S, Mve-Obiang A, Sparer T et al (2005) Mycobacterium ulcerans toxic macrolide, mycolactone modulates the host immune response and cellular location of M. ulcerans in vitro and in vivo. Cell Microbiol 7:1295–1304. https://doi.org/10.1111/j.1462-5822.2005.00557.x
13. George KM, Chatterjee D, Gunawardana G et al (1999) Mycolactone: a polyketide toxin from Mycobacterium ulcerans required for virulence. Science 283:854–857
14. Bieri R, Scherr N, Ruf M-T et al (2017) The macrolide toxin mycolactone promotes bim-dependent apoptosis in Buruli ulcer through inhibition of mTOR. ACS Chem Biol 12:1297–1307. https://doi.org/10.1021/acschembio.7b00053
15. Pahlevan AA, Wright DJ, Andrews C et al (1999) The inhibitory action of Mycobacterium ulcerans soluble factor on monocyte/T cell cytokine production and NF-kappa B function. J Immunol 163:3928–3935
16. Snyder DS, Small PLC (2003) Uptake and cellular actions of mycolactone, a virulence determinant for Mycobacterium ulcerans. Microb Pathog 34:91–101
17. Baron L, Paatero AO, Morel J-D et al (2016) Mycolactone subverts immunity by selectively blocking the Sec61 translocon. J Exp Med 213:2885–2896. https://doi.org/10.1084/jem.20160662
18. Hall BS, Hill K, McKenna M et al (2014) The pathogenic mechanism of the Mycobacterium ulcerans virulence factor, mycolactone, depends on blockade of protein translocation into the ER. PLoS Pathog 10:e1004061. https://doi.org/10.1371/journal.ppat.1004061
19. Gama JB, Ohlmeier S, Martins TG et al (2014) Proteomic analysis of the action of the Mycobacterium ulcerans toxin mycolactone: targeting host cells cytoskeleton and collagen. PLoS Negl Trop Dis 8:e3066. https://doi.org/10.1371/journal.pntd.0003066
20. Guenin-Macé L, Veyron-Churlet R, Thoulouze M-I et al (2013) Mycolactone activation of Wiskott-Aldrich syndrome proteins underpins Buruli ulcer formation. J Clin Invest 123:1501–1512. https://doi.org/10.1172/JCI66576
21. Marion E, Jarry U, Cano C et al (2016) FVB/N mice spontaneously heal ulcerative lesions induced by Mycobacterium ulcerans and switch M. ulcerans into a low mycolactone producer. J Immunol 196:2690–2698. https://doi.org/10.4049/jimmunol.1502194
22. Oliveira MS, Fraga AG, Torrado E et al (2005) Infection with Mycobacterium ulcerans induces persistent inflammatory responses in mice. Infect Immun 73:6299–6310. https://doi.org/10.1128/IAI.73.10.6299-6310.2005

23. Ortiz RH, Leon DA, Estevez HO et al (2009) Differences in virulence and immune response induced in a murine model by isolates of Mycobacterium ulcerans from different geographic areas. Clin Exp Immunol 157:271–281. https://doi.org/10.1111/j.1365-2249.2009.03941.x
24. Guenin-Macé L, Oldenburg R, Chrétien F, Demangel C (2014) Pathogenesis of skin ulcers: lessons from the Mycobacterium ulcerans and Leishmania spp. pathogens. Cell Mol Life Sci 71:2443–2450. https://doi.org/10.1007/s00018-014-1561-z
25. Merritt RW, Walker ED, Small PLC et al (2010) Ecology and transmission of Buruli ulcer disease: a systematic review. PLoS Negl Trop Dis 4:e911. https://doi.org/10.1371/journal.pntd.0000911
26. Wallace JR, Mangas KM, Porter JL et al (2017) Mycobacterium ulcerans low infectious dose and mechanical transmission support insect bites and puncturing injuries in the spread of Buruli ulcer. PLoS Negl Trop Dis 11:e0005553. https://doi.org/10.1371/journal.pntd.0005553
27. Lee H-M, Shin D-M, Choi D-K et al (2009) Innate immune responses to Mycobacterium ulcerans via toll-like receptors and dectin-1 in human keratinocytes. Cell Microbiol 11:678–692. https://doi.org/10.1111/j.1462-5822.2009.01285.x
28. Coutanceau E, Marsollier L, Brosch R et al (2005) Modulation of the host immune response by a transient intracellular stage of Mycobacterium ulcerans: the contribution of endogenous mycolactone toxin. Cell Microbiol 7:1187–1196. https://doi.org/10.1111/j.1462-5822.2005.00546.x
29. Roberts B, Hirst R (1996) Identification and characterisation of a superoxide dismutase and catalase from Mycobacterium ulcerans. J Med Microbiol 45:383–387. https://doi.org/10.1099/00222615-45-5-383
30. Ogbechi J, Ruf M-T, Hall BS et al (2015) Mycolactone-dependent depletion of endothelial cell thrombomodulin is strongly associated with fibrin deposition in Buruli ulcer lesions. PLoS Pathog 11:e1005011. https://doi.org/10.1371/journal.ppat.1005011
31. Phillips RO, Sarfo FS, Landier J et al (2014) Combined inflammatory and metabolic defects reflected by reduced serum protein levels in patients with Buruli ulcer disease. PLoS Negl Trop Dis 8:e2786. https://doi.org/10.1371/journal.pntd.0002786
32. Bénard A, Sala C, Pluschke G (2016) Mycobacterium ulcerans mouse model refinement for pre-clinical profiling of vaccine candidates. PLoS One 11:e0167059. https://doi.org/10.1371/journal.pone.0167059
33. Portaels F, Meyers WM, Ablordey A et al (2008) First cultivation and characterization of Mycobacterium ulcerans from the environment. PLoS Negl Trop Dis 2:e178. https://doi.org/10.1371/journal.pntd.0000178
34. Scherr N, Gersbach P, Dangy J-P et al (2013) Structure-activity relationship studies on the macrolide exotoxin mycolactone of Mycobacterium ulcerans. PLoS Negl Trop Dis 7:e2143. https://doi.org/10.1371/journal.pntd.0002143
35. Tafelmeyer P, Laurent C, Lenormand P et al (2008) Comprehensive proteome analysis of Mycobacterium ulcerans and quantitative comparison of mycolactone biosynthesis. Proteomics 8:3124–3138. https://doi.org/10.1002/pmic.200701018
36. Mve-Obiang A, Lee RE, Portaels F, Small PLC (2003) Heterogeneity of Mycolactones produced by clinical isolates of Mycobacterium ulcerans: implications for virulence. Infect Immun 71:774–783. https://doi.org/10.1128/IAI.71.2.774-783.2003
37. Huber CA, Ruf M-T, Pluschke G, Käser M (2008) Independent loss of immunogenic proteins in Mycobacterium ulcerans suggests immune evasion. Clin Vaccine Immunol 15:598–606. https://doi.org/10.1128/CVI.00472-07
38. Qi W, Käser M, Röltgen K et al (2009) Genomic diversity and evolution of Mycobacterium ulcerans revealed by next-generation sequencing. PLoS Pathog 5:e1000580. https://doi.org/10.1371/journal.ppat.1000580
39. Pimsler M, Sponsler TA, Meyers WM (1988) Immunosuppressive properties of the soluble toxin from Mycobacterium ulcerans. J Infect Dis 157:577–580
40. Sarfo FS, Phillips RO, Rangers B et al (2010) Detection of Mycolactone A/B in Mycobacterium ulcerans-infected human tissue. PLoS Negl Trop Dis 4:e577. https://doi.org/10.1371/journal.pntd.0000577

41. Phillips R, Adjei O, Lucas S et al (2004) Pilot randomized double-blind trial of treatment of Mycobacterium ulcerans disease (Buruli ulcer) with topical nitrogen oxides. Antimicrob Agents Chemother 48:2866–2870. https://doi.org/10.1128/AAC.48.8.2866-2870.2004
42. Phillips R, Kuijper S, Benjamin N et al (2004) In vitro killing of Mycobacterium ulcerans by acidified nitrite. Antimicrob Agents Chemother 48:3130–3132. https://doi.org/10.1128/AAC.48.8.3130-3132.2004
43. Torrado E, Fraga AG, Logarinho E et al (2010) IFN-gamma-dependent activation of macrophages during experimental infections by Mycobacterium ulcerans is impaired by the toxin mycolactone. J Immunol 184:947–955. https://doi.org/10.4049/jimmunol.0902717
44. Coutanceau E, Decalf J, Martino A et al (2007) Selective suppression of dendritic cell functions by Mycobacterium ulcerans toxin mycolactone. J Exp Med 204:1395–1403. https://doi.org/10.1084/jem.20070234
45. Grotzke JE, Kozik P, Morel J-D et al (2017) Sec61 blockade by mycolactone inhibits antigen cross-presentation independently of endosome-to-cytosol export. Proc Natl Acad Sci U S A 114:E5910–E5919. https://doi.org/10.1073/pnas.1705242114
46. Boulkroun S, Guenin-Macé L, Thoulouze M-I et al (2010) Mycolactone suppresses T cell responsiveness by altering both early signaling and posttranslational events. J Immunol 184:1436–1444. https://doi.org/10.4049/jimmunol.0902854
47. Bieri R, Bolz M, Ruf M-T, Pluschke G (2016) Interferon-γ is a crucial activator of early host immune defense against Mycobacterium ulcerans infection in mice. PLoS Negl Trop Dis 10:e0004450. https://doi.org/10.1371/journal.pntd.0004450
48. Fraga AG, Martins TG, Torrado E et al (2012) Cellular immunity confers transient protection in experimental Buruli ulcer following BCG or mycolactone-negative Mycobacterium ulcerans vaccination. PLoS One 7:e33406. https://doi.org/10.1371/journal.pone.0033406
49. Guenin-Macé L, Carrette F, Asperti-Boursin F et al (2011) Mycolactone impairs T cell homing by suppressing microRNA control of L-selectin expression. Proc Natl Acad Sci U S A 108:12833–12838. https://doi.org/10.1073/pnas.1016496108
50. Addo P, Owusu E, Adu-Addai B et al (2005) Findings from a Buruli ulcer mouse model study. Ghana Med J 39:86–93
51. Fenner F (1956) The pathogenic behavior of Mycobacterium ulcerans and Mycobacterium balnei in the mouse and the developing chick embryo. Am Rev Tuberc 73:650–673
52. Hong H, Coutanceau E, Leclerc M et al (2008) Mycolactone diffuses from Mycobacterium ulcerans-infected tissues and targets mononuclear cells in peripheral blood and lymphoid organs. PLoS Negl Trop Dis 2:e325. https://doi.org/10.1371/journal.pntd.0000325
53. Pattyn SR (1965) Bactériologie et pathologie humaine et expérimentale des ulcères à Mycobacterium ulcerans. Ann Soc Belges Méd Trop Parasitol Mycol 45:121–129
54. Sarfo FS, Le Chevalier F, Aka N et al (2011) Mycolactone diffuses into the peripheral blood of Buruli ulcer patients—implications for diagnosis and disease monitoring. PLoS Negl Trop Dis 5:e1237. https://doi.org/10.1371/journal.pntd.0001237
55. Fraga AG, Cruz A, Martins TG et al (2011) Mycobacterium ulcerans triggers T-cell immunity followed by local and regional but not systemic immunosuppression. Infect Immun 79:421–430. https://doi.org/10.1128/IAI.00820-10
56. Gooding TM, Johnson PDR, Smith M et al (2002) Cytokine profiles of patients infected with Mycobacterium ulcerans and unaffected household contacts. Infect Immun 70:5562–5567
57. Gooding TM, Kemp AS, Robins-Browne RM et al (2003) Acquired T-helper 1 lymphocyte anergy following infection with Mycobacterium ulcerans. Clin Infect Dis 36:1076–1077. https://doi.org/10.1086/368315
58. Phillips R, Horsfield C, Kuijper S et al (2006) Cytokine response to antigen stimulation of whole blood from patients with Mycobacterium ulcerans disease compared to that from patients with tuberculosis. Clin Vaccine Immunol 13:253–257. https://doi.org/10.1128/CVI.13.2.253-257.2006
59. Phillips R, Sarfo FS, Guenin-Macé L et al (2009) Immunosuppressive signature of cutaneous Mycobacterium ulcerans infection in the peripheral blood of patients with Buruli ulcer disease. J Infect Dis 200:1675–1684. https://doi.org/10.1086/646615

60. Prévot G, Bourreau E, Pascalis H et al (2004) Differential production of systemic and intralesional gamma interferon and interleukin-10 in nodular and ulcerative forms of Buruli disease. Infect Immun 72:958–965
61. Schipper HS, Rutgers B, Huitema MG et al (2007) Systemic and local interferon-gamma production following Mycobacterium ulcerans infection. Clin Exp Immunol 150:451–459. https://doi.org/10.1111/j.1365-2249.2007.03506.x
62. Westenbrink BD, Stienstra Y, Huitema MG et al (2005) Cytokine responses to stimulation of whole blood from patients with Buruli ulcer disease in Ghana. Clin Diagn Lab Immunol 12:125–129. https://doi.org/10.1128/CDLI.12.1.125-129.2005
63. Yeboah-Manu D, Peduzzi E, Mensah-Quainoo E et al (2006) Systemic suppression of interferon-gamma responses in Buruli ulcer patients resolves after surgical excision of the lesions caused by the extracellular pathogen Mycobacterium ulcerans. J Leukoc Biol 79:1150–1156. https://doi.org/10.1189/jlb.1005581
64. Gooding TM, Johnson PD, Campbell DE et al (2001) Immune response to infection with Mycobacterium ulcerans. Infect Immun 69:1704–1707. https://doi.org/10.1128/IAI.69.3.1704-1707.2001
65. Kiszewski AE, Becerril E, Aguilar LD et al (2006) The local immune response in ulcerative lesions of Buruli disease. Clin Exp Immunol 143:445–451. https://doi.org/10.1111/j.1365-2249.2006.03020.x
66. Niang F, Sarfo FS, Frimpong M et al (2015) Metabolomic profiles delineate mycolactone signature in Buruli ulcer disease. Sci Rep 5:17693. https://doi.org/10.1038/srep17693
67. O'Neill LAJ, Kishton RJ, Rathmell J (2016) A guide to immunometabolism for immunologists. Nat Rev Immunol 16:553–565. https://doi.org/10.1038/nri.2016.70
68. Barker DJ (1973) Epidemiology of Mycobacterium ulcerans infection. Trans R Soc Trop Med Hyg 67:43–50
69. Dobos KM, Spotts EA, Marston BJ et al (2000) Serologic response to culture filtrate antigens of Mycobacterium ulcerans during Buruli ulcer disease. Emerg Infect Dis 6:158–164. https://doi.org/10.3201/eid0602.000208
70. Stanford JL, Revill WD, Gunthorpe WJ, Grange JM (1975) The production and preliminary investigation of Burulin, a new skin test reagent for Mycobacterium ulcerans infection. J Hyg 74:7–16
71. Okenu DMN, Ofielu LO, Easley KA et al (2004) Immunoglobulin M antibody responses to Mycobacterium ulcerans allow discrimination between cases of active Buruli ulcer disease and matched family controls in areas where the disease is endemic. Clin Diagn Lab Immunol 11:387–391
72. Diaz D, Döbeli H, Yeboah-Manu D et al (2006) Use of the immunodominant 18-kiloDalton small heat shock protein as a serological marker for exposure to Mycobacterium ulcerans. Clin Vaccine Immunol 13:1314–1321. https://doi.org/10.1128/CVI.00254-06
73. Pidot SJ, Porter JL, Tobias NJ et al (2010) Regulation of the 18 kDa heat shock protein in Mycobacterium ulcerans: an alpha-crystallin orthologue that promotes biofilm formation. Mol Microbiol 78:1216–1231. https://doi.org/10.1111/j.1365-2958.2010.07401.x
74. Röltgen K, Bratschi MW, Ross A et al (2014) Late onset of the serological response against the 18 kDa small heat shock protein of Mycobacterium ulcerans in children. PLoS Negl Trop Dis 8:e2904. https://doi.org/10.1371/journal.pntd.0002904
75. Yeboah-Manu D, Röltgen K, Opare W et al (2012) Sero-epidemiology as a tool to screen populations for exposure to Mycobacterium ulcerans. PLoS Negl Trop Dis 6:e1460. https://doi.org/10.1371/journal.pntd.0001460
76. Dreyer A, Röltgen K, Dangy JP et al (2015) Identification of the Mycobacterium ulcerans protein MUL_3720 as a promising target for the development of a diagnostic test for Buruli ulcer. PLoS Negl Trop Dis 9:e0003477. https://doi.org/10.1371/journal.pntd.0003477
77. Pidot SJ, Porter JL, Marsollier L et al (2010) Serological evaluation of Mycobacterium ulcerans antigens identified by comparative genomics. PLoS Negl Trop Dis 4:e872. https://doi.org/10.1371/journal.pntd.0000872

78. Converse PJ, Nuermberger EL, Almeida DV, Grosset JH (2011) Treating Mycobacterium ulcerans disease (Buruli ulcer): from surgery to antibiotics, is the pill mightier than the knife? Future Microbiol 6:1185–1198. https://doi.org/10.2217/fmb.11.101
79. Ruf M-T, Schütte D, Chauffour A et al (2012) Chemotherapy-associated changes of histopathological features of Mycobacterium ulcerans lesions in a Buruli ulcer mouse model. Antimicrob Agents Chemother 56:687–696. https://doi.org/10.1128/AAC.05543-11
80. Sarfo FS, Phillips RO, Zhang J et al (2014) Kinetics of mycolactone in human subcutaneous tissue during antibiotic therapy for Mycobacterium ulcerans disease. BMC Infect Dis 14:202. https://doi.org/10.1186/1471-2334-14-202
81. Sarfo FS, Converse PJ, Almeida DV et al (2013) Microbiological, histological, immunological, and toxin response to antibiotic treatment in the mouse model of Mycobacterium ulcerans disease. PLoS Negl Trop Dis 7:e2101. https://doi.org/10.1371/journal.pntd.0002101
82. Ruf M-T, Sopoh GE, Brun LV et al (2011) Histopathological changes and clinical responses of Buruli ulcer plaque lesions during chemotherapy: a role for surgical removal of necrotic tissue? PLoS Negl Trop Dis 5:e1334. https://doi.org/10.1371/journal.pntd.0001334
83. Schütte D, Um-Boock A, Mensah-Quainoo E et al (2007) Development of highly organized lymphoid structures in Buruli ulcer lesions after treatment with rifampicin and streptomycin. PLoS Negl Trop Dis 1:e2. https://doi.org/10.1371/journal.pntd.0000002
84. Schütte D, Umboock A, Pluschke G (2009) Phagocytosis of Mycobacterium ulcerans in the course of rifampicin and streptomycin chemotherapy in Buruli ulcer lesions. Br J Dermatol 160:273–283. https://doi.org/10.1111/j.1365-2133.2008.08879.x
85. Martins TG, Gama JB, Fraga AG et al (2012) Local and regional re-establishment of cellular immunity during curative antibiotherapy of murine Mycobacterium ulcerans infection. PLoS One 7:e32740. https://doi.org/10.1371/journal.pone.0032740
86. Nienhuis WA, Stienstra Y, Abass KM et al (2012) Paradoxical responses after start of antimicrobial treatment in Mycobacterium ulcerans infection. Clin Infect Dis 54:519–526. https://doi.org/10.1093/cid/cir856
87. O'Brien DP, Robson M, Friedman ND et al (2013) Incidence, clinical spectrum, diagnostic features, treatment and predictors of paradoxical reactions during antibiotic treatment of Mycobacterium ulcerans infections. BMC Infect Dis 13:416. https://doi.org/10.1186/1471-2334-13-416
88. O'Brien DP, Robson ME, Callan PP, McDonald AH (2009) "Paradoxical" immune-mediated reactions to Mycobacterium ulcerans during antibiotic treatment: a result of treatment success, not failure. Med J Aust 191:564–566
89. Sarfo FS, Phillips R, Asiedu K et al (2010) Clinical efficacy of combination of rifampin and streptomycin for treatment of Mycobacterium ulcerans disease. Antimicrob Agents Chemother 54:3678–3685. https://doi.org/10.1128/AAC.00299-10
90. Ruf M-T, Chauty A, Adeye A et al (2011) Secondary Buruli ulcer skin lesions emerging several months after completion of chemotherapy: paradoxical reaction or evidence for immune protection? PLoS Negl Trop Dis 5:e1252. https://doi.org/10.1371/journal.pntd.0001252
91. Barogui YT, Klis S-A, Johnson RC et al (2016) Genetic susceptibility and predictors of paradoxical reactions in Buruli ulcer. PLoS Negl Trop Dis 10:e0004594. https://doi.org/10.1371/journal.pntd.0004594
92. O'Brien DP, Friedman ND, Cowan R et al (2015) Mycobacterium ulcerans in the elderly: more severe disease and suboptimal outcomes. PLoS Negl Trop Dis 9:e0004253. https://doi.org/10.1371/journal.pntd.0004253
93. de Zeeuw J, Duggirala S, Nienhuis WA et al (2013) Serum levels of neopterin during antimicrobial treatment for Mycobacterium ulcerans infection. Am J Trop Med Hyg 89:498–500. https://doi.org/10.4269/ajtmh.12-0599
94. Friedman ND, McDonald AH, Robson ME, O'Brien DP (2012) Corticosteroid use for paradoxical reactions during antibiotic treatment for Mycobacterium ulcerans. PLoS Negl Trop Dis 6:e1767. https://doi.org/10.1371/journal.pntd.0001767
95. Trevillyan JM, Johnson PDR (2013) Steroids control paradoxical worsening of Mycobacterium ulcerans infection following initiation of antibiotic therapy. Med J Aust 198:443–444

96. Tuffour J, Owusu-Mireku E, Ruf M-T et al (2015) Challenges associated with management of Buruli ulcer/human immunodeficiency virus coinfection in a treatment center in Ghana: a case series study. Am J Trop Med Hyg 93:216–223. https://doi.org/10.4269/ajtmh.14-0571
97. Wanda F, Nkemenang P, Ehounou G et al (2014) Clinical features and management of a severe paradoxical reaction associated with combined treatment of Buruli ulcer and HIV co-infection. BMC Infect Dis 14:423. https://doi.org/10.1186/1471-2334-14-423
98. Johnson RC, Nackers F, Glynn JR et al (2008) Association of HIV infection and Mycobacterium ulcerans disease in Benin. AIDS 22:901–903. https://doi.org/10.1097/QAD.0b013e3282f7690a
99. Toll A, Gallardo F, Ferran M et al (2005) Aggressive multifocal Buruli ulcer with associated osteomyelitis in an HIV-positive patient. Clin Exp Dermatol 30:649–651. https://doi.org/10.1111/j.1365-2230.2005.01892.x
100. Vincent QB, Ardant M-F, Marsollier L et al (2014) HIV infection and Buruli ulcer in Africa. Lancet Infect Dis 14:796–797. https://doi.org/10.1016/S1473-3099(14)70882-5
101. Prasad R (1993) Pulmonary sarcoidosis and chronic cutaneous atypical mycobacter ulcer. Aust Fam Physician 22:755–758
102. Martins TG, Trigo G, Fraga AG et al (2012) Corticosteroid-induced immunosuppression ultimately does not compromise the efficacy of antibiotherapy in murine Mycobacterium ulcerans infection. PLoS Negl Trop Dis 6:e1925. https://doi.org/10.1371/journal.pntd.0001925
103. Ogino S, King EE, Beck AH et al (2012) Interdisciplinary education to integrate pathology and epidemiology: towards molecular and population-level health science. Am J Epidemiol 176:659–667. https://doi.org/10.1093/aje/kws226
104. Jacobsen KH, Padgett JJ (2010) Risk factors for Mycobacterium ulcerans infection. Int J Infect Dis 14:e677–e681. https://doi.org/10.1016/j.ijid.2009.11.013
105. Amofah G, Bonsu F, Tetteh C et al (2002) Buruli ulcer in Ghana: results of a national case search. Emerg Infect Dis 8:167–170. https://doi.org/10.3201/eid0802.010119
106. Debacker M, Aguiar J, Steunou C et al (2004) Mycobacterium ulcerans disease: role of age and gender in incidence and morbidity. Trop Med Int Health 9:1297–1304. https://doi.org/10.1111/j.1365-3156.2004.01339.x
107. Debacker M, Portaels F, Aguiar J et al (2006) Risk factors for Buruli ulcer, Benin. Emerg Infect Dis 12:1325–1331. https://doi.org/10.3201/eid1209.050598
108. Landier J, Fontanet A, Texier G (2014) Defining and targeting high-risk populations in Buruli ulcer. Lancet Glob Health 2:e629. https://doi.org/10.1016/S2214-109X(14)70311-0
109. Tai AYC, Athan E, Friedman ND et al (2018) Increased severity and spread of Mycobacterium ulcerans, Southeastern Australia. Emerg Infect Dis 24:58. https://doi.org/10.3201/eid2401.171070
110. The Uganda Buruli Group (1971) Epidemiology of Mycobacterium ulcerans infection (Buruli ulcer) at Kinyara, Uganda. Trans R Soc Trop Med Hyg 65:763–775
111. Vincent QB, Ardant M-F, Adeye A et al (2014) Clinical epidemiology of laboratory-confirmed Buruli ulcer in Benin: a cohort study. Lancet Glob Health 2:e422–e430. https://doi.org/10.1016/S2214-109X(14)70223-2
112. Bratschi MW, Bolz M, Minyem JC et al (2013) Geographic distribution, age pattern and sites of lesions in a cohort of Buruli ulcer patients from the Mapé Basin of Cameroon. PLoS Negl Trop Dis 7:e2252. https://doi.org/10.1371/journal.pntd.0002252
113. N'krumah RTAS, Koné B, Cissé G et al (2017) Characteristics and epidemiological profile of Buruli ulcer in the district of Tiassalé, south Côte d'Ivoire. Acta Trop 175:138–144. https://doi.org/10.1016/j.actatropica.2016.12.023
114. Nhamoyebonde S, Leslie A (2014) Biological differences between the sexes and susceptibility to tuberculosis. J Infect Dis 209(Suppl 3):S100–S106. https://doi.org/10.1093/infdis/jiu147
115. Vesosky B, Turner J (2005) The influence of age on immunity to infection with Mycobacterium tuberculosis. Immunol Rev 205:229–243. https://doi.org/10.1111/j.0105-2896.2005.00257.x
116. Alcaïs A, Fieschi C, Abel L, Casanova J-L (2005) Tuberculosis in children and adults: two distinct genetic diseases. J Exp Med 202:1617–1621. https://doi.org/10.1084/jem.20052302

117. Dreesman A, Corbière V, Dirix V et al (2017) Age-stratified T cell responses in children infected with Mycobacterium tuberculosis. Front Immunol 8:1059. https://doi.org/10.3389/fimmu.2017.01059
118. Guerra-Laso JM, González-García S, González-Cortés C et al (2013) Macrophages from elders are more permissive to intracellular multiplication of Mycobacterium tuberculosis. Age (Dordr) 35:1235–1250. https://doi.org/10.1007/s11357-012-9451-5
119. Guzzetta G, Kirschner D (2013) The roles of immune memory and aging in protective immunity and endogenous reactivation of tuberculosis. PLoS One 8:e60425. https://doi.org/10.1371/journal.pone.0060425
120. Harari A, Rozot V, Bellutti Enders F et al (2011) Dominant TNF-α + Mycobacterium tuberculosis-specific CD4+ T cell responses discriminate between latent infection and active disease. Nat Med 17:372–376. https://doi.org/10.1038/nm.2299
121. Orme IM (1987) Aging and immunity to tuberculosis: increased susceptibility of old mice reflects a decreased capacity to generate mediator T lymphocytes. J Immunol 138:4414–4418
122. Klein SL, Flanagan KL (2016) Sex differences in immune responses. Nat Rev Immunol 16:626–638. https://doi.org/10.1038/nri.2016.90
123. Lamberts SW, van den Beld AW, van der Lely AJ (1997) The endocrinology of aging. Science 278:419–424
124. Giefing-Kröll C, Berger P, Lepperdinger G, Grubeck-Loebenstein B (2015) How sex and age affect immune responses, susceptibility to infections, and response to vaccination. Aging Cell 14:309–321. https://doi.org/10.1111/acel.12326
125. Yamamoto Y, Saito H, Setogawa T, Tomioka H (1991) Sex differences in host resistance to Mycobacterium marinum infection in mice. Infect Immun 59:4089–4096
126. Sanchez AM, Flamini MI, Baldacci C et al (2010) Estrogen receptor-alpha promotes breast cancer cell motility and invasion via focal adhesion kinase and N-WASP. Mol Endocrinol 24:2114–2125. https://doi.org/10.1210/me.2010-0252
127. Barker DJ (1972) The distribution of Buruli disease in Uganda. Trans R Soc Trop Med Hyg 66:867–874
128. Raghunathan PL, Whitney EAS, Asamoa K et al (2005) Risk factors for Buruli ulcer disease (Mycobacterium ulcerans infection): results from a case-control study in Ghana. Clin Infect Dis 40:1445–1453. https://doi.org/10.1086/429623
129. Ampah KA, Asare P, Binnah DD-G et al (2016) Burden and historical trend of Buruli ulcer prevalence in selected communities along the Offin River of Ghana. PLoS Negl Trop Dis 10:e0004603. https://doi.org/10.1371/journal.pntd.0004603
130. Johnson PDR, Lavender CJ (2009) Correlation between Buruli ulcer and vector-borne notifiable diseases, Victoria, Australia. Emerg Infect Dis 15:614–615. https://doi.org/10.3201/eid1504.081162
131. Marsollier L, Deniaux E, Brodin P et al (2007) Protection against Mycobacterium ulcerans lesion development by exposure to aquatic insect saliva. PLoS Med 4:e64. https://doi.org/10.1371/journal.pmed.0040064
132. Tian RDB, Lepidi H, Nappez C, Drancourt M (2016) Experimental survival of Mycobacterium ulcerans in watery soil, a potential source of Buruli ulcer. Am J Trop Med Hyg 94:89–92. https://doi.org/10.4269/ajtmh.15-0568
133. Tomczyk S, Deribe K, Brooker SJ et al (2014) Association between footwear use and neglected tropical diseases: a systematic review and meta-analysis. PLoS Negl Trop Dis 8:e3285. https://doi.org/10.1371/journal.pntd.0003285
134. Miyamoto Y, Komine M, Takatsuka Y et al (2014) Two cases of Buruli ulcer in Japanese brothers. J Dermatol 41:771–772. https://doi.org/10.1111/1346-8138.12540
135. O'Brien DP, Wynne JW, Buultjens AH et al (2017) Exposure risk for infection and lack of human-to-human transmission of Mycobacterium ulcerans disease, Australia. Emerg Infect Dis 23:837–840. https://doi.org/10.3201/eid2305.160809
136. Ohtsuka M, Kikuchi N, Yamamoto T et al (2014) Buruli ulcer caused by Mycobacterium ulcerans subsp shinshuense: a rare case of familial concurrent occurrence and detection of insertion sequence 2404 in Japan. JAMA Dermatol 150:64–67. https://doi.org/10.1001/jamadermatol.2013.6816

137. Sopoh GE, Barogui YT, Johnson RC et al (2010) Family relationship, water contact and occurrence of Buruli ulcer in Benin. PLoS Negl Trop Dis 4:e746. https://doi.org/10.1371/journal.pntd.0000746
138. Debacker M, Zinsou C, Aguiar J et al (2002) Mycobacterium ulcerans disease (Buruli ulcer) following human bite. Lancet 360:1830. https://doi.org/10.1016/S0140-6736(02)11771-5
139. Debacker M, Zinsou C, Aguiar J et al (2003) First case of Mycobacterium ulcerans disease (Buruli ulcer) following a human bite. Clin Infect Dis 36:e67–e68. https://doi.org/10.1086/367660
140. Capela C, Sopoh GE, Houezo JG et al (2015) Clinical epidemiology of Buruli ulcer from Benin (2005-2013): effect of time-delay to diagnosis on clinical forms and severe phenotypes. PLoS Negl Trop Dis 9:e0004005. https://doi.org/10.1371/journal.pntd.0004005
141. Bibert S, Bratschi MW, Aboagye SY et al (2017) Susceptibility to Mycobacterium ulcerans disease (Buruli ulcer) is associated with IFNG and iNOS gene polymorphisms. Front Microbiol 8:1903. https://doi.org/10.3389/fmicb.2017.01903
142. Goswami T, Bhattacharjee A, Babal P et al (2001) Natural-resistance-associated macrophage protein 1 is an H+/bivalent cation antiporter. Biochem J 354:511–519
143. Stienstra Y, van der Werf TS, Oosterom E et al (2006) Susceptibility to Buruli ulcer is associated with the SLC11A1 (NRAMP1) D543N polymorphism. Genes Immun 7:185–189. https://doi.org/10.1038/sj.gene.6364281
144. Capela C, Dossou AD, Silva-Gomes R et al (2016) Genetic variation in autophagy-related genes influences the risk and phenotype of Buruli ulcer. PLoS Negl Trop Dis 10:e0004671. https://doi.org/10.1371/journal.pntd.0004671
145. Juárez E, Carranza C, Hernández-Sánchez F et al (2012) NOD2 enhances the innate response of alveolar macrophages to Mycobacterium tuberculosis in humans. Eur J Immunol 42:880–889. https://doi.org/10.1002/eji.201142105
146. Manzanillo PS, Ayres JS, Watson RO et al (2013) The ubiquitin ligase parkin mediates resistance to intracellular pathogens. Nature 501:512–516. https://doi.org/10.1038/nature12566
147. Lewis CM (2002) Genetic association studies: design, analysis and interpretation. Brief Bioinform 3:146–153
148. Fletcher HA (2016) Sleeping beauty and the story of the Bacille Calmette-Guérin vaccine. MBio 7:e01370–e01316. https://doi.org/10.1128/mBio.01370-16
149. Nackers F, Dramaix M, Johnson RC et al (2006) BCG vaccine effectiveness against Buruli ulcer: a case-control study in Benin. Am J Trop Med Hyg 75:768–774
150. Phillips RO, Phanzu DM, Beissner M et al (2015) Effectiveness of routine BCG vaccination on Buruli ulcer disease: a case-control study in the Democratic Republic of Congo, Ghana and Togo. PLoS Negl Trop Dis 9:e3457. https://doi.org/10.1371/journal.pntd.0003457
151. Pommelet V, Vincent QB, Ardant M-F et al (2014) Findings in patients from Benin with osteomyelitis and polymerase chain reaction-confirmed Mycobacterium ulcerans infection. Clin Infect Dis 59:1256–1264. https://doi.org/10.1093/cid/ciu584
152. Portaels F, Aguiar J, Debacker M et al (2002) Prophylactic effect of Mycobacterium bovis BCG vaccination against osteomyelitis in children with Mycobacterium ulcerans disease (Buruli ulcer). Clin Diagn Lab Immunol 9:1389–1391
153. Portaels F, Aguiar J, Debacker M et al (2004) Mycobacterium bovis BCG vaccination as prophylaxis against Mycobacterium ulcerans osteomyelitis in Buruli ulcer disease. Infect Immun 72:62–65
154. Smith PG, Revill WD, Lukwago E, Rykushin YP (1976) The protective effect of BCG against Mycobacterium ulcerans disease: a controlled trial in an endemic area of Uganda. Trans R Soc Trop Med Hyg 70:449–457
155. The Uganda Buruli Group (1969) BCG vaccination against Mycobacterium ulcerans infection (Buruli ulcer). First results of a trial in Uganda. Lancet 1:111–115
156. Fenner F (1957) Homologous and heterologous immunity in infections of mice with Mycobacterium ulcerans and Mycobacterium balnei. Am Rev Tuberc 76:76–89
157. Converse PJ, Almeida DV, Nuermberger EL, Grosset JH (2011) BCG-mediated protection against Mycobacterium ulcerans infection in the mouse. PLoS Negl Trop Dis 5:e985. https://doi.org/10.1371/journal.pntd.0000985

158. Tanghe A, Adnet P-Y, Gartner T, Huygen K (2007) A booster vaccination with Mycobacterium bovis BCG does not increase the protective effect of the vaccine against experimental Mycobacterium ulcerans infection in mice. Infect Immun 75:2642–2644. https://doi.org/10.1128/IAI.01622-06
159. Watanabe M, Nakamura H, Nabekura R et al (2015) Protective effect of a dewaxed whole-cell vaccine against Mycobacterium ulcerans infection in mice. Vaccine 33:2232–2239. https://doi.org/10.1016/j.vaccine.2015.03.046
160. Bolz M, Bénard A, Dreyer AM et al (2016) Vaccination with the surface proteins MUL_2232 and MUL_3720 of Mycobacterium ulcerans induces antibodies but fails to provide protection against Buruli ulcer. PLoS Negl Trop Dis 10:e0004431. https://doi.org/10.1371/journal.pntd.0004431
161. Bolz M, Kerber S, Zimmer G, Pluschke G (2015) Use of recombinant virus replicon particles for vaccination against Mycobacterium ulcerans disease. PLoS Negl Trop Dis 9:e0004011. https://doi.org/10.1371/journal.pntd.0004011
162. Geroult S, Phillips RO, Demangel C (2014) Adhesion of the ulcerative pathogen Mycobacterium ulcerans to DACC-coated dressings. J Wound Care 23:417–418, 422–424. https://doi.org/10.12968/jowc.2014.23.8.417
163. Tanghe A, Content J, Van Vooren JP et al (2001) Protective efficacy of a DNA vaccine encoding antigen 85A from Mycobacterium bovis BCG against Buruli ulcer. Infect Immun 69:5403–5411
164. Tanghe A, Dangy J-P, Pluschke G, Huygen K (2008) Improved protective efficacy of a species-specific DNA vaccine encoding mycolyl-transferase Ag85A from Mycobacterium ulcerans by homologous protein boosting. PLoS Negl Trop Dis 2:e199. https://doi.org/10.1371/journal.pntd.0000199
165. Hart BE, Hale LP, Lee S (2015) Recombinant BCG expressing Mycobacterium ulcerans Ag85A imparts enhanced protection against experimental Buruli ulcer. PLoS Negl Trop Dis 9:e0004046. https://doi.org/10.1371/journal.pntd.0004046
166. Hart BE, Lee S (2016) Overexpression of a Mycobacterium ulcerans Ag85B-EsxH fusion protein in recombinant BCG improves experimental Buruli ulcer vaccine efficacy. PLoS Negl Trop Dis 10:e0005229. https://doi.org/10.1371/journal.pntd.0005229
167. Coutanceau E, Legras P, Marsollier L et al (2006) Immunogenicity of Mycobacterium ulcerans Hsp65 and protective efficacy of a Mycobacterium leprae Hsp65-based DNA vaccine against Buruli ulcer. Microbes Infect 8:2075–2081. https://doi.org/10.1016/j.micinf.2006.03.009
168. Roupie V, Pidot SJ, Einarsdottir T et al (2014) Analysis of the vaccine potential of plasmid DNA encoding nine mycolactone polyketide synthase domains in Mycobacterium ulcerans infected mice. PLoS Negl Trop Dis 8:e2604. https://doi.org/10.1371/journal.pntd.0002604
169. Butt AM, Nasrullah I, Tahir S, Tong Y (2012) Comparative genomics analysis of Mycobacterium ulcerans for the identification of putative essential genes and therapeutic candidates. PLoS One 7:e43080. https://doi.org/10.1371/journal.pone.0043080
170. Sarfo FS, Phillips RO, Ampadu E et al (2009) Dynamics of the cytokine response to Mycobacterium ulcerans during antibiotic treatment for M. ulcerans disease (Buruli ulcer) in humans. Clin Vaccine Immunol 16:61–65. https://doi.org/10.1128/CVI.00235-08
171. Trigo G, Martins TG, Fraga AG et al (2013) Phage therapy is effective against infection by Mycobacterium ulcerans in a murine footpad model. PLoS Negl Trop Dis 7:e2183. https://doi.org/10.1371/journal.pntd.0002183
172. Huygen K (2003) Prospects for vaccine development against Buruli disease. Expert Rev Vaccines 2:561–569. https://doi.org/10.1586/14760584.2.4.561
173. Dangy J-P, Scherr N, Gersbach P et al (2016) Antibody-mediated neutralization of the exotoxin mycolactone, the main virulence factor produced by Mycobacterium ulcerans. PLoS Negl Trop Dis 10:e0004808. https://doi.org/10.1371/journal.pntd.0004808
174. Bolz M, Ruggli N, Borel N et al (2016) Local cellular immune responses and pathogenesis of Buruli ulcer lesions in the experimental Mycobacterium ulcerans pig infection model. PLoS Negl Trop Dis 10:e0004678. https://doi.org/10.1371/journal.pntd.0004678

175. Bolz M, Ruggli N, Ruf M-T et al (2014) Experimental infection of the pig with Mycobacterium ulcerans: a novel model for studying the pathogenesis of Buruli ulcer disease. PLoS Negl Trop Dis 8:e2968. https://doi.org/10.1371/journal.pntd.0002968
176. Clancey JK (1964) Mycobacterial skin ulcers in Uganda: description of a new Mycobacterium (Mycobacterium Buruli). J Pathol Bacteriol 88:175–187
177. Walsh DS, Cruz ECD, Abalos RM et al (2007) Clinical and histologic features of skin lesions in a cynomolgus monkey experimentally infected with Mycobacterium ulcerans (Buruli ulcer) by intradermal inoculation. Am J Trop Med Hyg 76:132–134
178. Manna V, Bem J, Marks R (1982) An animal model for chronic ulceration. Br J Dermatol 106:169–181
179. Summerfield A, Meurens F, Ricklin ME (2015) The immunology of the porcine skin and its value as a model for human skin. Mol Immunol 66:14–21. https://doi.org/10.1016/j.molimm.2014.10.023
180. Beissner M, Piten E, Maman I et al (2012) Spontaneous clearance of a secondary Buruli ulcer lesion emerging ten months after completion of chemotherapy—a case report from Togo. PLoS Negl Trop Dis 6:e1747. https://doi.org/10.1371/journal.pntd.0001747
181. Gordon CL, Buntine JA, Hayman JA et al (2011) Spontaneous clearance of Mycobacterium ulcerans in a case of Buruli ulcer. PLoS Negl Trop Dis 5:e1290. https://doi.org/10.1371/journal.pntd.0001290
182. Marion E, Chauty A, Kempf M et al (2016) Clinical features of spontaneous partial healing during Mycobacterium ulcerans infection. Open Forum Infect Dis 3:ofw013. https://doi.org/10.1093/ofid/ofw013
183. Revill WD, Morrow RH, Pike MC, Ateng J (1973) A controlled trial of the treatment of Mycobacterium ulcerans infection with clofazimine. Lancet 2:873–877
184. Silva-Gomes R, Marcq E, Trigo G et al (2015) Spontaneous healing of Mycobacterium ulcerans lesions in the Guinea Pig model. PLoS Negl Trop Dis 9:e0004265. https://doi.org/10.1371/journal.pntd.0004265
185. Deshayes C, Angala SK, Marion E et al (2013) Regulation of mycolactone, the Mycobacterium ulcerans toxin, depends on nutrient source. PLoS Negl Trop Dis 7:e2502. https://doi.org/10.1371/journal.pntd.0002502

Buruli Ulcer in Animals and Experimental Infection Models

Miriam Bolz and Marie-Thérèse Ruf

1 Naturally Infected Animals

Naturally infected animals presenting with typical BU lesions have so far only been described in Australia and there only in one major endemic focus, the central coastal Victoria close to Melbourne. Between 1980 and 1985, 11 *M. ulcerans* positive koalas (*Phascolarctos cinereus*) in a population of approximately 200 koalas on Raymond Island were described as having ulcers on the face, forearm, rump, groin or foot pad [1, 2]. More recently, examination of common ringtail (*Pseudocheirus peregrinus*) and common brushtail (*Trichosurus vulpecula*) possums from Point Lonesdale, a small BU endemic region with a recent human outbreak, revealed that 38% of the analyzed ringtail possums and 24% of the brushtail possums had laboratory-confirmed *M. ulcerans* skin lesions and/or their feces tested positive for *M. ulcerans* DNA by PCR. Lesions were found on tails, toes, feet and noses with the majority occurring on the tail [3]. Another study by O'Brien et al. showed that *M. ulcerans* bacteria were present in possum lesions and from 90% of the animals with lesions, positive cultures could be obtained [4]. The high number of possums with skin lesions suggests that possums may represent an animal reservoir for *M. ulcerans* in south-eastern Australia [3–5].

Sporadically, *M. ulcerans* positive lesions were also diagnosed in domesticated animals like dogs [6], a cat [7], horses [8] and alpacas [9]. In the dogs, lesions were found on the feet, legs and the rump and diagnosis was done by IS*2404* real-time PCR (qPCR). Molecular typing in three animals confirmed that the infection was caused by disease-causing human strains [6]. The cat presented with a lesion on the nose and acid fast bacilli (AFB) staining as well as molecular methods confirmed the infection with *M. ulcerans* [7].

Despite considerable effort, several studies conducted in Africa were not able to corroborate the findings from Australia [10–12].

M. Bolz (✉) · M.-T. Ruf
Swiss Tropical and Public Health Institute, Basel, Switzerland

G. Pluschke, K. Röltgen (eds.), *Buruli Ulcer*,
https://doi.org/10.1007/978-3-030-11114-4_9

2 The Mouse (*Mus musculus*) Model

2.1 History of the BU Mouse Model

Already with the first documentation of BU and its causative agent *M. ulcerans,* the description of potential animal models was published. Besides the report of the clinical picture and the cultivation of the new mycobacterial species, MacCallum et al. dedicated two separate chapters of their publication in 1957 to "Experimental investigations in laboratory animals" and the description of the "pathology of the experimental lesions in the rat" [13]. In their attempt to grow the unknown mycobacteria, the researchers conducted a series of experiments in guinea pigs, rats and mice, mostly injecting ground-up tissue of a patient lesion or exudate from a patient's ulcer. As a next step they serially passaged a saline extract of lung tissue from infected animals that had died. None of the rats showed any signs of disease. Curiously, the rat of the fourth passage was forgotten and 16 months after intraperitoneal (i.p.) injection developed edematous and ulcerated limbs and the tail sloughed off. Microscopic examination of ulcers revealed high numbers of AFB, thus the rat became the first animal to sustain *M. ulcerans* and characterize the bacteria while optimal in vitro growth conditions still had to be found [13]. Beside the rats, the scientists also reported the infection of 16 white mice, 12 guinea-pigs, two rabbits, one fowl and three lizards and made observations that would be re-confirmed in many more animal studies to follow: mice seemed more susceptible to infection than rats and guinea-pigs.

The actual development of the most commonly used animal model in BU research today, the mouse (*Mus musculus*), traces back to Fenner [14], who used it primarily for the evaluation of available antimicrobial compounds against the newly discovered *M. ulcerans* bacteria [14, 15]. In those early days of BU research, the mouse model was used to study all the major aspects of the disease that would later regain interest: chemotherapy testing [16], characterization of the immune response against *M. ulcerans* [17] and pathogenesis [18].

During the next 20 years, animal models in BU research were only occasionally used, but importantly, the knowledge gained from previous studies served as a basis for the development of the mouse model for *Mycobacterium leprae* infection, the cause of leprosy [19]. A first grading system for assessing disease severity in the mouse model was described in 1972 in the context of drug testing [20] and further refinement of the mouse model to how it is mainly used today (foot pad injection) was then described in 1975 [21], a publication that reconfirmed the relative resistance of guinea pigs to *M. ulcerans* infection once more.

For another 20–30 years BU animal models were nearly forgotten again, before an increasing number of publications reflected the regained interest in the topic. The animal model that emerged as most commonly used was the mouse. Its applications can be roughly split into four different topics as further elaborated in this chapter: testing of antimicrobial compounds, vaccine development, study of the pathogenesis of BU and studies on the toxin mycolactone.

2.2 Experimental Infection of the Mouse

2.2.1 Infection Sites

Very early in BU research and the use of the mouse model it was recognized that *M. ulcerans* needs a relatively cold location to grow in the mouse body [16, 18]. Even when animals were inoculated intracerebrally, the most vulnerable site of infection remained to be the tail [18]. Conveniently, low temperatures in mice are found in places that are not extensively covered with fur: the tail, the hind foot pads and the ears. Each of the three sites have their practical advantages and disadvantages; for example, the ears are very thin structures, which makes injection of bacteria challenging but on the other hand processing for histopathology is easy as there are no bones that have to be softened by decalcification, like it is the case in mouse foot pads. However, mouse ears do not contain a lot of fat tissue; hence they are very rarely used as site of infection, as *M. ulcerans* is mostly located close to fat cell areas in the human skin [22, 23]. It appears that nowadays, the few laboratories that use mice as animal models for BU have a preference for either infecting the tail [24–26] or the foot pads [27–33]. A recent study by Bénard et al. compared a new infection site, the hock, to the foot pad and the ear, concluding that in their hands the foot pad is the site of choice in the mouse for a well-defined, consistent and reproducible infection [34].

2.2.2 Mouse Strains

The mouse strains used for infection with *M. ulcerans* in different laboratories are not overly diverse. While earlier studies had used outbred Swiss mice [28], the vast majority of newer studies used immunocompetent inbred strains that are very frequently used in research in general: BALB/c [24, 28, 30, 35–37] and C57BL/6 mice [22, 32, 38] provided by different suppliers. Furthermore, a lot of studies did not use mixed sex mice but were only conducted in female animals. Again, depending on the laboratory, there seems to be a certain preference, rendering female BALB/c the most used mice for the BU mouse foot pad model by far. The first publication directly comparing the two most commonly used inbred strains, with each other and to infection in FVB/N mice was published in 2016 [25]. While infection led to ulceration of the tails and subsequent death of the animals in BALB/c and C57BL/6 mice, FVB/N mice were able to spontaneously heal the developed ulcers and did not succumb to the infection, regardless of the different *M. ulcerans* strains used in the study [25]. The only other study published on the specific topic of mouse strain comparability reported that C57BL/6 might be slightly more susceptible to the infection, displaying increased leukocyte infiltration and bacterial growth compared to BALB/c mice [34].

Immunocompromised or transgenic mice were occasionally used to study the role of host factors in *M. ulcerans* infection. TNF receptor P55-deficient mice were shown to be equally susceptible to infection with highly virulent *M. ulcerans* strains as wild-type C57BL/6 mice, but showed increased susceptibility to non- and intermediately virulent strains [39]. Bieri et al. documented the infection in interferon-γ (IFNγ)-deficient mice (B6.129S7-Ifng^{tm1Ts}/J), identifying the cytokine as critical regulator of early host defense mechanisms against *M. ulcerans* infection. Compared to wildtype mice, IFNγ-deficient mice displayed a faster progression of the infection with increased tissue necrosis, increased edema formation and a higher bacterial burden [40]. These results were somewhat comparable with an earlier study that reported an increased susceptibility of IFNγ-deficient mice to an avirulent (mycolactone-negative) and an intermediately virulent *M. ulcerans* strain [41]. Studies in Rag2$^{-/-}$ mice and nude mice done in the same laboratory in Portugal showed a year later that lymphocytes are protective against infection with low-virulence strains but not against infection with highly virulent *M. ulcerans* [42].

Three more specialized transgenic mouse lines were used to study very specific questions in the mouse model for BU: Type 2 angiotensin II receptor knock-out mice (AT_2R, [43]) were used for studying nerve damage caused by mycolactone but the infection in those mice was not further characterized [44]. Similarly, Fas- and Bim-knock out (KO) mice were used to establish the mode of action of the toxin mycolactone in vivo [45]. Mice lacking the proapoptotic Bcl-2 family member Bim did not develop necrotic BU lesions, but were able to control the mycobacterial multiplication. Recently a study used ICR mice, an outbred strain, for infection with *M. ulcerans* in order to study questions of BU transmission [46].

2.2.3 Mycobacterial Strains Used for Experimental Infection

Compared to the mouse strains used in the foot pad model of BU, the *M. ulcerans* strains used to infect the mice are much more diverse. Again, preferences appear to be present in different laboratories, probably mostly dependent on which strains were available to the laboratory at some point in time and which strains could be maintained without loss of virulence. As repeated passage by in vitro culture prompts some of the *M. ulcerans* strains to lose their virulence plasmid [47, 48], some laboratories opted to passage their strains trough mouse foot pads in order to preserve their virulence [49–51]. Alternatively, low passage primary isolates have been used for productive infection of mice [52].

The most comprehensive study on how the use of *M. ulcerans* strains from different geographical origin and with different cultivation history influences virulence and immune responses induced in experimentally infected mice was published in 2009 by Ortiz et al. Eleven different *M. ulcerans* strains isolated from different parts of the world over a time span of 47 years were used to infect 6–8 weeks old male BALB/c mice in the foot pad. Subsequently, their ability to cause a productive infection was reported as well as a detailed characterization of the inflammation induced by the different strains [53]. The bacterial strains used in these experiments partially overlap with the set of strains which was regularly used in the laboratory of Jorge Pedrosa, the only other research group that systematically characterized murine infection caused by different strains of *M. ulcerans* [30, 39, 41, 42, 48, 54]. The vast majority of BU mouse model studies were conducted with a single *M. ulcerans* strain. A few of those *M. ulcerans* isolates warrant mentioning as they are especially popular: **Cu001** was originally isolated from a BU patient in Adzopé, Ivory Coast, in 1996 and has been successively passaged in foot pads of BALB/c mice ever since [28, 35]. It is probably the strain most often used for in vivo antibiotic treatment efficacy testing. **Mu1615** is among the oldest isolates frequently used in mouse infection studies. It was originally obtained from a Malaysian BU patient in the 1960s and produces mycolactone A/B like the African isolates [55, 56]. A spontaneous mycolactone negative mutant derived from this strain [47] has been used as avirulent *M. ulcerans* strain in the mouse model. **S1013** was originally isolated in 2010 from a swab taken from the undermined edges of the ulcerative lesion of a Cameroonian BU patient [57]. Its passage number has been kept to an absolute minimum in order to preserve its virulence for studies in the mouse footpad model and the pig model of BU [40, 45, 52, 58–60].

2.2.4 Dose and Preparation of the Bacterial Inocula

Similarly to all the other experimental details of the BU mouse model described above, preparation of the inoculum, as well as how the number of bacteria in the inoculum is quantified varies from laboratory to laboratory. The waxy extracellular matrix of *M. ulcerans* together with the extreme slow growth of the bacteria are the two major factors that make preparation and dosing of an inoculum of *M. ulcerans* rather difficult. One strategy that some laboratories employ for dispersing clumps in their culture is to vortex the bacteria with glass beads prior to quantification of the inoculum [30, 32]. Sonication of the culture is another possible way of reducing clumps. However, both methods pose the risk of substantially altering the surface structure of the bacilli and with it potentially their interaction with the host after inoculation.

For estimating the number of bacteria in the inoculum, colony forming unit (CFU) counting or counting of AFB by microscopy are the most commonly used methods. The AFB counting method according to Shepard and McRae [61] offers the advantage that the actual dose is quantified at the time of injection, while 2–3 months waiting time for colonies to grow is necessary after plating. On the other hand, CFU plating reveals the number of live *M. ulcerans* (or microclumps), which cannot be determined by microscopic enumeration. Hydrophobicity and the strong tendency of *M. ulcerans* to clump affect both methods, as well as dosing with the help of optical density (OD) or the so called McFarland standard [62]. Because an immediate estimate of how many bacteria are in a solution proved to be so difficult, some laboratories opted to produce cell banks of ready to go bacterial inocula that were frozen down for later use [63]. While this strategy allows to standardize inoculation, the freezing may affect to some extent relevant properties of the bacteria. In our hands, dosing the infection based on the wet weight of the bacterial pellet resulted in the most reproducible infection of mouse foot pads [34, 40, 52, 58]. A wide range of inoculum sizes has been used. In some studies as few as a hundred or a thousand CFU [45, 50, 58] were injected, while others used up to 10^5 or 10^6 CFU [54, 64].

2.2.5 Infection Outcomes and What to Measure

With an estimated generation time of 3–4 days in the mouse foot pad [14], animal experiments with *M. ulcerans* often last for a long time. Depending on the infection dose and the virulence of the strain injected, development of a visible pathology only starts a few weeks after infection [14, 53], with a typical lag time of 3–4 weeks. When foot pads are inoculated, the first visible signs are usually edema on the top of the foot pad, followed by reddening of the foot and ankle and swelling to the point where foot pads appear to be "leaky", so that cage bedding material starts to stick to the foot pad (Fig. 1). Infected mouse ears do not tolerate high bacterial inocula and infection often results in rapid loss of tissue [34]. Similarly, mouse tails do not comprise a lot of fat, hence the ulceration described in publications where mice were infected in the tail might have limited comparability to human BU lesions [24].

Several approaches to monitor progression of the infection exist. Measuring foot pad thickness with a caliper allows for repeated measurements while assessment of bacterial proliferation by CFU plating, AFB counting, or qRT-PCR require the euthanasia of the animal. For CFU determination the infected tissue is usually ground up, sometimes de-contaminated to avoid overgrowth with faster growing microorganisms and further processed [22, 52]. Very recently an optimized procedure for extraction and quantification of bacterial RNA from infected mouse tissue was published, which provides a surrogate marker for viability of the bacteria [66, 67]. Parallel to measuring foot pad thickness, a grading system can be applied to estimate the severity of the disease (Fig. 1). The grading system as it is used in many laboratories today was originally described in 1972 and has not substantially changed since then [20]. Both, grading system as well as foot pad thickness measurements have mostly replaced the practice of measuring time to death of the animals. Instead, animals are monitored on a regular basis and animals are euthanized in compliance with laws for protection of experimental animals. If enumeration of AFB or quantification of bacterial DNA is not the primary objective of the analysis, all infected tissues can be processed for histopathology instead, and a description of inflammatory cellular infiltrates, tissue damage, etc. can be done.

For in vivo monitoring of the infection, Zhang et al. engineered recombinant bioluminescent *M. ulcerans* strains expressing luxAB from *Vibrio harveyi* [68]. The

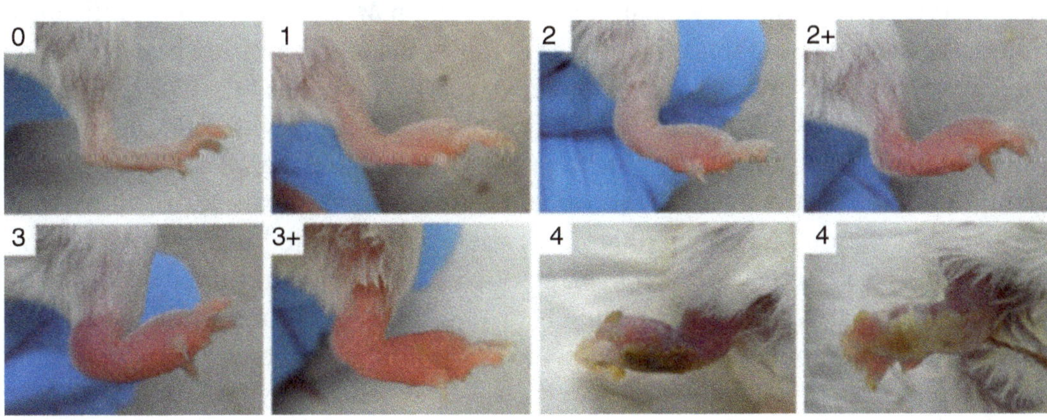

Fig. 1 Pathogenesis after subcutanous *M. ulcerans* injection in the mouse foot pad. Subcutaneous (s.c.) injection of acid fast bacilli into the healthy mouse foot pad (0) leads to progressive swelling and inflammation. (1) Grade 1, slight swelling; (2 and 2+) Grade 2, swelling with inflammation; (3 and 3+) Grade 3, swelling with inflammation of the leg; (4) Grade 4, swelling with inflammation and possible ulceration, cage bedding sticking to the sole of the foot [65]

relative light unit count measured for these strains in real-time correlated with the CFU counts [68]. The measured bioluminescence represents a suitable real-time surrogate marker for viable bacteria in mouse foot pads [33]. While these first constructs required exogenous substrate, limiting their utility in vivo, Zhang et al. went on engineering alternative bacterial strains that are autoluminescent and showed that they could reduce time, effort, animal numbers and costs of BU animal model experiments [69].

Another rather new approach proposed for monitoring antibiotic treatment success is the detection of mycolactone by thin layer chromatography (TLC) or fluorescent TLC [70].

2.3 Research Applications for the Mouse Model

2.3.1 Antimicrobial Compound Testing

Antimicrobial compound testing is the application for which the BU mouse model was most frequently used since the discovery of the disease. Early on, researchers were interested in which antimicrobials would work against the newly described bacterium, which was the major incentive to have an animal model in the first place [15, 71, 72]. When the mouse model saw a revival in the early 2000s, testing new drugs in vivo was still its main use.

Systematic testing of different antimicrobial treatments in the mouse foot pad model with Rifampin, Rifabutin, Amikacin (AMK), Clarithromycin (CLR) and Streptomycin (STR) being the most intensively evaluated actives, showed that a combination therapy with Rifampicin (RIF) and AMK or STR might be effective for the treatment of human BU [28, 35] and led to the conduction of a human treatment trial using RIF and STR [73]. Mice were usually treated right away from the day of infection and only when a potential effective combination of drugs was found, later treatment starts were evaluated [29, 74]. When doses were administered in different time intervals, it became very clear, that the effect on *M. ulcerans* was significantly stronger when mice were treated daily [75]. Therapy durations varied between studies, but were mostly between 4 and 8 weeks [28, 29, 35, 36, 74, 76, 77]. Finally, Lefrançois et al. showed in 2007 that combination treatment with RIF and an aminoglycoside (STR or AMK) was able to prevent BU in mice with 8 weeks being the optimal duration of treatment [78]. Since the provisional guidelines of the World Health Organization (WHO) in 2004, BU patients were treated with a combination of RIF and STR [79]. This proved to be an effective treatment for the disease; however, the fact that STR had to be injected daily was less than optimal. Hence, researchers started to look for a cure that could be administered as a fully oral regimen and tested them in the mouse model [80–82]. As shown by Ji et al. in 2007 the combination of RIF and CLR has the highest potential of replacing the current standard treatment [82]. The WHO Technical Advisory Group on BU decided in 2017 that the WHO recommendation for treatment of the disease should be changed to RIF and CLR, pending the availability of full results of a treatment trial with this drug combination.

On a few occasions, treatment alternatives to drug administration were also tested in the BU mouse model: The effect of corticosteroid-induced immunosuppression on therapy with antibiotics was examined [83], phage therapy was shown to be effective in mice [54] and even the efficacy of hydrated clays on *M. ulcerans* growth was assessed in vivo on mouse tails [84]. However, as ideal the mouse model might be for drug testing and every therapy that can be applied by injection or orally, for the evaluation of topically applied therapies, including thermotherapy, the mouse model is highly impractical.

2.3.2 Vaccine Development

Including a vaccination phase prior to infection of mice with *M. ulcerans* further prolongs the overall time of the animal experiments, resulting in experiments that often last for several months. For example a single experiment including two immunizations with an adjuvanted formulation of recombinant protein followed by infection with a low dose of bacteria results in almost five months of experiment time, not counting the time needed for specimen processing and follow up analyses [58]. An important factor determining the time span is the time researchers allow between the last immunization and the infection with *M. ulcerans*. Three weeks is generally agreed on as the minimum waiting time, although some research laboratories have published longer intervals, lasting up to 16 weeks [32, 85]. Mostly, protective efficacy of the tested vaccines is assessed by comparing the speed and severity of the infection between mice receiving the candidate vaccine and a placebo group, very similarly to testing of antimicrobial compounds, where non- or mock-treated animals are used as negative controls.

After a first attempt of protecting rats and mice with *Mycobacterium fortuitum* against experimental *M. ulcerans* infections in 1985 [86], vaccination against BU was not experimentally addressed for another 15 years. Evidence in the literature of a partial protective effect of Bacillus Calmette-Guérin (BCG) vaccination in humans against BU prompted the research group of Kris Huygen to experimentally address this question in the mouse model and to become the major driver of the search for a BU vaccine in the following years [87, 88]. Tanghe et al. showed that intravenous vaccination with BCG (10^6 CFU) or intra-muscular vaccination with plasmid DNA encoding the BCG version of antigen 85A (Ag85A) both reduced the bacterial burden in foot pads of infected mice [32]. Although BCG vaccination has been repeatedly shown to reduce CFU numbers in experimentally infected mice and/or slow down pathogenesis [24, 48, 51], the protection conferred by BCG was always only partial, of only relatively short duration and could not be prolonged by a booster vaccination [85]. Furthermore, a comparison between BALB/c mice and C57BL/6 mice revealed that the efficacy of BCG vaccination against *M. ulcerans* may vary with both choice of host and pathogen strain [27]. A single sub-cutaneous vaccination with a mycolactone-negative *M. ulcerans* strain two months prior to foot pad infection with a virulent isolate was found to lead to a similar delay in foot pad swelling as a single BCG immunization [48], countering arguments that a homologous immunization would be more efficient than a heterologous with the *M. bovis* derived BCG. Nevertheless, Hart et al. further pursued this approach with some success, as they were able to show that a prime vaccination with recombinant BCG expressing *M. ulcerans* Ag85A followed by a similar recombinant *Mycobacterium smegmatis* boost conferred superior protection against in vivo challenge than BCG alone [89]. While in this study immunizations were done intra-venously, the follow up study showed that a single sub-cutaneous injection with a BCG strain overexpressing two *M. ulcerans* antigens significantly prolonged survival times of mice compared to vaccination with the previously tested vaccination strains and application routes [63]. Even though the observed differences in bacterial burden at some point during the experiment were correlated with longer survival of the mice, none of the vaccination strategies tested so far was able to completely protect against experimental infection with *M. ulcerans*.

Besides DNA-vaccines [24, 32, 51, 90] and immunization with live BCG and/or mycolactone negative *M. ulcerans*, a few other attempts have been made at developing a vaccine against BU: Mice were exposed to Naucoris aquatic insect bites or sensitized to Naucoris salivary gland homogenates [26], immunized with recombinant virus replicon particles [52], adjuvanted recombinant proteins [58] or dewaxed whole-cell vaccines [64], all with limited protective efficacy at best.

2.3.3 Study of the Pathogenesis of Buruli Ulcer and the Immune Response to the Disease

While BU in humans results in different clinical presentations, pathogenesis in experimentally infected mice usually follows a defined pattern that is shaped by the virulence of the *M. ulcerans* strain used for infection. Two histopathological hallmarks of the human disease are the presence of large extracellular clusters of AFB and the almost complete absence of inflammatory infiltrates in the center of the lesions [91, 92], but not necessarily in the periphery of lesions [93]. While acquiring human tissue specimens from early BU cases is challenging due to practical and ethical reasons [93], the mouse model of *M. ulcerans* infection allowed to describe the early events in the pathogenesis of *M. ulcerans* infection in detail. In the mouse, at least some of the inoculated *M. ulcerans* bacteria are initially captured by phagocytes at the injection site and transported to the draining lymph node. The importance of initial intracellular stages prior to the emergence of extracellular clusters [22] is not entirely clear. Infection with virulent strains in mice then leads to the fast influx of neutrophils and to some degree of monocytes/macrophages, thus an early acute inflammatory response is induced [22, 30, 31]. Even though cellular responses are initiated in the draining lymph nodes, the progression from this initial acute to a more chronic inflammation with a predominance of mononuclear cells and/or lymphocytes is obstructed [30] because mycolactone is driving the cellular infiltrates into apoptosis. As a consequence AFB become extracellular and proliferate in the developing necrotic and acellular subcutaneous lesion. At the boundary between the expanding necrotic areas and the surrounding healthy tissue, acute inflammatory infiltrates and intracellular bacteria are found [30]. This belt of immune cells remains a site of constant interaction between the pathogen and the immune system. As the infection progresses in infected mouse foot pads, edema becomes a prominent feature in histopathology, which is macroscopically reflected by the onset of foot pad swelling [30, 31]. Swelling then extends to the ankle and the lower part of the leg, at which point mice are usually euthanized for ethical reasons [31]. Treatment of infected mice with RIF-STR rapidly leads to changes in the histopathological appearance of the infected tissue. The predominantly neutrophilic immune response changes to an infiltration dominated by lymphocytes and macrophages with prominent B-lymphocyte clusters and macrophage accumulations surrounding the bacteria, which loose solid Ziehl-Neelsen staining and viability [31, 94, 95].

Both, immune response and pathogenesis observed in *M. ulcerans* mouse infections are strongly influenced by the toxin mycolactone. Mycolactone-negative bacteria are faster phagocytosed, no prominent necrosis is developing and the initial neutrophilic response becomes readily replaced by a predominantly lymphocytic and macrophagic infiltrate [22, 30]. Goto et al. examined whether nerve invasion occurred after infection of mouse foot pads with *M. ulcerans* and concluded, that the reported painlessness of BU might partially be attributed to intraneural invasion of bacilli [96]. Another study used injection of mycolactone rather than bacilli to study the effects of the toxin more directly in vivo [97].

3 Other Animal Models

3.1 Guinea Pig (*Cavia porcellus*)

Guinea pigs have been used as model for *M. ulcerans* infection as early as in 1974, when the presence of an exogenous diffusible toxin, nowadays known as mycolactone, was proposed [98, 99]. Culture filtrate intradermally inoculated into guinea pig skin caused focal necrosis and inflammation, which closely resembled the skin lesions developing after injection of viable organisms as well as human BU skin lesions [98, 99]. In later experiments the effect of extracted mycolactone was compared to the effects caused by infection of the skin with wild type and mycolactone-deficient *M. ulcerans* bacteria. It became clear that the histopathological effects observed were largely attributable to the polyketide toxin mycolactone, the main virulence factor of *M. ulcerans*, which induced apoptosis as demonstrated by TUNEL staining [100–102]. Macroscopic changes in the guinea pig after an intradermal injection of 10^6 and 10^7 *M. ulcerans* bacteria included the occurrence of a nodule after three days and the presence of ulcers after 8–12 days. Histopathological hallmarks were described as the occurrence of central necrotic areas with extracellular AFB, which enlarged during the six weeks observation time and were surrounded by inflammatory infiltrates [100]. Recovery and re-growth of bacteria from the skin lesions was possible until week six after infection, but viability decreased substantially during the observation period [100]. In a time-lapse study conducted by Silva-Gomes et al. in which the ear and the back of the guinea pigs were subcutaneously inoculated, the early nodule-like structures progressed to ulcers with overlying scabs which healed between day 18–26 post-inoculation and were accompanied by bacterial clearance regardless of the strain or the infectious dose used [103].

One proposed potential way of contracting BU is the contamination of existing open wounds with *M. ulcerans* through exposure to a contaminated environment. In order to test this hypothesis, the skin of guinea pigs has been abraded and cultured *M. ulcerans* cells were applied on the open superficial wounds [104]. However, topical application failed to establish an infection during the 90 days of observation and all abrasion sites healed during the first week post exposure. In contrast, from all the sites intradermally injected with 10^6 *M. ulcerans* bacteria as positive control, re-cultivation was possible 90 days post infection [104].

Overall, guinea pigs seem to be quite resistant to *M. ulcerans* infections, and are far less frequently used as BU research model than mice.

3.2 Pig (*Sus scrofa*)

The pig has only recently been evaluated as a model to study BU [59, 60]. Pig and human skin share many similarities like thickness, general structure of the epidermis, dermis and the subcutaneous tissue, blood supply and the adnexal structures [105–109]. Moreover, the porcine immune system resembles the human immune system much more than the murine immune system [110]. This has made the porcine skin a preferred model for burn and wound healing studies [111–113]. In the studies conducted by Bolz et al. that represent the first description of the pig (*Sus scrofa*) as a model to study *M. ulcerans* infection, eight week old piglets were infected subcutaneously with 100 µl of different numbers (2×10^3 to 2×10^7) of *M. ulcerans* bacteria [60]. For injections, both flanks [59, 60] as well as the upper and lower legs (unpublished results) were used to study up to 24 individual inoculation sites on a single pig, which greatly reduces the number of animals needed for individual experiments. After 2.5 weeks macroscopic examinations revealed elevated, movable and firm nodular structures with the highest inoculation doses. After 6.5 weeks these lesions presented either as indurated plaques or had ulcerated. Histopathological analysis showed that all key features typically found in early human BU lesions [93] were also present in the pig 2.5 weeks after inoculation (Fig. 2) [60]. At the inoculation site a necrotic core structure containing mainly extracellular clusters of AFB (Fig. 2a) and fat cell ghosts developed. The core as well as some infiltrating cells were strongly stained by the TUNEL method (Fig. 2b), indicating the presence of large numbers of apoptotic cells. The necrotic core was surrounded by a dense belt of infiltrating cells, which were mainly composed of macrophages and CD3 positive T-cells (Fig. 2c) [59]. Inside the necrotic core debris of neutrophils was still detectable [59]. At week 6.5 the infiltration was even more organized, with the necrotic core being surrounded by a belt of neutrophils followed by a belt of macrophages and the outermost belt being mainly composed of CD3 positive T-cells. Some of the pig lesions had ulcerated during this observation time. Although the majority of bacteria and necrotic slough were expelled during that process, small necrotic areas with AFB remained and undermined edges started to form. A slight epidermal hyperplasia was already observed after 2.5 weeks, however at later time points it was much more pronounced [60].

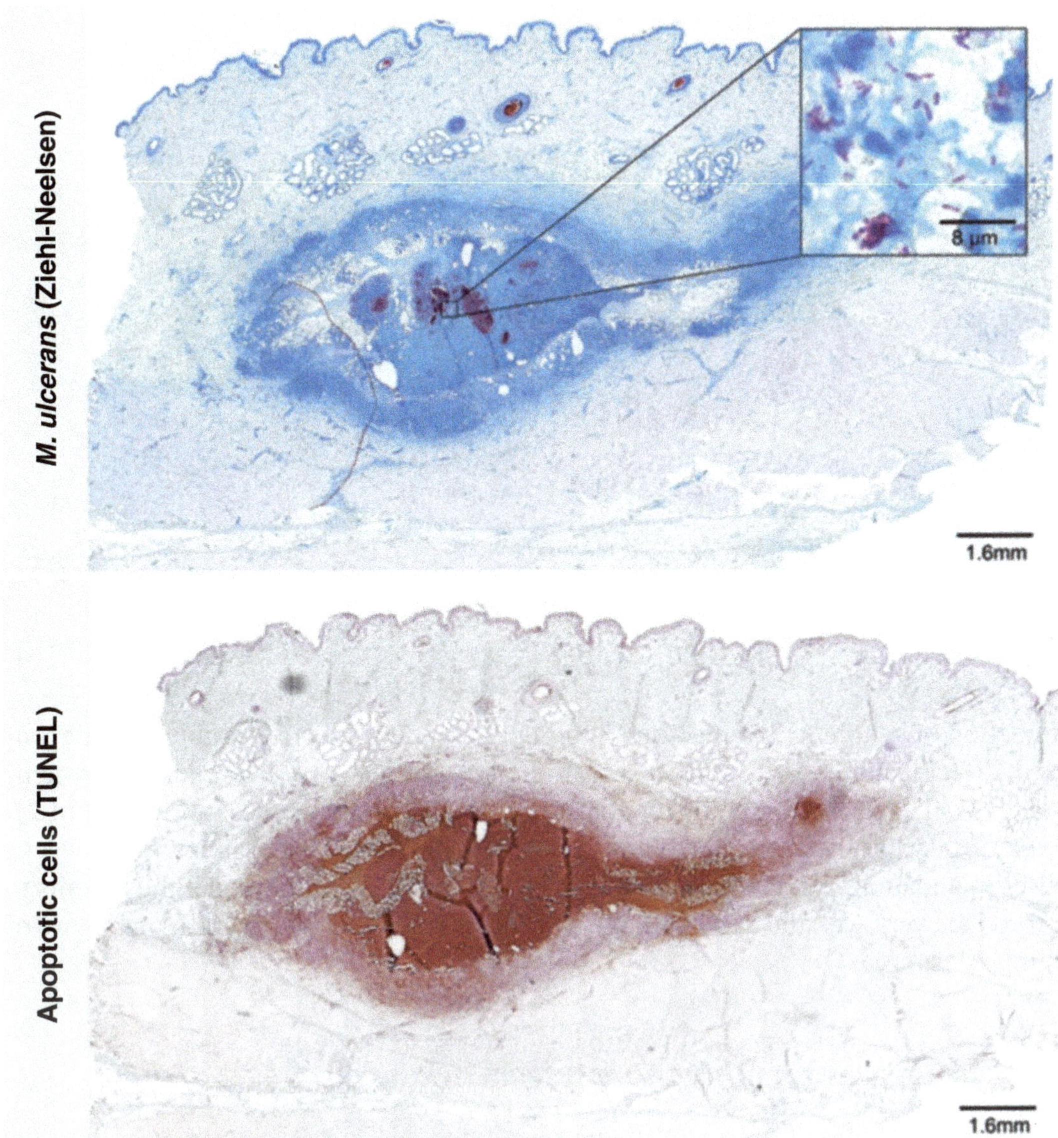

Fig. 2 Histopathological appearance of a typical pig BU lesion two weeks after subcutaneous injection of 7×10^7 *M. ulcerans* bacteria. (**a**) Ziehl-Neelsen staining revealed the presence of large extracellular clusters, globi like structures and single intra- and extracellular AFB (pink rods). (**b**) The TUNEL staining method for apoptotic cells revealed a strong staining of the necrotic core as well as some TUNEL positive cells in the surrounding infiltration belt. (**c**) CD3 positive T-cells are only present in the outermost area of the infiltration and start to form a belt around the lesion core, which develops further until week 6

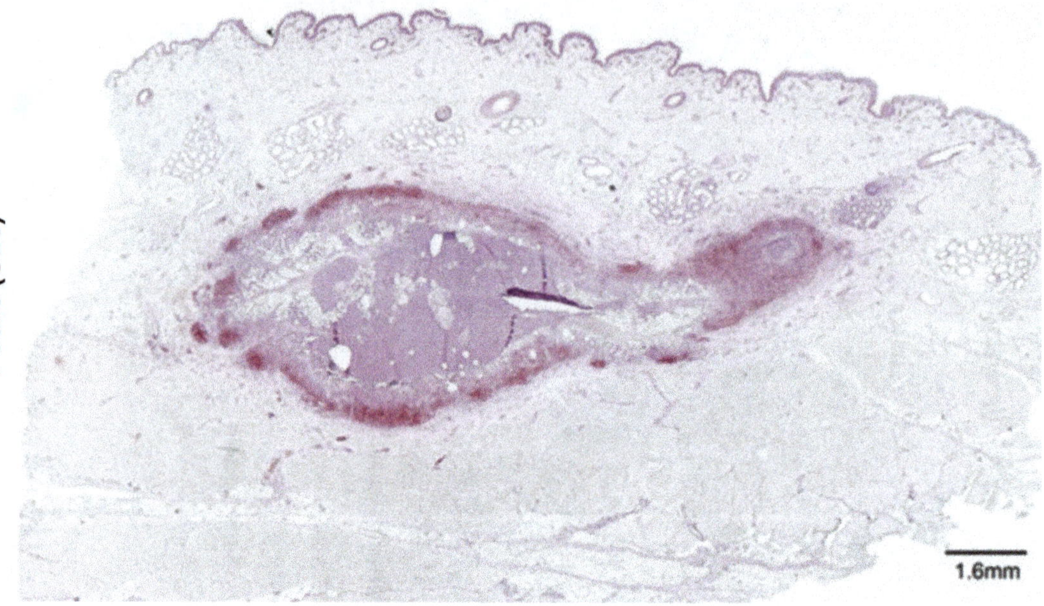

Fig. 2 (continued)

Injection of synthetic mycolactone into pig skin induced a comparable dose dependent effect in the tissue, with the formation of a necrotic core surrounded by inflammatory cells ([60] and unpublished results). In follow up studies cultivation of *M. ulcerans* bacteria re-isolated from infected pig skin was only possible up to 3 weeks after inoculation, and no chronic infection developed. The pig model thus seems to reproduce very well the early pathogenesis but not the chronic, necrotizing nature of human lesions.

3.3 Grasscutter (*Thryonomys swinderianus*)

Addo et al. [62] used the small hystricomorph rodent *Thryonomys swinderianus,* commonly called grasscutter, as a model for BU. The abundance of these animals in sub-Saharan Africa, the already existent establishment as laboratory animals and the substantially larger body size than mice (1.5–5 kg) were major reasons for choosing these animals. Grasscutters were subcutaneously inoculated into the shaved right thigh with 200 µl *M. ulcerans* suspension (1.0 and 5.0 McFarland standards) [62]. Progressive skin lesions (erythema, papule, nodule, edema, and undermined ulcer) developed in all inoculated animals with onset and severity of the lesions correlating with the infection dose. After the development of a nodule and a scab (40 to 154 days after inoculation), which was demonstrated to contain AFB, infection stalled and an inactive phase (265 to 371 days) occurred. Relapse occurred in 90% of the animals and a fulminant progression of the disease with the formation of undermined ulcers was observed 441 to 540 days post inoculation. No systemic spread of bacteria was observed, with blood, urine and feces consistently being AFB-negative whilst the exudate and pus of the lesions was consistently AFB-positive. In histopathological analyses typical features of an *M. ulcerans* infection in the skin were observed together with rather untypically strong inflammation, neurogenic atrophy and osteomyelitis [62]. Re-cultivation of AFB from the infected animals however was not done and therefore viability of the AFB detected in exudate could not be determined.

3.4 Anole Lizard (*Anolis carolinensis*)

Based on the fact that *M. ulcerans* grows best at temperatures of 30–32 °C, and poikilothermic animals have a preferred body temperature of 32 °C, Marcus et al. tested the anole lizard, *Anolis carolinensis,* as a model to study BU [114, 115]. Lizards were inoculated with an *M. ulcerans* solution containing 10^3 CFU into the right lateral thorax, caudal to the edge of the right scapula. Five times a week lizards were incubated at 32 °C for 6–8 h. Six weeks post inoculation three different kind of lesions could be distinguished by histopathology [114]: a diffuse non-encapsulated, non-necrotizing granulomatous reaction with intracellular AFB (67% of the animals), a diffuse necrotizing granulomatous myositis of the thoracic wall with large numbers of extracellular AFB (24% of the animals) and the presence of several small, discrete, non-necrotizing encapsulated granulomas on the superficial surface of the supraspinatus nerve containing few intracellular AFB (4.8% of the animals). Lizards inoculated with dead *M. ulcerans* bacteria also developed non-necrotizing, non-encapsulated lesions, however, nearly all bacteria were found intracellularly. In contrast to the human disease, lesions did not develop inside the subcutaneous tissuc but on thc thoracic wall. Re cultivated bacteria after single passage in the anole lizard showed a more rapid growth in vitro and were more virulent when introduced into mouse foot pads [114].

3.5 Nine-Banded Armadillo (*Dasypus novemcinctus*)

The nine-banded armadillo is a well-studied and characterized animal model in leprosy research [116] with a body temperature of 30–35.8 °C. Walsh et al. showed that it can also be used to study BU in vivo [117]. In this study animals were intradermally infected with 3×10^3 to 3×10^8 *M. ulcerans* into the lower abdominal and medial thigh regions. Injection sites were observed every 2–4 weeks for up to six months or death of the animal. 37% of the armadillos developed progressive cutaneous lesions, 15.5% developed evanescent papule-nodular lesions, 15.5% died within one week after inoculation and 32% of the animals developed no clinical signs [117]. Progressive cutaneous lesions presented with clinical signs comparable to the human disease like undermined edges and extensive necrosis. All animals which developed ulcers died between four and six months after inoculation due to unidentified reasons. Histopathologically, lesions showed coagulating necrosis of the deep dermis, panniculitis, edema, sparse inflammation, fat cell ghosts and AFB, mainly present as extracellular clumps. Three months after inoculation, cultures of one animal on Loewenstein-Jensen media were negative, although AFB were still present [117].

3.6 Cynomologus Monkey (*Maca fascicularis*)

In a study with only one monkey it was shown that intradermal inoculation of *M. ulcerans* (2.2×10^8) induced the formation of a papule which ulcerated within 2–4 weeks after inoculation [118]. In biopsies, infiltration, edema, inflammatory infiltrates and extracellular bacteria were observed. Ulcers healed spontaneously between eight and 12 weeks post infection. At week six, cultures on Loewenstein-Jensen medium were negative, indicating that the cynomolgus monkey is only modestly susceptible to *M. ulcerans* [118].

3.7 African Rat (*Mastomys natalensis*)

The common Africa rat was used by Singh and colleagues because of its abundance in sub-Saharan Africa. Inoculation of different doses into the tail vein led to ulceration and eventually the loss of the tail [119]. Supposedly due to the naturally suppressed immune system, internal organs (lung, spleen, liver) as well as the footpads were involved in a small number of individuals [119].

3.8 Common Brushtail Possum (*Trichosurus vulpecula*)

Due to the lower body temperature and a high susceptibility of *Trichosurus vulpecula* to *M. tuberculosis* and *M. bovis,* researchers infected these animals in two experimental series with *M. ulcerans*. Inoculation with patient material into the hind leg of one possum led to the formation of a deep ulcer. Material from this lesion was used to inoculate four more animals, subcutaneously or intraperitoneally. Cutaneous ulcers and lesions at peripheral sites developed and AFB were recovered from them [120, 121]. In the second series it was shown that also non-inoculated *T. vulpecula* which were housed together with *M. ulcerans* inoculated animals developed ulcers from which AFB could be isolated [120]. Recent reports showed that wild possum species (*Pseudocheirus peregrinus, Trichosurus vulpecula, Trichosurus cunninghami)* in BU-endemic areas of Victoria, Australia are infected with *M. ulcerans*. Most clinically apparent cases were adults with ulcerative cutaneous lesions, generally confined to body areas without fur [4].

4 Conclusions

Overall, none of the currently used animal models fully reflects the spectrum of human BU disease. The mouse model is the most widely used animal model for BU in research because it is cheaper than the other models, bacteria are multiplying well and the onset of the disease is relatively fast. Especially for drug and vaccine studies, that both require large numbers of animals, this model is most advantageous. However, due to the lack of subcutaneous fat tissue and limited tissue depth, the clinical and histopathological appearance of the lesions in mouse foot pads and mouse tails are quite different from human BU lesions. In contrast, in the pig and guinea pig models both clinical and histopathological appearances of BU resemble the human situation, making both species a good model for the characterization of the early pathogenesis of BU. However both species are only modestly susceptible to *M. ulcerans* infection and do not develop chronic infections. All other animal models described in this chapter have not been studied well enough to conclude for which purposes they would be more suitable than the well-established mouse footpad model.

References

1. Mitchell PJ, Jerrett IV, Slee KJ (1984) Skin ulcers caused by *Mycobacterium ulcerans* in koalas near Bairnsdale, Australia. Pathology (Phila) 16:256–260
2. Mitchell PJ, McOrist S, Bilney R (1987) Epidemiology of *Mycobacterium ulcerans* infection in koalas (*Phascolarctos cinereus*) on Raymond Island, southeastern Australia. J Wildl Dis 23:386–390
3. Fyfe JAM, Lavender CJ, Handasyde KA, Legione AR, O'Brien CR, Stinear TP, Pidot SJ, Seemann T, Benbow ME, Wallace JR, McCowan C, Johnson PDR (2010) A major role for mammals in the ecology of *Mycobacterium ulcerans*. PLoS Negl Trop Dis 4:e791. https://doi.org/10.1371/journal.pntd.0000791
4. O'Brien CR, Handasyde KA, Hibble J, Lavender CJ, Legione AR, McCowan C, Globan M, Mitchell AT, McCracken HE, Johnson PDR, Fyfe JAM (2014) Clinical, microbiological and pathological findings of *Mycobacterium ulcerans* infection in three Australian possum species. PLoS Negl Trop Dis 8:e2666. https://doi.org/10.1371/journal.pntd.0002666
5. Carson C, Lavender CJ, Handasyde KA, O'Brien CR, Hewitt N, Johnson PDR, Fyfe JAM (2014) Potential wildlife sentinels for monitoring the endemic spread of human Buruli ulcer in South-East Australia. PLoS Negl Trop Dis 8:e2668. https://doi.org/10.1371/journal.pntd.0002668
6. O'Brien CR, McMillan E, Harris O, O'Brien DP, Lavender CJ, Globan M, Legione AR, Fyfe JA (2011) Localised *Mycobacterium ulcerans* infection in four dogs. Aust Vet J 89:506–510. https://doi.org/10.1111/j.1751-0813.2011.00850.x
7. Elsner L, Wayne J, O'Brien CR, McCowan C, Malik R, Hayman JA, Globan M, Lavender CJ, Fyfe JA (2008) Localised *Mycobacterium ulcerans* infection in a cat in Australia. J Feline Med Surg 10:407–412. https://doi.org/10.1016/j.jfms.2008.03.003
8. van Zyl A, Daniel J, Wayne J, McCowan C, Malik R, Jelfs P, Lavender CJ, Fyfe JA (2010) *Mycobacterium ulcerans* infections in two horses in South-Eastern Australia. Aust Vet J 88:101–106. https://doi.org/10.1111/j.1751-0813.2009.00544.x
9. O'Brien C, Kuseff G, McMillan E, McCowan C, Lavender C, Globan M, Jerrett I, Oppedisano F, Johnson P, Fyfe J (2013) *Mycobacterium ulcerans* infection in two alpacas. Aust Vet J 91:296–300. https://doi.org/10.1111/avj.12071

10. Durnez L, Suykerbuyk P, Nicolas V, Barrière P, Verheyen E, Johnson CR, Leirs H, Portaels F (2010) Terrestrial small mammals as reservoirs of *Mycobacterium ulcerans* in Benin. Appl Environ Microbiol 76:4574–4577. https://doi.org/10.1128/AEM.00199-10
11. Portaels F, Chemlal K, Elsen P, Johnson PD, Hayman JA, Hibble J, Kirkwood R, Meyers WM (2001) *Mycobacterium ulcerans* in wild animals. Rev Sci Tech Int Off Epizoot 20: 252–264
12. Vandelannoote K, Durnez L, Amissah D, Gryseels S, Dodoo A, Yeboah S, Addo P, Eddyani M, Leirs H, Ablordey A, Portaels F (2010) Application of real-time PCR in Ghana, a Buruli ulcer-endemic country, confirms the presence of *Mycobacterium ulcerans* in the environment. FEMS Microbiol Lett 304:191–194. https://doi.org/10.1111/j.1574-6968.2010.01902.x
13. MacCallum P, Tolhurst JC (1948) A new mycobacterial infection in man. J Pathol Bacteriol 60:93–122
14. Fenner F (1956) The pathogenic behavior of *Mycobacterium ulcerans* and *Mycobacterium balnei* in the mouse and the developing chick embryo. Am Rev Tuberc 73:650–673
15. Leach RH, Fenner F (1954) Studies on *Mycobacterium ulcerans* and *Mycobacterium balnei* III. Growth in the semi-synthetic culture media of Dubos and drug sensitivity in vitro and in vivo. Aust J Exp Biol Med Sci 32:835–852
16. Feldman WH, Karlson AG (1957) Mycobacterium ulcerans infections; response to chemotherapy in mice. Am Rev Tuberc 75:266–279
17. Fenner F (1957) Homologous and heterologous immunity in infections of mice with *Mycobacterium ulcerans* and *Mycobacterium balnei*. Am Rev Tuberc 76:76–89
18. Feldman WH, Karlson AG, Herrick JF (1957) Mycobacterium ulcerans; pathogenesis of infection in mice, including determinations of dermal temperatures. Am J Pathol 33:1163–1179
19. Shepard CC (1960) The experimental disease that follows the injection of human leprosy bacilli into foot-pads of mice. J Exp Med 112:445–454
20. Stanford JL, Phillips I (1972) Rifampicin in experimental *Mycobacterium ulcerans* infection. J Med Microbiol 5:39–45. https://doi.org/10.1099/00222615-5-1-39
21. Ullmann U, Schubert GE, Kieninger G (1975) Bacteriological investigations and animal experiments with Mycobacterium ulcerans (Tübingen 1971) (author's transl). Zentralblatt Bakteriol Parasitenkd Infekt Hyg Erste Abt Orig Reihe Med Mikrobiol Parasitol 232: 318–327
22. Coutanceau E, Marsollier L, Brosch R, Perret E, Goossens P, Tanguy M, Cole ST, Small PLC, Demangel C (2005) Modulation of the host immune response by a transient intracellular stage of *Mycobacterium ulcerans*: the contribution of endogenous mycolactone toxin. Cell Microbiol 7:1187–1196. https://doi.org/10.1111/j.1462-5822.2005.00546.x
23. Ruf M-T, Bolz M, Vogel M, Bayi PF, Bratschi MW, Sopho GE, Yeboah-Manu D, Um Boock A, Junghanss T, Pluschke G (2016) Spatial distribution of *Mycobacterium ulcerans* in Buruli ulcer lesions: implications for laboratory diagnosis. PLoS Negl Trop Dis 10:e0004767. https://doi.org/10.1371/journal.pntd.0004767
24. Coutanceau E, Legras P, Marsollier L, Reysset G, Cole ST, Demangel C (2006) Immunogenicity of *Mycobacterium ulcerans* Hsp65 and protective efficacy of a *Mycobacterium leprae* Hsp65-based DNA vaccine against Buruli ulcer. Microbes Infect 8:2075–2081. https://doi.org/10.1016/j.micinf.2006.03.009
25. Marion E, Jarry U, Cano C, Savary C, Beauvillain C, Robbe-Saule M, Preisser L, Altare F, Delneste Y, Jeannin P, Marsollier L (2016) FVB/N mice spontaneously heal ulcerative lesions induced by *Mycobacterium ulcerans* and switch *M. Ulcerans* into a low Mycolactone producer. J Immunol 196:2690–2698. https://doi.org/10.4049/jimmunol.1502194
26. Marsollier L, Deniaux E, Brodin P, Marot A, Wondje CM, Saint-André J-P, Chauty A, Johnson C, Tekaia F, Yeramian E, Legras P, Carbonnelle B, Reysset G, Eyangoh S, Milon G, Cole ST, Aubry J (2007) Protection against Mycobacterium ulcerans lesion development by exposure to aquatic insect saliva. PLoS Med 4:e64. https://doi.org/10.1371/journal.pmed.0040064
27. Converse PJ, Almeida DV, Nuermberger EL, Grosset JH (2011) BCG-mediated protection against *Mycobacterium ulcerans* infection in the mouse. PLoS Negl Trop Dis 5:e985. https://doi.org/10.1371/journal.pntd.0000985

28. Dega H, Robert J, Bonnafous P, Jarlier V, Grosset J (2000) Activities of several antimicrobials against *Mycobacterium ulcerans* infection in mice. Antimicrob Agents Chemother 44:2367–2372
29. Ji B, Lefrançois S, Robert J, Chauffour A, Truffot C, Jarlier V (2006) In vitro and in vivo activities of rifampin, streptomycin, amikacin, moxifloxacin, R207910, linezolid, and PA-824 against *Mycobacterium ulcerans*. Antimicrob Agents Chemother 50:1921–1926. https://doi.org/10.1128/AAC.00052-06
30. Oliveira MS, Fraga AG, Torrado E, Castro AG, Pereira JP, Filho AL, Milanezi F, Schmitt FC, Meyers WM, Portaels F, Silva MT, Pedrosa J (2005) Infection with *Mycobacterium ulcerans* induces persistent inflammatory responses in mice. Infect Immun 73:6299–6310. https://doi.org/10.1128/IAI.73.10.6299-6310.2005
31. Ruf M-T, Schütte D, Chauffour A, Jarlier V, Ji B, Pluschke G (2012) Chemotherapy-associated changes of histopathological features of *Mycobacterium ulcerans* lesions in a Buruli ulcer mouse model. Antimicrob Agents Chemother 56:687–696. https://doi.org/10.1128/AAC.05543-11
32. Tanghe A, Content J, Van Vooren JP, Portaels F, Huygen K (2001) Protective efficacy of a DNA vaccine encoding antigen 85A from *Mycobacterium bovis* BCG against Buruli ulcer. Infect Immun 69:5403–5411
33. Zhang T, Li S-Y, Converse PJ, Almeida DV, Grosset JH, Nuermberger EL (2011) Using bioluminescence to monitor treatment response in real time in mice with *Mycobacterium ulcerans* infection. Antimicrob Agents Chemother 55:56–61. https://doi.org/10.1128/AAC.01260-10
34. Bénard A, Sala C, Pluschke G (2016) Mycobacterium ulcerans mouse model refinement for pre-clinical profiling of vaccine candidates. PLoS One 11:e0167059. https://doi.org/10.1371/journal.pone.0167059
35. Bentoucha A, Robert J, Dega H, Lounis N, Jarlier V, Grosset J (2001) Activities of new macrolides and fluoroquinolones against *Mycobacterium ulcerans* infection in mice. Antimicrob Agents Chemother 45:3109–3112. https://doi.org/10.1128/AAC.45.11.3109-3112.2001
36. Dhople AM (2001) In vivo susceptibility of *Mycobacterium ulcerans* to KRM-1648, a new benzoxazinorifamycin, in comparison with rifampicin. Anti-mycobacterial activity of KRM-1648. Arzneimittelforschung 51:501–505. https://doi.org/10.1055/s-0031-1300070
37. Marsollier L, Honoré N, Legras P, Manceau AL, Kouakou H, Carbonnelle B, Cole ST (2003a) Isolation of three *Mycobacterium ulcerans* strains resistant to rifampin after experimental chemotherapy of mice. Antimicrob Agents Chemother 47:1228–1232
38. Houngbédji MG, Boissinot M, Bergeron GM, Frenette J (2008) Subcutaneous injection of *Mycobacterium ulcerans* causes necrosis, chronic inflammatory response and fibrosis in skeletal muscle. Microbes Infect 10:1236–1243. https://doi.org/10.1016/j.micinf.2008.07.041
39. Torrado E, Adusumilli S, Fraga AG, Small PLC, Castro AG, Pedrosa J (2007) Mycolactone-mediated inhibition of tumor necrosis factor production by macrophages infected with *Mycobacterium ulcerans* has implications for the control of infection. Infect Immun 75:3979–3988. https://doi.org/10.1128/IAI.00290-07
40. Bieri R, Bolz M, Ruf M-T, Pluschke G (2016) Interferon-γ is a crucial activator of early host immune defense against *Mycobacterium ulcerans* infection in mice. PLoS Negl Trop Dis 10:e0004450. https://doi.org/10.1371/journal.pntd.0004450
41. Torrado E, Fraga AG, Logarinho E, Martins TG, Carmona JA, Gama JB, Carvalho MA, Proença F, Castro AG, Pedrosa J (2010) IFN-gamma-dependent activation of macrophages during experimental infections by *Mycobacterium ulcerans* is impaired by the toxin mycolactone. J Immunol 184:947–955. https://doi.org/10.4049/jimmunol.0902717
42. Fraga AG, Cruz A, Martins TG, Torrado E, Saraiva M, Pereira DR, Meyers WM, Portaels F, Silva MT, Castro AG, Pedrosa J (2011) *Mycobacterium ulcerans* triggers T-cell immunity followed by local and regional but not systemic immunosuppression. Infect Immun 79:421–430. https://doi.org/10.1128/IAI.00820-10
43. Hein L, Barsh GS, Pratt RE, Dzau VJ, Kobilka BK (1995) Behavioural and cardiovascular effects of disrupting the angiotensin II type-2 receptor in mice. Nature 377:744–747. https://doi.org/10.1038/377744a0

44. Marion E, Song O-R, Christophe T, Babonneau J, Fenistein D, Eyer J, Letournel F, Henrion D, Clere N, Paille V, Guérineau NC, Saint André J-P, Gersbach P, Altmann K-H, Stinear TP, Comoglio Y, Sandoz G, Preisser L, Delneste Y, Yeramian E, Marsollier L, Brodin P (2014) Mycobacterial toxin induces analgesia in Buruli ulcer by targeting the angiotensin pathways. Cell 157:1565–1576. https://doi.org/10.1016/j.cell.2014.04.040
45. Bieri R, Scherr N, Ruf M-T, Dangy J-P, Gersbach P, Gehringer M, Altmann K-H, Pluschke G (2017) The macrolide toxin mycolactone promotes Bim-dependent apoptosis in Buruli ulcer through inhibition of mTOR. ACS Chem Biol 12:1297–1307. https://doi.org/10.1021/acschembio.7b00053
46. Azumah BK, Addo PG, Dodoo A, Awandare G, Mosi L, Boakye DA, Wilson MD (2017) Experimental demonstration of the possible role of *Acanthamoeba polyphaga* in the infection and disease progression in Buruli ulcer (BU) using ICR mice. PLoS One 12:e0172843. https://doi.org/10.1371/journal.pone.0172843
47. Dufresne SS, Frenette J (2013) Investigation of wild-type and mycolactone-negative mutant *Mycobacterium ulcerans* on skeletal muscle: IGF-1 protects against mycolactone-induced muscle catabolism. Am J Physiol Regul Integr Comp Physiol 304:R753–R762. https://doi.org/10.1152/ajpregu.00587.2012
48. Fraga AG, Martins TG, Torrado E, Huygen K, Portaels F, Silva MT, Castro AG, Pedrosa J (2012) Cellular immunity confers transient protection in experimental Buruli ulcer following BCG or mycolactone-negative *Mycobacterium ulcerans* vaccination. PLoS One 7:e33406. https://doi.org/10.1371/journal.pone.0033406
49. Chauffour A, Robert J, Veziris N, Aubry A, Jarlier V (2016) Sterilizing activity of fully oral intermittent regimens against *Mycobacterium ulcerans* infection in mice. PLoS Negl Trop Dis 10:e0005066. https://doi.org/10.1371/journal.pntd.0005066
50. Converse PJ, Tyagi S, Xing Y, Li S-Y, Kishi Y, Adamson J, Nuermberger EL, Grosset JH (2015) Efficacy of rifampin plus clofazimine in a murine model of *Mycobacterium ulcerans* disease. PLoS Negl Trop Dis 9:e0003823. https://doi.org/10.1371/journal.pntd.0003823
51. Tanghe A, Dangy J-P, Pluschke G, Huygen K (2008) Improved protective efficacy of a species-specific DNA vaccine encoding mycolyl-transferase Ag85A from *Mycobacterium ulcerans* by homologous protein boosting. PLoS Negl Trop Dis 2:e199. https://doi.org/10.1371/journal.pntd.0000199
52. Bolz M, Kerber S, Zimmer G, Pluschke G (2015) Use of recombinant virus replicon particles for vaccination against Mycobacterium ulcerans disease. PLoS Negl Trop Dis 9:e0004011. https://doi.org/10.1371/journal.pntd.0004011
53. Ortiz RH, Leon DA, Estevez HO, Martin A, Herrera JL, Romo LF, Portaels F, Pando RH (2009) Differences in virulence and immune response induced in a murine model by isolates of *Mycobacterium ulcerans* from different geographic areas. Clin Exp Immunol 157:271–281. https://doi.org/10.1111/j.1365-2249.2009.03941.x
54. Trigo G, Martins TG, Fraga AG, Longatto-Filho A, Castro AG, Azeredo J, Pedrosa J (2013) Phage therapy is effective against infection by *Mycobacterium ulcerans* in a murine footpad model. PLoS Negl Trop Dis 7:e2183. https://doi.org/10.1371/journal.pntd.0002183
55. Mve-Obiang A, Lee RE, Umstot ES, Trott KA, Grammer TC, Parker JM, Ranger BS, Grainger R, Mahrous EA, Small PLC (2005) A newly discovered mycobacterial pathogen isolated from laboratory colonies of Xenopus species with lethal infections produces a novel form of mycolactone, the *Mycobacterium ulcerans* macrolide toxin. Infect Immun 73:3307–3312. https://doi.org/10.1128/IAI.73.6.3307-3312.2005
56. Pettit JH, Marchette NJ, Rees RJ (1966) *Mycobacterium ulcerans* infection. Clinical and bacteriological study of the first cases recognized in South East Asia. Br J Dermatol 78:187–197
57. Bratschi MW, Bolz M, Minyem JC, Grize L, Wantong FG, Kerber S, Njih Tabah E, Ruf M-T, Mou F, Noumen D, Um Boock A, Pluschke G (2013) Geographic distribution, age pattern and sites of lesions in a cohort of Buruli ulcer patients from the Mapé Basin of Cameroon. PLoS Negl Trop Dis 7:e2252. https://doi.org/10.1371/journal.pntd.0002252
58. Bolz M, Bénard A, Dreyer AM, Kerber S, Vettiger A, Oehlmann W, Singh M, Duthie MS, Pluschke G (2016a) Vaccination with the surface proteins MUL_2232 and MUL_3720 of

Mycobacterium ulcerans induces antibodies but fails to provide protection against Buruli ulcer. PLoS Negl Trop Dis 10:e0004431. https://doi.org/10.1371/journal.pntd.0004431
59. Bolz M, Ruggli N, Borel N, Pluschke G, Ruf M-T (2016b) Local cellular immune responses and pathogenesis of Buruli ulcer lesions in the experimental *Mycobacterium ulcerans* pig infection model. PLoS Negl Trop Dis 10:e0004678. https://doi.org/10.1371/journal.pntd.0004678
60. Bolz M, Ruggli N, Ruf M-T, Ricklin ME, Zimmer G, Pluschke G (2014) Experimental infection of the pig with *Mycobacterium ulcerans*: a novel model for studying the pathogenesis of Buruli ulcer disease. PLoS Negl Trop Dis 8:e2968. https://doi.org/10.1371/journal.pntd.0002968
61. Shepard CC, McRae DH (1968) A method for counting acid-fast bacteria. Int J Lepr Mycobact Dis Off Organ Int Lepr Assoc 36:78–82
62. Addo P, Adu-Addai B, Quartey M, Abbas M, Okang I, Owusu E, Ofori-Adjei D, Awumbila B (2006) Clinical and histopathological presentation of Buruli ulcer in experimentally infected Grasscutters (*Thryonomys swinderianus*). Internet J Trop Med 3:e2
63. Hart BE, Lee S (2016) Overexpression of a *Mycobacterium ulcerans* Ag85B-EsxH fusion protein in recombinant BCG improves experimental Buruli ulcer vaccine efficacy. PLoS Negl Trop Dis 10:e0005229. https://doi.org/10.1371/journal.pntd.0005229
64. Watanabe M, Nakamura H, Nabekura R, Shinoda N, Suzuki E, Saito H (2015) Protective effect of a dewaxed whole-cell vaccine against *Mycobacterium ulcerans* infection in mice. Vaccine 33:2232–2239. https://doi.org/10.1016/j.vaccine.2015.03.046
65. Bolz M (2014) Prospects for the development of a subunit vaccine against *Mycobacterium ulcerans* disease (Buruli ulcer). PhD, University of Basel
66. Robbe-Saule M, Babonneau J, Sismeiro O, Marsollier L, Marion E (2017) An optimized method for extracting bacterial RNA from mouse skin tissue colonized by *Mycobacterium ulcerans*. Front Microbiol 8:512. https://doi.org/10.3389/fmicb.2017.00512
67. Sarpong-Duah M, Frimpong M, Beissner M, Saar M, Laing K, Sarpong F, Loglo AD, Abass KM, Frempong M, Sarfo FS, Bretzel G, Wansbrough-Jones M, Phillips RO (2017) Clearance of viable *Mycobacterium ulcerans* from Buruli ulcer lesions during antibiotic treatment as determined by combined 16S rRNA reverse transcriptase/IS 2404 qPCR assay. PLoS Negl Trop Dis 11:e0005695. https://doi.org/10.1371/journal.pntd.0005695
68. Zhang T, Bishai WR, Grosset JH, Nuermberger EL (2010) Rapid assessment of antibacterial activity against *Mycobacterium ulcerans* by using recombinant luminescent strains. Antimicrob Agents Chemother 54:2806–2813. https://doi.org/10.1128/AAC.00400-10
69. Zhang T, Li S-Y, Converse PJ, Grosset JH, Nuermberger EL (2013) Rapid, serial, non-invasive assessment of drug efficacy in mice with autoluminescent *Mycobacterium ulcerans* infection. PLoS Negl Trop Dis 7:e2598. https://doi.org/10.1371/journal.pntd.0002598
70. Converse PJ, Xing Y, Kim KH, Tyagi S, Li S-Y, Almeida DV, Nuermberger EL, Grosset JH, Kishi Y (2014) Accelerated detection of mycolactone production and response to antibiotic treatment in a mouse model of *Mycobacterium ulcerans* disease. PLoS Negl Trop Dis 8:e2618. https://doi.org/10.1371/journal.pntd.0002618
71. Lunn HF, Rees RJW (1964) Treatment of mycobacterial skin ulcers in Uganda with a riminophenazine derivative (B.663). Lancet 283:247–249. https://doi.org/10.1016/S0140-6736(64)92351-7
72. Pattyn SR, Royackers J (1965) Treatment of experimental infection by *Mycobacterium ulcerans* and *Mycobacterium balnei* in mice. Ann Soc Belg Med Trop Parasitol Mycol 45: 31–38
73. Etuaful S, Carbonnelle B, Grosset J, Lucas S, Horsfield C, Phillips R, Evans M, Ofori-Adjei D, Klustse E, Owusu-Boateng J, Amedofu GK, Awuah P, Ampadu E, Amofah G, Asiedu K, Wansbrough-Jones M (2005) Efficacy of the combination rifampin-streptomycin in preventing growth of *Mycobacterium ulcerans* in early lesions of Buruli ulcer in humans. Antimicrob Agents Chemother 49:3182–3186. https://doi.org/10.1128/AAC.49.8.3182-3186.2005
74. Marsollier L, Prévot G, Honoré N, Legras P, Manceau AL, Payan C, Kouakou H, Carbonnelle B (2003b) Susceptibility of *Mycobacterium ulcerans* to a combination of amikacin/rifampicin. Int J Antimicrob Agents 22:562–566

75. Dega H, Bentoucha A, Robert J, Jarlier V, Grosset J (2002) Bactericidal activity of rifampin-amikacin against *Mycobacterium ulcerans* in mice. Antimicrob Agents Chemother 46:3193–3196
76. Dhople AM, Namba K (2003) Activities of sitafloxacin (DU-6859a), either singly or in combination with rifampin, against *Mycobacterium ulcerans* infection in mice. J Chemother 15:47–52. https://doi.org/10.1179/joc.2003.15.1.47
77. Nakanaga K, Saito H, Ishii N, Goto M (2004) Comparison of inhibitory effect of rifalazil and rifampicin against *Mycobacterium ulcerans* infection induced in mice. Kekkaku 79:333–339
78. Lefrançois S, Robert J, Chauffour A, Ji B, Jarlier V (2007) Curing *Mycobacterium ulcerans* infection in mice with a combination of rifampin-streptomycin or rifampin-amikacin. Antimicrob Agents Chemother 51:645–650. https://doi.org/10.1128/AAC.00821-06
79. WHO and GBU Initiative (2004) Provisional guidance on the role of specific antibiotics in the management of *Mycobacterium ulcerans* disease (Buruli ulcer). WHO and GBU Initiative, Geneva, Switzerland
80. Almeida D, Converse PJ, Ahmad Z, Dooley KE, Nuermberger EL, Grosset JH (2011) Activities of rifampin, rifapentine and clarithromycin alone and in combination against mycobacterium ulcerans disease in mice. PLoS Negl Trop Dis 5:e933. https://doi.org/10.1371/journal.pntd.0000933
81. Ji B, Chauffour A, Robert J, Jarlier V (2008) Bactericidal and sterilizing activities of several orally administered combined regimens against *Mycobacterium ulcerans* in mice. Antimicrob Agents Chemother 52:1912–1916. https://doi.org/10.1128/AAC.00193-08
82. Ji B, Chauffour A, Robert J, Lefrançois S, Jarlier V (2007) Orally administered combined regimens for treatment of *Mycobacterium ulcerans* infection in mice. Antimicrob Agents Chemother 51:3737–3739. https://doi.org/10.1128/AAC.00730-07
83. Martins TG, Trigo G, Fraga AG, Gama JB, Longatto-Filho A, Saraiva M, Silva MT, Castro AG, Pedrosa J (2012b) Corticosteroid-induced immunosuppression ultimately does not compromise the efficacy of antibiotherapy in murine *Mycobacterium ulcerans* infection. PLoS Negl Trop Dis 6:e1925. https://doi.org/10.1371/journal.pntd.0001925
84. Adusumilli S, Haydel SE (2016) In vitro antibacterial activity and in vivo efficacy of hydrated clays on *Mycobacterium ulcerans* growth. BMC Complement Altern Med 16:40. https://doi.org/10.1186/s12906-016-1020-5
85. Tanghe A, Adnet P-Y, Gartner T, Huygen K (2007) A booster vaccination with *Mycobacterium bovis* BCG does not increase the protective effect of the vaccine against experimental *Mycobacterium ulcerans* infection in mice. Infect Immun 75:2642–2644. https://doi.org/10.1128/IAI.01622-06
86. Singh NB, Srivastava A (1985) *Mycobacterium fortuitum* (TMC 1529): potentially immunogenic strain against experimental infections of *Mycobacterium ulcerans* in rats and mice. Indian J Exp Biol 23:408–409
87. Einarsdottir T, Huygen K (2011) Buruli ulcer. Hum Vaccin 7:1198–1203. https://doi.org/10.4161/hv.7.11.17751
88. Huygen K, Adjei O, Affolabi D, Bretzel G, Demangel C, Fleischer B, Johnson RC, Pedrosa J, Phanzu DM, Phillips RO, Pluschke G, Siegmund V, Singh M, van der Werf TS, Wansbrough-Jones M, Portaels F (2009) Buruli ulcer disease: prospects for a vaccine. Med Microbiol Immunol 198:69–77. https://doi.org/10.1007/s00430-009-0109-6
89. Hart BE, Hale LP, Lee S (2015) Recombinant BCG expressing *Mycobacterium ulcerans* Ag85A imparts enhanced protection against experimental Buruli ulcer. PLoS Negl Trop Dis 9:e0004046. https://doi.org/10.1371/journal.pntd.0004046
90. Roupie V, Pidot SJ, Einarsdottir T, Van Den Poel C, Jurion F, Stinear TP, Huygen K (2014) Analysis of the vaccine potential of plasmid DNA encoding nine mycolactone polyketide synthase domains in *Mycobacterium ulcerans* infected mice. PLoS Negl Trop Dis 8:e2604. https://doi.org/10.1371/journal.pntd.0002604
91. Guarner J, Bartlett J, Whitney EAS, Raghunathan PL, Stienstra Y, Asamoa K, Etuaful S, Klutse E, Quarshie E, van der Werf TS, van der Graaf WTA, King CH, Ashford DA (2003) Histopathologic features of *Mycobacterium ulcerans* infection. Emerg Infect Dis 9:651–656

92. Hayman J (1993) Out of Africa: observations on the histopathology of *Mycobacterium ulcerans* infection. J Clin Pathol 46:5–9
93. Ruf M-T, Steffen C, Bolz M, Schmid P, Pluschke G (2017) Infiltrating leukocytes surround early Buruli ulcer lesions, but are unable to reach the mycolactone producing mycobacteria. Virulence 8:1918–1926. https://doi.org/10.1080/21505594.2017.1370530
94. Martins TG, Gama JB, Fraga AG, Saraiva M, Silva MT, Castro AG, Pedrosa J (2012a) Local and regional re-establishment of cellular immunity during curative antibiotherapy of murine *Mycobacterium ulcerans* infection. PLoS One 7:e32740. https://doi.org/10.1371/journal.pone.0032740
95. Sarfo FS, Converse PJ, Almeida DV, Zhang J, Robinson C, Wansbrough-Jones M, Grosset JH (2013) Microbiological, histological, immunological, and toxin response to antibiotic treatment in the mouse model of *Mycobacterium ulcerans* disease. PLoS Negl Trop Dis 7:e2101. https://doi.org/10.1371/journal.pntd.0002101
96. Goto M, Nakanaga K, Aung T, Hamada T, Yamada N, Nomoto M, Kitajima S, Ishii N, Yonezawa S, Saito H (2006) Nerve damage in *Mycobacterium ulcerans*-infected mice: probable cause of painlessness in buruli ulcer. Am J Pathol 168:805–811. https://doi.org/10.2353/ajpath.2006.050375
97. En J, Goto M, Nakanaga K, Higashi M, Ishii N, Saito H, Yonezawa S, Hamada H, Small PLC (2008) Mycolactone is responsible for the painlessness of *Mycobacterium ulcerans* infection (Buruli ulcer) in a murine study. Infect Immun 76:2002–2007. https://doi.org/10.1128/IAI.01588-07
98. Krieg RE, Hockmeyer WT, Connor DH (1974) Toxin of *Mycobacterium ulcerans*. Production and effects in Guinea pig skin. Arch Dermatol 110:783–788
99. Read JK, Heggie CM, Meyers WM, Connor DH (1974) Cytotoxic activity of *Mycobacterium ulcerans*. Infect Immun 9:1114–1122
100. Adusumilli S, Mve-Obiang A, Sparer T, Meyers W, Hayman J, Small PLC (2005) Mycobacterium ulcerans toxic macrolide, mycolactone modulates the host immune response and cellular location of *M. ulcerans* in vitro and in vivo. Cell Microbiol 7:1295–1304. https://doi.org/10.1111/j.1462-5822.2005.00557.x
101. George KM, Chatterjee D, Gunawardana G, Welty D, Hayman J, Lee R, Small PL (1999) Mycolactone: a polyketide toxin from *Mycobacterium ulcerans* required for virulence. Science 283:854–857
102. George KM, Pascopella L, Welty DM, Small PL (2000) A *Mycobacterium ulcerans* toxin, mycolactone, causes apoptosis in Guinea pig ulcers and tissue culture cells. Infect Immun 68:877–883
103. Silva-Gomes R, Marcq E, Trigo G, Gonçalves CM, Longatto-Filho A, Castro AG, Pedrosa J, Fraga AG (2015) Spontaneous healing of *Mycobacterium ulcerans* lesions in the Guinea pig model. PLoS Negl Trop Dis 9:e0004265. https://doi.org/10.1371/journal.pntd.0004265
104. Williamson HR, Mosi L, Donnell R, Aqqad M, Merritt RW, Small PLC (2014) *Mycobacterium ulcerans* fails to infect through skin abrasions in a Guinea pig infection model: implications for transmission. PLoS Negl Trop Dis 8:e2770. https://doi.org/10.1371/journal.pntd.0002770
105. Hammond SA, Tsonis C, Sellins K, Rushlow K, Scharton-Kersten T, Colditz I, Glenn GM (2000) Transcutaneous immunization of domestic animals: opportunities and challenges. Adv Drug Deliv Rev 43:45–55
106. Liu Y, Chen J, Shang H, Liu C, Wang Y, Niu R, Wu J, Wei H (2010) Light microscopic, electron microscopic, and immunohistochemical comparison of Bama minipig (*Sus scrofa* domestica) and human skin. Comp Med 60:142–148
107. Mahl JA, Vogel BE, Court M, Kolopp M, Roman D, Nogués V (2006) The minipig in dermatotoxicology: methods and challenges. Exp Toxicol Pathol Off J Ges Toxikol Pathol 57:341–345. https://doi.org/10.1016/j.etp.2006.03.004
108. Meyer W, Schwarz R, Neurand K (1978) The skin of domestic mammals as a model for the human skin, with special reference to the domestic pig. Curr Probl Dermatol 7:39–52
109. Montagna W, Yun JS (1964) The skin of the domestic pig. J Invest Dermatol 42:11–21

110. Summerfield A, Meurens F, Ricklin ME (2015) The immunology of the porcine skin and its value as a model for human skin. Mol Immunol 66:14–21. https://doi.org/10.1016/j.molimm.2014.10.023
111. Jung Y, Son D, Kwon S, Kim J, Han K (2013) Experimental pig model of clinically relevant wound healing delay by intrinsic factors. Int Wound J 10:295–305. https://doi.org/10.1111/j.1742-481X.2012.00976.x
112. Sheu S-Y, Wang W-L, Fu Y-T, Lin S-C, Lei Y-C, Liao J-H, Tang N-Y, Kuo T-F, Yao C-H (2014) The pig as an experimental model for mid-dermal burns research. Burns J Int Soc Burn Inj 40:1679–1688. https://doi.org/10.1016/j.burns.2014.04.023
113. Sullivan TP, Eaglstein WH, Davis SC, Mertz P (2001) The pig as a model for human wound healing. Wound Repair Regen Off Publ Wound Heal Soc Eur Tissue Repair Soc 9:66–76
114. Marcus LC, Stottmeier KD, Morrow RH (1975) Experimental infection of anole lizards (*Anolis carolinensis*) with Mycobacterium ulcerans by the subcutaneous route. Am J Trop Med Hyg 24:649–655
115. Marcus LC, Stottmeier KD, Morrow RH (1976) Experimental alimentary infection of anole lizards (*Anolis carolinensis*) with *Mycobacterium ulcerans*. Am J Trop Med Hyg 25:630–632
116. Walsh GP, Meyers WM, Binford CH (1986) Naturally acquired leprosy in the nine-banded armadillo: a decade of experience 1975-1985. J Leukoc Biol 40:645–656
117. Walsh DS, Meyers WM, Krieg RE, Walsh GP (1999) Transmission of *Mycobacterium ulcerans* to the nine-banded armadillo. Am J Trop Med Hyg 61:694–697
118. Walsh DS, Dela Cruz EC, Abalos RM, Tan EV, Walsh GP, Portaels F, Meyers WM (2007) Clinical and histologic features of skin lesions in a cynomolgus monkey experimentally infected with *Mycobacterium ulcerans* (Buruli ulcer) by intradermal inoculation. Am J Trop Med Hyg 76:132–134
119. Singh NB, Srivastava A, Verma VK, Kumar A, Gupta SK (1984) *Mastomys natalensis*: a new animal model for *Mycobacterium ulcerans* research. Indian J Exp Biol 7:393–394
120. Bolliger A, Forbes BRV, Kirkland WB (1950) Transmission of a recently isolated mycobacterium to phalangers (*Tichosurus vulpecula*). Aust J Sci 12:146–147
121. Forbes BR, Wannan JS, Kirkland WB (1954) Indolent cutaneous ulceration due to infection with *Mycobacterium ulcerans*. Med J Aust 41:475–479

Laboratory Diagnosis of Buruli Ulcer: Challenges and Future Perspectives

Katharina Röltgen, Israel Cruz, Joseph M. Ndung'u, and Gerd Pluschke

1 Introduction

The clinical presentation of Buruli ulcer (BU) is manifold and includes relatively unspecific, non-ulcerative manifestations such as nodules, papules, plaques, and edema, which may eventually progress to necrotic ulcers [1]. Early case detection and adequate treatment are essential to prevent the formation of large cutaneous lesions that are often associated with serious morbidity and permanent disability. While surgical resection of BU lesions has long been the only treatment option, demonstration of the efficacy of rifampicin against *M. ulcerans* in a mouse footpad model [2, 3] and of a combined regimen of rifampicin and streptomycin in a clinical trial [4], have shifted first-line treatment recommendations to antibiotic therapy. Routine implementation of the drug regimen consisting of rifampicin and streptomycin administered daily for 8 weeks, has greatly improved specific therapy and reduced the frequency of relapses [5–7]. However, considering the history of antibiotic resistance in other bacterial pathogens such as *M. tuberculosis* [8], concerns have arisen that inappropriate use of these antibiotics may lead to similar patterns of resistance in *M. ulcerans*. These concerns appear justified, as rifampicin, currently the only highly effective drug available for the treatment of BU, is also a major component in the treatment of other often co-prevalent mycobacterial infections such as tuberculosis and leprosy. Furthermore, streptomycin should be prescribed with caution, as long-term streptomycin toxicity in the form of a high incidence of persistent hearing loss has been documented in a follow up study of former BU patients who had received the combination therapy [9]. A new combination

K. Röltgen · G. Pluschke (✉)
Swiss Tropical and Public Health Institute, Basel, Switzerland

University of Basel, Basel, Switzerland
e-mail: gerd.pluschke@swisstph.ch

I. Cruz · J. M. Ndung'u
Foundation for Innovative New Diagnostics (FIND), Geneva, Switzerland

G. Pluschke, K. Röltgen (eds.), *Buruli Ulcer*,
https://doi.org/10.1007/978-3-030-11114-4_10

therapy with rifampicin and clarithromycin is currently being evaluated as first-line treatment of BU. Pre-treatment laboratory confirmation of clinically suspected BU cases has thus gained in importance; particularly because the differential diagnosis of skin conditions with similar manifestations is broad ([1, 10], http://www.who.int/neglected_diseases/resources/9789241513531/en/) (Fig. 1) and misclassification of clinically suspected cases seems to be more common than previously assumed [11].

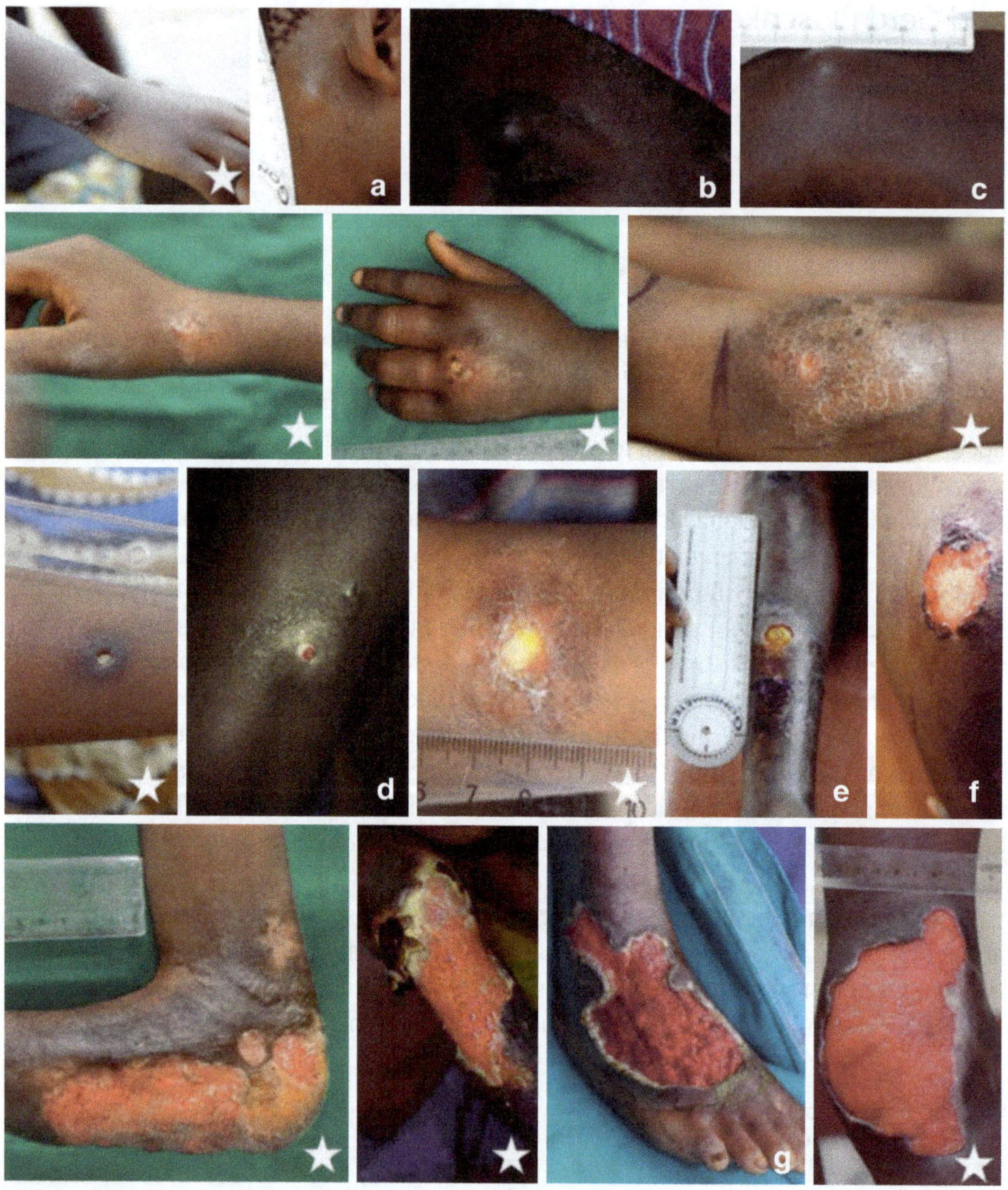

Fig. 1 Differential diagnosis of BU. Infection with *M. ulcerans* can cause a variety of clinical manifestations, including both non-ulcerative and ulcerative forms. The differential diagnosis of BU thus comprises a wide spectrum of other skin conditions with a similar appearance. Shown are BU lesions (marked by a star) and non-BU lesions ((**a**) swollen lymph node; (**b**) lipoma; (**c**) ganglion; (**d**) cutaneous tuberculosis; (**e**) sickle cell necrosis; (**f**) burn; (**g**) snake bite). Pictures provided by Markus Schindler, Thomas Junghanss and Moritz Vogel are included

For a detailed description of the currently available laboratory techniques and procedures for the detection of *M. ulcerans* the reader is referred to the WHO manual *"Laboratory Diagnosis of Buruli ulcer"* edited by Françoise Portaels [12]. Laboratory tests routinely used for the detection of *M. ulcerans* infections include microscopic detection of acid fast bacilli (AFBs) in stained smears from clinical specimens and DNA detection by PCR targeting the *M. ulcerans*-specific insertion sequence (IS) element IS*2404*. Histopathological analysis of sections from the affected tissue and primary cultivation of the mycobacteria require sophisticated infrastructure and can only be performed by specifically trained personnel [12]. Furthermore, cultivation of the extremely slow growing mycobacteria takes several weeks to months and is thus impractical to aid pre-treatment diagnosis.

Of the established tests for the detection of *M. ulcerans*, IS*2404* PCR has proven to be the most sensitive and specific, if performed according to demanding quality assurance schemes [13, 14]. Therefore, it is currently considered the diagnostic gold standard. However, data on the sensitivity and specificity of IS*2404* PCR as well as of the other reconfirmatory tests are difficult to interpret from available comparative studies. Indeed, many studies reporting on the test performance of *M. ulcerans* detection evaluated sensitivity and specificity by comparing the respective tests to clinical diagnosis, which has limited accuracy, even when performed by experienced health staff. The sensitivity of microscopy—the only test that can be performed at district hospital level—has been reported to be relatively low. In BU endemic countries with resource-rich healthcare systems and good laboratory infrastructure, such as Australia and Japan, PCR-based diagnosis is routinely done, whereas in resource-poor settings, logistical challenges and high costs often impede rapid PCR-based laboratory diagnosis at centralized reference laboratories. As a consequence, the diagnosis of BU at remote health facilities is often based on clinical judgment only. A simple and rapid point-of-care diagnostic test for BU is therefore of urgent need. Following recommendations of WHO, a diagnostic test suitable for application in developing countries should meet the so called ASSURED [15] criteria of being Affordable, Sensitive, Specific, User-friendly, Robust and rapid, Equipment free, and Deliverable to the end user. This chapter highlights current approaches as well as future prospects for the diagnosis of BU at district hospitals and the primary healthcare level, where the majority of BU patients are diagnosed to date.

2 Currently Available Laboratory Diagnostic Tests

2.1 Specimen Collection and Reference Standards

Before the introduction of antibiotic therapy in 2004 [16], the only treatment option for BU was surgical excision of the lesions, with or without subsequent skin grafting. Laboratory reconfirmation was often done retrospectively by analyzing specimens excised during surgery [17–19]. After 2004, when the importance of pre-treatment reconfirmatory laboratory diagnosis became more broadly recognized, alternative sources for diagnostic specimens included punch biopsies obtained from non-ulcerative lesions and swab samples taken from the undermined edges of ulcerative lesions, where the bacterial load is typically higher than in the core of a lesion. However, due to the invasiveness of punch biopsies, a consensus has been reached that in the interest of the patients, the method should be limited to special circumstances such as the establishment of differential diagnosis, investigations on paradoxical reactions or the reconfirmation of suspected treatment failure [12, 20]. Since 2007, a less invasive technique referred to as fine-needle aspiration (FNA), which can be performed at all healthcare levels and on both non-ulcerative lesions and ulcers with scarred edges that hinder the collection of swabs, has gradually replaced the use of punch biopsies for routine laboratory confirmation. The current recommendation for the collection of diagnostic specimens is thus to take swab and FNA samples from ulcers and non-ulcerative lesions, respectively. Recommendations for storage and transport conditions (e.g. media and optimal temperatures) for different types of samples are detailed in the aforementioned WHO manual "*Laboratory diagnosis of Buruli ulcer*" [12].

IS*2404* PCR-based tests performed at reference centres have become a gold standard for the diagnosis of BU. While the accuracy of new diagnostic tests for *M. ulcerans* infection should therefore be assessed by a comparison to results obtained by PCR, the true accuracy of PCR assays is in turn difficult to evaluate, as it can only be compared to imperfect reference standards that have both limited specificity such as clinical diagnosis [10], or limited sensitivity, such as microscopy [21] and culture [22].

2.2 IS*2404* PCR: The Current Gold Standard

In many mycobacterial species so called insertion sequence (IS) elements have been identified, representing suitable targets for PCR-based detection assays [23–26]. IS sequences are typically characterized by species-specificity and the presence of multiple copies within one genome, facilitating a highly specific and sensitive detection of the respective pathogens. In 1997, Ross et al. provided a milestone for the PCR-based diagnosis of BU by identifying an *M. ulcerans*-specific repetitive DNA fragment [27], which was later characterized in detail and designated IS*2404* [28]. The high copy number of IS*2404* (between 150 and more than 200 copies per genome in *M. ulcerans* isolates from Australia and Africa) and of another IS, referred to as IS*2606* (63 to 98 copies per genome in *M. ulcerans* isolates from Australia and Africa) [29], predestine their application as targets for sensitive PCR amplification tests for *M. ulcerans*. High specificity of developed PCR assays targeting IS*2404* was indicated by the lack of IS*2404* PCR positivity among a wide range of other mycobacterial species [28, 30, 31]. The design of real-time quantitative PCR (qPCR) assays targeting IS*2404* [32, 33] has been another milestone in BU diagnostics development and has replaced conventional gel-based PCR as the routine method for laboratory confirmation of BU in many diagnostic and research laboratories [14]. Besides an increased sensitivity of *M. ulcerans* detection, major advantages of the qPCR assay are the reduced risk of contamination, and an improved turnaround time.

Nevertheless, like any molecular test, the highly sensitive IS*2404* PCR—whether performed as gel-based test or as a qPCR assay—is prone to contamination leading to false positive test results. Cross-contamination of samples with *M. ulcerans* genomic DNA may occur during sample collection or processing. However the most common problem is carryover of PCR product from previous amplification reactions. For instance, in a study describing histopathological features of BU, specimens obtained from cases with filarial nodules and a keratin cyst without any histopathological indication for BU, tested positive in a nested IS*2404* PCR [34]. To ensure accuracy of PCR assay results, it is thus essential to strictly adhere to the three-room principle necessitating one room for preparing the PCR mix, a second room for processing the samples and addition of template DNA in a PCR cabinet, and a third room for PCR amplification. Moreover, the accuracy of PCR is also endangered by false negative test results. In the study mentioned above, IS*2404* PCR was negative for a number of histology-confirmed BU patients [34]. If clinical diagnosis of BU appears convincing, but PCR results are negative, it is recommended to collect and test a new set of samples to verify laboratory test results. False negative testing can result from a low concentration of *M. ulcerans* DNA in lesion specimens, a poor DNA extraction efficiency, low PCR sensitivity, and/or the presence of PCR inhibitors.

Thus strict adherence to good clinical laboratory practices and implementation of quality assurance protocols are necessary to generate reliable IS*2404* PCR results. In 2008, the Technical Advisory Group of the WHO Global BU Initiative therefore recommended the establishment of an external quality assessment program (EQAP) for the PCR-based detection of *M. ulcerans* in clinical and also in environmental samples. This system was implemented and coordinated by WHO Collaborating Centres for BU (the Institute of Tropical Medicine (ITM) in Belgium for clinical samples and the Victorian Infectious Diseases Reference Laboratory in Australia for environmental samples) [14]. For the proficiency testing of clinical samples, coded specimens with known content were distributed by ITM to national reference and research laboratories that were asked to process the samples using DNA extraction and PCR procedures they usually apply for the detection of *M. ulcerans* DNA. Two rounds of clinical EQAP revealed a marked diversity in the quality of *M. ulcerans* DNA detection between laboratories. In the two assessment rounds, only 36% and 31% of the participating reference laboratories had more than 90% concordant results with the proficiency panels. 64% and 38% reported false positive and 55% and 81% false negative results, respectively. These data demonstrate the need for continued internal and external quality assurance [14]. Transport of samples from remote endemic areas to central reference laboratories, ensuring sample integrity and timely return of results, represents another major challenge for the African BU control programs.

In summary, PCR targeting IS*2404* is a sensitive and specific diagnostic laboratory test for *M. ulcerans*, but necessitates a well-equipped laboratory infrastructure, specifically trained laboratory staff and strict quality control. Recent advances in isothermal amplification and PCR product detection technologies have reduced equipment needs, but not staff requirements.

2.3 Detection of AFBs by Microscopy: A Test for the Primary Healthcare Level with Limited Sensitivity

At the primary healthcare level, where most of the BU patients are diagnosed to date, microscopic detection of AFBs in direct smears from lesion specimens using a conventional light microscope is the only available confirmatory test for *M. ulcerans* infection. This method relies on one of the characteristic properties of mycobacteria, namely their ability to form deeply coloured complexes with arylmethan dyes in phenol water such as carbol fuchsin, carbol crystal violet or carbol auramine O, that usually resist de-colorization by acidic ethanol (acid-fastness) [35]. For tuberculosis, sputum smear microscopy has for decades been the most widely used tool for the laboratory diagnosis in low- and middle-income countries. The most commonly used staining technique for the diagnosis of different mycobacterial infections is the so called Ziehl-Neelsen (ZN) method based on carbol fuchsin. When compared with IS*2404* PCR, the reported sensitivity rates of direct smear microscopy for the detection of *M. ulcerans* ranged between 26% and 67%.[1] The efficacy of the detection of small numbers of AFBs depends strongly on technical skills of the microscopist and on quality of the microscopic equipment. While being a simple and cost-effective first-line test for BU in resource-constrained settings, limited sensitivity even in the hands of well-trained personnel is a serious drawback. Furthermore, misdiagnosis of cutaneous tuberculosis as BU based on microscopy may occur [40], as staining is not specific for *M. ulcerans*.

Although fluorescence microscopy based on auramine O staining has several advantages over light microscopy using ZN staining, its widespread use has long been hindered by the need for a more expensive fluorescence microscope. However, the recent advent of low-cost ultrabright light emitting diodes (LED) has enabled the development of simple and affordable fluorescence microscopes [41, 42]. Advantages of fluorescence microscopy include not only a simpler, quicker, and cheaper staining procedure, but also the possibility to screen slides with a lower power objective lens, which may improve sensitivity. For the diagnosis of tuberculosis, fluorescence microscopy seems to be as specific and more sensitive than light microscopy [43]. However, in a study on the detection of *M. ulcerans*, no significant difference in sensitivity between the two microscopic detection methods was observed [38].

2.4 Considerations on the Accuracy of IS*2404* qPCR and AFB Detection by Microscopy: Direct Comparison of the Two Techniques

A comparative analysis of IS*2404* qPCR and smear microscopy conducted under optimal laboratory conditions using well-characterized lesion specimens from BU patients, showed that the amount of *M. ulcerans* DNA in extracts from the samples

[1] Twenty-six percent (11/43 punch biopsies) [36], 34% (344/1020 swab, FNA and tissue specimens) [11], 56.1% (69/123 swabs) [21], 58.4% (45/77 swab and FNA specimens) [37], 59.4% (66/111 tissue specimens) [38], 64.6% (822/1273 tissue specimens) [39], 67% (83/124 swab and FNA specimens) [22].

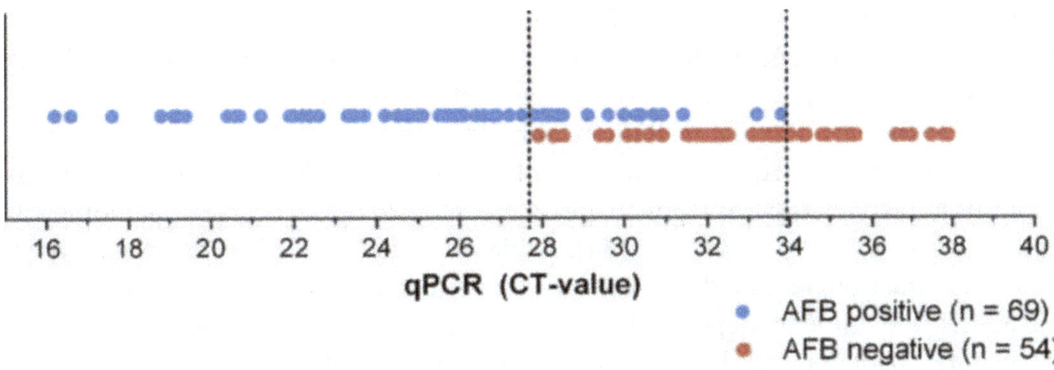

Fig. 2 Correlation between IS*2404* qPCR Ct values and microscopic AFB detection. A total of 123 IS*2404* qPCR-positive swab samples from BU lesions were analyzed by smear microscopy after ZN staining. While AFBs were detected in 69 (blue dots) of the 123 samples (56%), no AFBs were found in 54 (red dots) of the 123 samples (44%). AFBs were detected in all specimens having a qPCR Ct <27.8, whereas all specimens having a qPCR Ct >33.8 were microscopy-negative [21]

correlated well with the probability of finding AFBs [21]. While AFBs were detected in all IS*2404* qPCR highly positive samples, the probability of microscopy-positivity decreased for specimens with lower DNA content. Samples with qPCR cycle threshold (Ct) values above 34 were consistently negative (Fig. 2). A reasonable explanation for these results is certainly the higher sensitivity of qPCR as compared to smear microscopy. However, failure to detect even a single AFB in 54 IS*2404*-positive swab smears after careful screening of entire microscopy slides using a high power objective lens [21], may also raise concerns about false positive qPCR results. Even if qPCR assays are performed under strict quality control, false positive results with high Ct values may emerge through cross-contamination of samples with minimal amounts of *M. ulcerans* DNA in the process of sample collection (such as healthcare workers or other personnel handling several patients and samples without changing their gloves). In this context, the number of qPCR cycles that should be performed and which Ct values should be considered meaningful may have to be reconsidered.

2.5 Histopathology and Cultivation of *M. ulcerans*: Research Tools Rather than Diagnostic Tests

The pathogenesis of BU is primarily driven by the unique polyketide-derived macrolide exotoxin mycolactone of *M. ulcerans*, which causes apoptosis in mammalian cells [44]. In early non-ulcerative nodular stages of the infection the epidermis remains intact, but contiguous coagulative necrosis is found in the lower dermis. If mycolactone-mediated destruction of the subcutaneous tissue extends, the dermis and epidermis overlying foci of early *M. ulcerans* infection eventually degenerate. This leads to the formation of ulcers with undermined edges and a necrotic slough in the base. If mycolactone-mediated progression of tissue necrosis spreads mainly laterally, patients may develop extended non-ulcerative plaque or edematous forms of the disease [1].

Macroscopic features are thus changing as BU lesions evolve, whereas the progressive contiguous coagulative necrosis of the dermal and subcutaneous adipose tissue with dead adipocytes appearing as characteristic "ghost" outlines [45, 46] are considered BU-specific histopathological hallmarks that can be found in both pre-ulcerative (nodules, plaques, edema) and ulcerative stages. Necrosis of the subcutaneous tissue and necrotic collagen in the dermis seem to be the best histopathological predictors of *M. ulcerans* disease [34], while other features, like epidermal hyperplasia, destruction of blood vessels and interstitial edema complete the picture. Vasculitis and mineral deposits have been observed mainly in specimens from African BU patients [46]. In active lesions, extracellular clusters of AFBs are typically located in deep layers of the necrotic subcutaneous fat tissue [21]. Due to an uneven distribution of the clusters [21], AFBs may not be found in all parts of an active BU lesion. While there is a remarkable lack of inflammatory leukocytes in the necrotic centre of the lesions, a belt of inflammatory cells surrounding the necrotic core can be observed already in early lesions [47].

For routine diagnosis of BU in the endemic African countries, histopathological analysis of tissue samples is impractical, as it necessitates sophisticated technology as well as specifically trained and highly experienced personnel. Moreover, adequate tissue specimens obtained from the centre of a non-ulcerative lesion or from the edge of an ulcer by invasive sampling techniques are required. Therefore, histopathological analyses are not routinely used for diagnosing BU in Africa, but rather for the establishment of a differential diagnosis such as for suspected cases of squamous carcinoma secondary to BU [48, 49], or as a research tool to monitor treatment success [50] or paradoxical reactions [51].

Culture of the causative organism of an infectious disease is a mainstay of bacteriological diagnosis, not least because it provides information on the viability of the pathogen and allows for drug susceptibility testing. However, the extremely long generation time of *M. ulcerans*—colonies appear only after more than 2 months of incubation [18, 39]—excludes the application of cultivation for a pre-treatment confirmation of the clinical diagnosis. Moreover, successful cultivation of *M. ulcerans* depends on well-trained personnel and a complex laboratory infrastructure, only available at a few reference centres in the BU endemic African countries. Primary cultivation of *M. ulcerans* is thus performed mainly for research purposes, such as the monitoring of the efficacy of new treatment modalities [4, 6] as well as the distinction between paradoxical reactions and treatment failures [52, 53]. In addition, *M. ulcerans* isolates have been used for molecular epidemiological studies [29, 54, 55] and are used for the surveillance of the potential emergence of drug resistance.

2.6 From Theory into Practice: Diagnosis of BU in Resource-Constrained Endemic Countries

BU mainly affects impoverished populations living in remote, rural areas of West and Central Africa with only limited access to health facilities. Patients often have to travel long distances to reach a BU treatment facility. Primary or district level health facilities, where the majority of BU patients present are usually not capable of performing PCR-based analyses. At best, direct microscopy of ZN-stained swab smears is performed, a test with limited sensitivity yielding a high proportion of false negative results. Specimens for PCR reconfirmation are usually stored and transported in bulk to reference laboratories, leading to delayed delivery of diagnostic test results. Major delays in diagnosis and late initiation of treatment may on the other hand distract patients from returning to facilities of the formal health system. To avoid the dropout of patients, antibiotic therapy is often initiated upon the clinical diagnosis.

As the routine use of IS*2404* PCR in resource-constrained countries is also limited by high costs, more cost-effective stepwise approaches starting with the local microscopic analysis of swab smears for the presence of AFBs and—if this initial test is negative—subsequent testing by PCR at a reference laboratory have been proposed [37, 56]. However, the right balance between saving costs and taking the risk of false positive microscopy results has yet to be evaluated, as patients with cutaneous tuberculosis may receive the short 8 week course of antibiotics, if the treatment decision is made only on the grounds of a positive microscopy result [40].

Until the early 2000s clinical specimens from suspected BU patients were primarily analyzed at international reference laboratories for retrospective confirmation of the clinical diagnosis. Although several BU endemic African countries including Benin, Cameroon, Central African Republic, Côte d'Ivoire, Democratic Republic of the Congo, Ghana and Togo have installed their own reference laboratories for the diagnosis of BU [12], the situation has remained very challenging with respect to both timely sample transport and quality assurance. Therefore, the development of a simple point-of-care diagnostic test for BU remains a major research priority [57].

3 Development of BU Diagnostics for District Hospital or Primary Healthcare Facility Level

In 2013, WHO together with the Foundation for Innovative New Diagnostics (FIND) convened a meeting of BU experts to review the need for BU diagnostics in the form of low-tech assays to be used in district hospitals and of simple, instrument-free rapid diagnostic tests (RDTs) that can be performed at the primary healthcare level. RDTs should be low-cost, simple to operate and read, stable, and yield results in a short period of time. They are expected to reduce the need for multiple healthcare visits, improve the chances that patients receive appropriate treatment, and can be highly cost-effective. Most available RDTs for neglected tropical diseases (NTDs) are based on immunoassays, including lateral flow, flow-through, agglutination, and dipstick test formats for antibody or antigen detection. Current strategies for the development of a BU RDT rely on antigen detection assays. For the diagnosis at district hospital level, options to detect *M. ulcerans* DNA by loop-mediated isothermal amplification (LAMP), antigens by enzyme-linked immunosorbent assay (ELISA) and mycolactone by fluorescence-based thin layer chromatography (f-TLC) are being evaluated.

3.1 LAMP: An Alternative for the Detection of *M. ulcerans* DNA

After its first description in 2000 [58], LAMP has gained attention as a rapid and cost-effective nucleic acid amplification method for the diagnosis of infectious diseases. Advantages of the LAMP technique over conventional PCR-based tests are lower technology requirements and simple read-out, as results can be read by the naked eye. However, this simple read-out does not allow for a distinction between the specific products and products of nonspecific amplification, posing a risk of false positive results [59], particularly when performed under suboptimal conditions [60]. Major improvements in LAMP test specificity have recently been reported by combining the technique with molecular beacons, targeting an internal sequence of the amplicon, thus allowing for a direct, specific detection of the expected product [61]. Nevertheless, significant training and infrastructure development is required to achieve acceptable performance of LAMP assays [60], making it suitable for district hospital settings, but not for the primary healthcare level.

In 2012, three different LAMP assays for the detection of *M. ulcerans* DNA were published. Sensitivity rates targeting either a sequence of the *M. ulcerans* virulence plasmid [62] or the multi-copy sequence IS*2404* [63, 64] were determined to be close to those of conventional IS*2404* PCR. No cross-reactivity was recorded with DNA of other closely related mycobacterial species [62–64]. However, the accuracy of these assays was assessed by testing only a very limited number of clinical specimens. As the potential occurrence of false positive results—compared to the conventional IS*2404* PCR gold standard—was reported [63] even when performed in well-equipped specialized laboratories, LAMP assays will have to be evaluated in decentralized settings, where the technique is intended to be applied. Initial results have shown that the extraction and purification of *M. ulcerans* DNA and the generation of isothermal conditions during the LAMP reaction remain major challenges for the application of LAMP assays under field conditions. Use of crude, non-purified DNA extracts and pocket warmers as heat source were shown to significantly decrease test sensitivity [64]. In order to overcome another shortcoming of the technique—the requirement of cold-chains for transport and storage of reagents—a dry-reagent LAMP assay targeting IS*2404* was developed. Validation of this assay with a limited number of qPCR-confirmed lesion specimens in a laboratory environment indicated a sensitivity and specificity comparable to that of conventional gel-based PCR [65].

Taken together, the LAMP method has the potential to be developed into a sensitive and simple test for BU at district hospital level. As the technique requires basic laboratory infrastructure for template preparation and generation of isothermal conditions, it is at its current stage of development not suitable for an application at the primary healthcare level.

3.2 Detection of Mycolactone by f-TLC: Struggling with the Complexity of Lipid Extracts

As the production of mycolactone appears to be restricted to clinically relevant *M. ulcerans* lineages and some other closely related *M. ulcerans* ecovars [66], the toxin is considered an ideal target for the diagnosis of BU. In addition to their diagnostic application, mycolactone-based tests also have the potential to be used for the monitoring of treatment success by measuring mycolactone levels as an indicator for the presence of viable *M. ulcerans* bacteria in treated BU lesions [67].

One strategy to detect mycolactone is by staining of mycolactone bands after TLC-based separation of lipids extracted from clinical specimens. After demonstration by TLC and mass spectrometry that mycolactone can principally be detected in extracts from tissue samples of patients with ulcerative and non-ulcerative BU lesions [68], the TLC method has gradually been optimized to facilitate its application in more peripheral laboratory settings. In order to overcome shortcomings of the technique such as low sensitivity and limited specificity due to the presence of other lipids with similar chromatographic behaviour in the lipid extracts, a boronate-assisted f-TLC was developed. Staining is based on the formation of cyclic boronates with the 1,3-diols present in the mycolactone variants produced by the human pathogenic *M. ulcerans* lineages [69]. It has been reported that with this technique, mycolactone could selectively be visualized using an ultraviolet lamp with a detection limit of 2 ng [69]. In a preliminary study, f-TLC was applied to a panel of IS*2404* PCR-positive samples from BU patients, returning 73.2% sensitivity and 85.7% specificity [70]. Implementation of the f-TLC method in hospital laboratories appears to be non-trivial and specific detection of mycolactone can be challenging due to background staining of co-extracted human lipids [71].

3.3 Serological Tests: Only Suitable for Seroepidemiological Studies

For serology-based diagnostic approaches, the identification of appropriate target antigens that are *M. ulcerans*-specific and at the same time capable of mounting strong immune responses in BU patients, but not in healthy individuals exposed to the pathogen is essential. Different strategies were used to find suitable antigens, but none of them led to the development of a diagnostic serological assay. While initial studies have reported that BU patients develop antibodies to antigens present in *M. ulcerans* culture filtrate, broad antigenic cross-reactivity among mycobacterial species complicated the design of a serological test specific for *M. ulcerans* infections. Thus it is not surprising that the same studies have also shown that sera from healthy control individuals living in BU endemic areas, as well as from tuberculosis patients, contained antibodies that recognized the *M. ulcerans* target antigens [72, 73].

Moreover, among *M. ulcerans*-specific antigens identified by comparative genomics, none enabled a distinction between BU patients and healthy control subjects living in the same BU endemic area [74]. Furthermore, from a panel of monoclonal antibodies (mAbs) generated against immunodominant *M. ulcerans* proteins, only those specific for the 18 kDa small heat shock protein (shsp)—revealed sufficiently limited interspecies cross-reactivity. However, this antigen is not suitable as target for a BU-specific serological test, as anti-shsp antibodies are also frequently found in sera of healthy individuals living in BU endemic areas [75]. This circumstance however has opened up the opportunity for seroepidemiological studies assessing the exposure of populations to *M. ulcerans* [76–78].

3.4 Detection of Mycolactone and *M. ulcerans* Proteins by Antigen Detection Assays: Prospects for the Development of an RDT

As the development of sero-diagnostic tests for BU has failed, current strategies to develop RDTs for BU rely on the detection of *M. ulcerans* protein antigens or mycolactone. The main advantage of assays utilizing antigen-antibody interactions is their potential to be converted into test formats that can be deployed at the primary healthcare level. In fact, for numerous infectious diseases developed ELISAs in the form of antigen capture assays have been converted or have the potential to be converted into point-of-care (lateral flow) diagnostic assays [79–82].

In the case of mycolactone as a capture assay target antigen, the lipid-like nature of this macrolide toxin, as well as its cytotoxic and immunosuppressive properties have long hampered the generation of specific antibodies that can be used for its detection. Recently, a new strategy to immunize mice with a protein conjugate of a non-toxic synthetic truncated mycolactone derivative facilitated the generation of mAbs specific for the upper side chain and part of the core structure of mycolactone [83]. By using the generated mAbs, a first prototype competition assay was designed, in which a mycolactone-specific mAb is used in combination with a mycolactone derivative as signaling molecule, to quantify the amount of mycolactone in a sample. While the assay showed excellent specificity, its sensitivity must be optimized to allow for the detection of mycolactone in clinical specimens. If non-competing pairs of anti-mycolactone mAbs can be generated, development of an antigen capture assay may become possible.

An ideal protein target antigen for the development of a diagnostic test for BU should (1) be highly expressed by *M. ulcerans*, (2) have no orthologs in other prevalent pathogenic mycobacteria and (3) be easily accessible through a cell surface location. These predefined criteria were shown to be met by the *M. ulcerans* protein MUL_3720 [84]. Immunization of mice with the recombinantly expressed MUL_3720 facilitated the generation of a panel of high affinity mAbs against this antigen. Tandems of non-competing MUL_3720-specific mAbs recognizing different epitopes were selected to enable the development of a highly specific MUL_3720 detection assay in a sandwich-ELISA format [84]. Preliminary analyses comparing qPCR and ELISA results indicate that the MUL_3720 capture assay is highly specific. Optimization of the ELISA format potentially suitable for district hospitals is on-going to reach a sufficiently high sensitivity of antigen detection. As a next step towards an RDT, the application of the generated mAbs in a lateral flow assay format is being evaluated.

4 Discussion

In spite of the availability of sensitive and specific IS*2404* PCR assays routinely performed in resource-rich BU endemic countries and implemented in several national reference laboratories in BU endemic African countries, there is still an urgent need for simple and accurate point-of-care tests for the diagnosis of BU at district hospital and at primary healthcare level. In addition to logistical challenges and delays in the transport of clinical specimens to reference laboratories, outcomes of external quality assessment programs have demonstrated major shortcomings associated with the routine application of the PCR tests [14]. Ideally, the diagnosis of BU should furthermore not depend on the availability of laboratory infrastructure or of specifically trained laboratory personnel and should be performed directly at the point-of-care, so that treatment can be started without delay.

Immunochromatographic detection of antigens or antibodies in a dipstick or lateral flow format constitute currently the core of commercially available point-of-care RDTs for infectious diseases [85]. Whereas attempts to design an antibody-based sero-diagnostic assay for BU have been equivocal [74–76], the recent identification of a target protein suitable for antigen-capture test formats, shows great promise [84]. In addition, mycolactone—the lipid-like molecule secreted by *M. ulcerans*—may represent an optimal target molecule for the development of a sensitive and specific antigen detection test [83]. The advantage of antigen-based detection assays is that they can be converted to technically simple, robust test formats, easily applicable even at the primary healthcare level. In contrast, prerequisites for most molecular amplification techniques are incompatible with the ASSURED guidelines for point-of-care diagnostics [15].

In early 2011, WHO endorsed the first "sample in—answer out" qPCR platform with fully integrated sample processing for the diagnosis of tuberculosis in low-resource settings [86]. The GeneXpert MTB/RIF, a molecular test that can detect both *M. tuberculosis* DNA in sputum samples and rifampicin resistance mutations, has been developed for use at district and sub-district levels in tuberculosis endemic countries. However, in view of the limited resources available for the control of BU, costs for the production and broad introduction of a GeneXpert diagnostic test for *M. ulcerans* would be disproportionate. Several studies have reported attempts to develop nucleic acid-based point-of-care tests for the detection of *M. ulcerans* by the LAMP procedure [62–64]. While LAMP amplifies DNA with high sensitivity and specificity, its application under field conditions has so far been limited. This is mainly due to technology requirements associated with LAMP, such as template preparation in field settings, production of kits with dried-down reagents, and methods for unambiguous detection of amplification products [87]. If these prerequisites are met, the LAMP platform can potentially be progressed into a format for the detection of *M. ulcerans* at district hospital level.

Point-of-care tests are key components to improve global health, but only if they are rigorously evaluated, and effectively regulated [88]. Careful pre-implementation evaluation of the accuracy of novel diagnostic tests for BU is absolutely essential and a consensus has to be reached on the reference standard to which the performance of new diagnostic tests is compared. In view of the broad differential diagnosis of BU and reported misclassification of cases based on the clinical presentation of patients, a comparison to clinical diagnosis—as done in many previous studies—is not satisfactory. Instead, test performance should be evaluated by a comparison to the current qPCR reference standard. However, strict quality assurance of the qPCR assay used for evaluation of the tests has to be ensured. Clearly, technological innovation is not sufficient. After successful development and evaluation of a new point-of-care test for BU, decentralized implementation of the test will involve training programs and monitoring of the effectiveness of the new tests in decentralized settings, to ensure the accuracy of test results.

5 Outlook

In the past years, the number of BU cases reported annually in many African endemic countries has declined. This may partly be attributed to the establishment of effective national BU control programs and a reduction in transmission intensity. On the other hand, the intensity of disease surveillance and case search activities may have declined, resulting in an underestimation of the true disease burden. Limited funding for the control of BU can lead to a lack of awareness and a loss of local competencies among health workers. For instance, in a recent retrospective assessment of the diagnosis of BU in Ghana between 2008 and 2016, a gradual decline in the annual laboratory confirmation rate of clinically suspected cases from 52% in 2008 and 76% in 2009 to only 15% in 2016 was revealed, reflecting both the decline in BU incidence and the loss of clinical expertise [11]. Considering this scenario, as well as the broad differential diagnosis of BU, the availability of a reliable point-of-care test for the diagnosis of BU is crucially important. Apart from BU, several other NTDs present with skin manifestations, either as the primary or as an associated clinical condition. Many skin diseases have similarities in terms of their ability to cause long-term disabilities, reinforcement of poverty and geographical distribution. Collectively, these highly disabling and stigmatizing diseases constitute a great burden on the affected populations. Major deficiencies in our understanding of many of these diseases and a lack of tools to combat them necessitate further investment in research and development of control strategies. Integration provides an opportunity to leverage funding and to spearhead efforts for the development, optimization, and implementation of new diagnostic, therapeutic, and preventive tools for the simultaneous control of several co-endemic skin NTDs [89].

References

1. Junghanss T, Johnson RC, Pluschke G (2014) Mycobacterium ulcerans disease. In: Farrar J, Hotez PJ, Junghanss T, Kang G, Lalloo D, White NJ (eds) Manson's tropical diseases, 23rd edn. Saunders, Edinburgh, pp 519–531
2. Dega H, Bentoucha A, Robert J, Jarlier V, Grosset J (2002) Bactericidal activity of rifampin-amikacin against Mycobacterium ulcerans in mice. Antimicrob Agents Chemother 46(10):3193–3196
3. Marsollier L, Prevot G, Honore N, Legras P, Manceau AL, Payan C et al (2003) Susceptibility of Mycobacterium ulcerans to a combination of amikacin/rifampicin. Int J Antimicrob Agents 22(6):562–566
4. Etuaful S, Carbonnelle B, Grosset J, Lucas S, Horsfield C, Phillips R et al (2005) Efficacy of the combination rifampin-streptomycin in preventing growth of Mycobacterium ulcerans in early lesions of Buruli ulcer in humans. Antimicrob Agents Chemother 49(8):3182–3186
5. Chauty A, Ardant MF, Adeye A, Euverte H, Guedenon A, Johnson C et al (2007) Promising clinical efficacy of streptomycin-rifampin combination for treatment of buruli ulcer (Mycobacterium ulcerans disease). Antimicrob Agents Chemother 51(11):4029–4035
6. Sarfo FS, Phillips R, Asiedu K, Ampadu E, Bobi N, Adentwe E et al (2010) Clinical efficacy of combination of rifampin and streptomycin for treatment of Mycobacterium ulcerans disease. Antimicrob Agents Chemother 54(9):3678–3685
7. Nienhuis WA, Stienstra Y, Thompson WA, Awuah PC, Abass KM, Tuah W et al (2010) Antimicrobial treatment for early, limited Mycobacterium ulcerans infection: a randomised controlled trial. Lancet 375(9715):664–672
8. Borrell S, Gagneux S (2011) Strain diversity, epistasis and the evolution of drug resistance in Mycobacterium tuberculosis. Clin Microbiol Infect 17(6):815–820
9. Klis S, Stienstra Y, Phillips RO, Abass KM, Tuah W, van der Werf TS (2014) Long term streptomycin toxicity in the treatment of Buruli ulcer: follow-up of participants in the BURULICO drug trial. PLoS Negl Trop Dis 8(3):e2739
10. Toutous Trellu L, Nkemenang P, Comte E, Ehounou G, Atangana P, Mboua DJ et al (2016) Differential diagnosis of skin ulcers in a Mycobacterium ulcerans endemic area: data from a prospective study in Cameroon. PLoS Negl Trop Dis 10(4):e0004385
11. Yeboah-Manu D, Aboagye SY, Asare P, Asante-Poku A, Ampah K, Danso E et al (2018) Laboratory confirmation of Buruli ulcer cases in Ghana, 2008-2016. PLoS Negl Trop Dis 12(6):e0006560
12. Portaels F (2014) Laboratory diagnosis of buruli ulcer: a manual for health care providers. WHO, Geneva
13. Lavender CJ, Fyfe JA (2013) Direct detection of Mycobacterium ulcerans in clinical specimens and environmental samples. Methods Mol Biol 943:201–216
14. Eddyani M, Lavender C, de Rijk WB, Bomans P, Fyfe J, de Jong B et al (2014) Multicenter external quality assessment program for PCR detection of Mycobacterium ulcerans in clinical and environmental specimens. PLoS One 9(2):e89407
15. Mabey D, Peeling RW, Ustianowski A, Perkins MD (2004) Diagnostics for the developing world. Nat Rev Microbiol 2(3):231–240
16. World Health Organization (2004) Provisional guidance on the role of specific antibiotics in the management of Mycobacterium ulcerans disease (Buruli ulcer). World Health Organization, Geneva
17. Mensah-Quainoo E, Yeboah-Manu D, Asebi C, Patafuor F, Ofori-Adjei D, Junghanss T et al (2008) Diagnosis of Mycobacterium ulcerans infection (Buruli ulcer) at a treatment centre in Ghana: a retrospective analysis of laboratory results of clinically diagnosed cases. Tropical Med Int Health 13(2):191–198
18. Yeboah-Manu D, Bodmer T, Mensah-Quainoo E, Owusu S, Ofori-Adjei D, Pluschke G (2004) Evaluation of decontamination methods and growth media for primary isolation of Mycobacterium ulcerans from surgical specimens. J Clin Microbiol 42(12):5875–5876

19. Phillips R, Horsfield C, Kuijper S, Lartey A, Tetteh I, Etuaful S et al (2005) Sensitivity of PCR targeting the IS2404 insertion sequence of Mycobacterium ulcerans in an assay using punch biopsy specimens for diagnosis of Buruli ulcer. J Clin Microbiol 43(8):3650–3656
20. World Health Organization (2010) Guidance on sampling techniques for laboratory-confirmation of Mycobacterium ulcerans infection (Buruli ulcer disease). World Health Organization, Geneva
21. Ruf MT, Bolz M, Vogel M, Bayi PF, Bratschi MW, Sopho GE et al (2016) Spatial distribution of Mycobacterium ulcerans in Buruli ulcer lesions: implications for laboratory diagnosis. PLoS Negl Trop Dis 10(6):e0004767
22. Yeboah-Manu D, Danso E, Ampah K, Asante-Poku A, Nakobu Z, Pluschke G (2011) Isolation of Mycobacterium ulcerans from swab and fine-needle-aspiration specimens. J Clin Microbiol 49(5):1997–1999
23. Thierry D, Brisson-Noel A, Vincent-Levy-Frebault V, Nguyen S, Guesdon JL, Gicquel B (1990) Characterization of a Mycobacterium tuberculosis insertion sequence, IS6110, and its application in diagnosis. J Clin Microbiol 28(12):2668–2673
24. Hermans PW, van Soolingen D, Dale JW, Schuitema AR, McAdam RA, Catty D et al (1990) Insertion element IS986 from Mycobacterium tuberculosis: a useful tool for diagnosis and epidemiology of tuberculosis. J Clin Microbiol 28(9):2051–2058
25. Poulet S, Cole ST (1995) Repeated DNA sequences in mycobacteria. Arch Microbiol 163(2):79–86
26. Kang TJ, Kim SK, Lee SB, Chae GT, Kim JP (2003) Comparison of two different PCR amplification products (the 18-kDa protein gene vs. RLEP repetitive sequence) in the diagnosis of Mycobacterium leprae. Clin Exp Dermatol 28(4):420–424
27. Ross BC, Marino L, Oppedisano F, Edwards R, Robins-Browne RM, Johnson PD (1997) Development of a PCR assay for rapid diagnosis of Mycobacterium ulcerans infection. J Clin Microbiol 35(7):1696–1700
28. Stinear T, Ross BC, Davies JK, Marino L, Robins-Browne RM, Oppedisano F et al (1999) Identification and characterization of IS2404 and IS2606: two distinct repeated sequences for detection of Mycobacterium ulcerans by PCR. J Clin Microbiol 37(4):1018–1023
29. Doig KD, Holt KE, Fyfe JA, Lavender CJ, Eddyani M, Portaels F et al (2012) On the origin of Mycobacterium ulcerans, the causative agent of Buruli ulcer. BMC Genomics 13:258
30. Guimaraes-Peres A, Portaels F, de Rijk P, Fissette K, Pattyn SR, van Vooren J et al (1999) Comparison of two PCRs for detection of Mycobacterium ulcerans. J Clin Microbiol 37(1):206–208
31. Siegmund V, Adjei O, Nitschke J, Thompson W, Klutse E, Herbinger KH et al (2007) Dry reagent-based polymerase chain reaction compared with other laboratory methods available for the diagnosis of Buruli ulcer disease. Clin Infect Dis 45(1):68–75
32. Fyfe JA, Lavender CJ, Johnson PD, Globan M, Sievers A, Azuolas J et al (2007) Development and application of two multiplex real-time PCR assays for the detection of Mycobacterium ulcerans in clinical and environmental samples. Appl Environ Microbiol 73(15):4733–4740
33. Rondini S, Mensah-Quainoo E, Troll H, Bodmer T, Pluschke G (2003) Development and application of real-time PCR assay for quantification of Mycobacterium ulcerans DNA. J Clin Microbiol 41(9):4231–4237
34. Guarner J, Bartlett J, Whitney EA, Raghunathan PL, Stienstra Y, Asamoa K et al (2003) Histopathologic features of Mycobacterium ulcerans infection. Emerg Infect Dis 9(6):651–656
35. Barksdale L, Kim KS (1977) Mycobacterium. Bacteriol Rev 41(1):217–372
36. Phillips RO, Sarfo FS, Osei-Sarpong F, Boateng A, Tetteh I, Lartey A et al (2009) Sensitivity of PCR targeting Mycobacterium ulcerans by use of fine-needle aspirates for diagnosis of Buruli ulcer. J Clin Microbiol 47(4):924–926
37. Yeboah-Manu D, Asante-Poku A, Asan-Ampah K, Ampadu ED, Pluschke G (2011) Combining PCR with microscopy to reduce costs of laboratory diagnosis of Buruli ulcer. Am J Trop Med Hyg 85(5):900–904

38. Affolabi D, Bankole H, Ablordey A, Hounnouga J, Koutchakpo P, Sopoh G et al (2008) Effects of grinding surgical tissue specimens and smear staining methods on Buruli ulcer microscopic diagnosis. Tropical Med Int Health 13(2):187–190
39. Eddyani M, Debacker M, Martin A, Aguiar J, Johnson CR, Uwizeye C et al (2008) Primary culture of Mycobacterium ulcerans from human tissue specimens after storage in semisolid transport medium. J Clin Microbiol 46(1):69–72
40. Bratschi MW, Njih Tabah E, Bolz M, Stucki D, Borrell S, Gagneux S et al (2012) A case of cutaneous tuberculosis in a Buruli ulcer-endemic area. PLoS Negl Trop Dis 6(8):e1751
41. Minion J, Sohn H, Pai M (2009) Light-emitting diode technologies for TB diagnosis: what is on the market? Expert Rev Med Devices 6(4):341–345
42. Albert H, Manabe Y, Lukyamuzi G, Ademun P, Mukkada S, Nyesiga B et al (2010) Performance of three LED-based fluorescence microscopy systems for detection of tuberculosis in Uganda. PLoS One 5(12):e15206
43. Steingart KR, Henry M, Ng V, Hopewell PC, Ramsay A, Cunningham J et al (2006) Fluorescence versus conventional sputum smear microscopy for tuberculosis: a systematic review. Lancet Infect Dis 6(9):570–581
44. Bieri R, Scherr N, Ruf MT, Dangy JP, Gersbach P, Gehringer M et al (2017) The Macrolide Toxin Mycolactone Promotes Bim-Dependent Apoptosis in Buruli Ulcer through Inhibition of mTOR. ACS Chem Biol 12(5):1297–1307
45. Connor DH, Lunn HF (1965) Mycobacterium ulcerans infection (with comments on pathogenesis). Int J Lepr 33(3):S698–S709
46. Hayman J (1993) Out of Africa: observations on the histopathology of Mycobacterium ulcerans infection. J Clin Pathol 46(1):5–9
47. Ruf MT, Steffen C, Bolz M, Schmid P, Pluschke G (2017) Infiltrating leukocytes surround early Buruli ulcer lesions, but are unable to reach the mycolactone producing mycobacteria. Virulence 8(8):1918–1926
48. Evans MR, Etuaful SN, Amofah G, Adjei O, Lucas S, Wansbrough-Jones MH (1999) Squamous cell carcinoma secondary to Buruli ulcer. Trans R Soc Trop Med Hyg 93(1):63–64
49. Kassi K, Kouame K, Allen W, Kouassi LA, Ance W, Kanga JM (2010) Squamous cell carcinoma secondary to Buruli ulcer: a clinical case report in a young girl. Bacteriol Virusol Parazitol Epidemiol 55(1):25–28
50. Ruf MT, Sopoh GE, Brun LV, Dossou AD, Barogui YT, Johnson RC et al (2011) Histopathological changes and clinical responses of Buruli ulcer plaque lesions during chemotherapy: a role for surgical removal of necrotic tissue? PLoS Negl Trop Dis 5(9):e1334
51. Ruf MT, Chauty A, Adeye A, Ardant MF, Koussemou H, Johnson RC et al (2011) Secondary Buruli ulcer skin lesions emerging several months after completion of chemotherapy: paradoxical reaction or evidence for immune protection? PLoS Negl Trop Dis 5(8):e1252
52. Friedman ND, McDonald AH, Robson ME, O'Brien DP (2012) Corticosteroid use for paradoxical reactions during antibiotic treatment for Mycobacterium ulcerans. PLoS Negl Trop Dis 6(9):e1767
53. O'Brien DP, Robson M, Friedman ND, Walton A, McDonald A, Callan P et al (2013) Incidence, clinical spectrum, diagnostic features, treatment and predictors of paradoxical reactions during antibiotic treatment of Mycobacterium ulcerans infections. BMC Infect Dis 13:416
54. Roltgen K, Qi W, Ruf MT, Mensah-Quainoo E, Pidot SJ, Seemann T et al (2010) Single nucleotide polymorphism typing of Mycobacterium ulcerans reveals focal transmission of buruli ulcer in a highly endemic region of Ghana. PLoS Negl Trop Dis 4(7):e751
55. Vandelannoote K, Meehan CJ, Eddyani M, Affolabi D, Phanzu DM, Eyangoh S et al (2017) Multiple introductions and recent spread of the emerging human pathogen Mycobacterium ulcerans across Africa. Genome Biol Evol 9(3):414–426
56. Bretzel G, Siegmund V, Nitschke J, Herbinger KH, Thompson W, Klutse E et al (2007) A stepwise approach to the laboratory diagnosis of Buruli ulcer disease. Tropical Med Int Health 12(1):89–96

57. Report from the Meeting of the Buruli ulcer Technical Advisory Group, World Health Organization, Headquarters, Geneva, Switzerland, March 2017. Available from http://www.who.int/neglected_diseases/events/WHO_BU_TAG_2017_report.pdf
58. Notomi T, Okayama H, Masubuchi H, Yonekawa T, Watanabe K, Amino N et al (2000) Loop-mediated isothermal amplification of DNA. Nucleic Acids Res 28(12):E63
59. Kimura Y, de Hoon MJ, Aoki S, Ishizu Y, Kawai Y, Kogo Y et al (2011) Optimization of turn-back primers in isothermal amplification. Nucleic Acids Res 39(9):e59
60. Gray CM, Katamba A, Narang P, Giraldo J, Zamudio C, Joloba M et al (2016) Feasibility and operational performance of tuberculosis detection by loop-mediated isothermal amplification platform in decentralized settings: results from a multicenter study. J Clin Microbiol 54(8):1984–1991
61. Liu W, Huang S, Liu N, Dong D, Yang Z, Tang Y et al (2017) Establishment of an accurate and fast detection method using molecular beacons in loop-mediated isothermal amplification assay. Sci Rep 7:40125
62. de Souza DK, Quaye C, Mosi L, Addo P, Boakye DA (2012) A quick and cost effective method for the diagnosis of Mycobacterium ulcerans infection. BMC Infect Dis 12:8
63. Njiru ZK, Yeboah-Manu D, Stinear TP, Fyfe JA (2012) Rapid and sensitive detection of Mycobacterium ulcerans by use of a loop-mediated isothermal amplification test. J Clin Microbiol 50(5):1737–1741
64. Ablordey A, Amissah DA, Aboagye IF, Hatano B, Yamazaki T, Sata T et al (2012) Detection of Mycobacterium ulcerans by the loop mediated isothermal amplification method. PLoS Negl Trop Dis 6(4):e1590
65. Beissner M, Phillips RO, Battke F, Bauer M, Badziklou K, Sarfo FS et al (2015) Loop-mediated isothermal amplification for laboratory confirmation of Buruli ulcer disease-towards a point-of-care test. PLoS Negl Trop Dis 9(11):e0004219
66. Pidot SJ, Asiedu K, Kaser M, Fyfe JA, Stinear TP (2010) Mycobacterium ulcerans and other mycolactone-producing mycobacteria should be considered a single species. PLoS Negl Trop Dis 4(7):e663
67. Converse PJ, Xing Y, Kim KH, Tyagi S, Li SY, Almeida DV et al (2014) Accelerated detection of mycolactone production and response to antibiotic treatment in a mouse model of Mycobacterium ulcerans disease. PLoS Negl Trop Dis 8(1):e2618
68. Sarfo FS, Phillips RO, Rangers B, Mahrous EA, Lee RE, Tarelli E et al (2010) Detection of mycolactone A/B in Mycobacterium ulcerans-infected human tissue. PLoS Negl Trop Dis 4(1):e577
69. Spangenberg T, Kishi Y (2010) Highly sensitive, operationally simple, cost/time effective detection of the mycolactones from the human pathogen Mycobacterium ulcerans. Chem Commun (Camb) 46(9):1410–1412
70. Wadagni A, Frimpong M, Phanzu DM, Ablordey A, Kacou E, Gbedevi M et al (2015) Simple, rapid Mycobacterium ulcerans disease diagnosis from clinical samples by fluorescence of mycolactone on thin layer chromatography. PLoS Negl Trop Dis 9(11):e0004247
71. Sarfo FS, Le Chevalier F, Aka N, Phillips RO, Amoako Y, Boneca IG et al (2011) Mycolactone diffuses into the peripheral blood of Buruli ulcer patients--implications for diagnosis and disease monitoring. PLoS Negl Trop Dis 5(7):e1237
72. Dobos KM, Spotts EA, Marston BJ, Horsburgh CR Jr, King CH (2000) Serologic response to culture filtrate antigens of Mycobacterium ulcerans during Buruli ulcer disease. Emerg Infect Dis 6(2):158–164
73. Okenu DM, Ofielu LO, Easley KA, Guarner J, Spotts Whitney EA, Raghunathan PL et al (2004) Immunoglobulin M antibody responses to Mycobacterium ulcerans allow discrimination between cases of active Buruli ulcer disease and matched family controls in areas where the disease is endemic. Clin Diagn Lab Immunol 11(2):387–391

74. Pidot SJ, Porter JL, Marsollier L, Chauty A, Migot-Nabias F, Badaut C et al (2010) Serological evaluation of Mycobacterium ulcerans antigens identified by comparative genomics. PLoS Negl Trop Dis 4(11):e872
75. Diaz D, Dobeli H, Yeboah-Manu D, Mensah-Quainoo E, Friedlein A, Soder N et al (2006) Use of the immunodominant 18-kiloDalton small heat shock protein as a serological marker for exposure to Mycobacterium ulcerans. Clin Vaccine Immunol 13(12):1314–1321
76. Yeboah-Manu D, Roltgen K, Opare W, Asan-Ampah K, Quenin-Fosu K, Asante-Poku A et al (2012) Sero-epidemiology as a tool to screen populations for exposure to Mycobacterium ulcerans. PLoS Negl Trop Dis 6(1):e1460
77. Roltgen K, Bratschi MW, Ross A, Aboagye SY, Ampah KA, Bolz M et al (2014) Late onset of the serological response against the 18 kDa small heat shock protein of Mycobacterium ulcerans in children. PLoS Negl Trop Dis 8(5):e2904
78. Ampah KA, Nickel B, Asare P, Ross A, De-Graft D, Kerber S et al (2016) A sero-epidemiological approach to explore transmission of Mycobacterium ulcerans. PLoS Negl Trop Dis 10(1):e0004387
79. da Costa VG, Marques-Silva AC, Moreli ML (2014) A meta-analysis of the diagnostic accuracy of two commercial NS1 antigen ELISA tests for early dengue virus detection. PLoS One 9(4):e94655
80. Vallur AC, Tutterrow YL, Mohamath R, Pattabhi S, Hailu A, Abdoun AO et al (2015) Development and comparative evaluation of two antigen detection tests for Visceral Leishmaniasis. BMC Infect Dis 15:384
81. Cruz HM, Scalioni Lde P, de Paula VS, da Silva EF, do OK, Milagres FA et al (2015) Evaluating HBsAg rapid test performance for different biological samples from low and high infection rate settings & populations. BMC Infect Dis 15:548
82. Li B, Sun Z, Li X, Li X, Wang H, Chen W et al (2017) Performance of pfHRP2 versus pLDH antigen rapid diagnostic tests for the detection of Plasmodium falciparum: a systematic review and meta-analysis. Arch Med Sci 13(3):541–549
83. Dangy JP, Scherr N, Gersbach P, Hug MN, Bieri R, Bomio C et al (2016) Antibody-mediated neutralization of the exotoxin mycolactone, the main virulence factor produced by Mycobacterium ulcerans. PLoS Negl Trop Dis 10(6):e0004808
84. Dreyer A, Roltgen K, Dangy JP, Ruf MT, Scherr N, Bolz M et al (2015) Identification of the Mycobacterium ulcerans protein MUL_3720 as a promising target for the development of a diagnostic test for Buruli ulcer. PLoS Negl Trop Dis 9(2):e0003477
85. Bissonnette L, Bergeron MG (2010) Diagnosing infections--current and anticipated technologies for point-of-care diagnostics and home-based testing. Clin Microbiol Infect 16(8):1044–1053
86. Boehme CC, Nicol MP, Nabeta P, Michael JS, Gotuzzo E, Tahirli R et al (2011) Feasibility, diagnostic accuracy, and effectiveness of decentralised use of the Xpert MTB/RIF test for diagnosis of tuberculosis and multidrug resistance: a multicentre implementation study. Lancet 377(9776):1495–1505
87. Njiru ZK (2012) Loop-mediated isothermal amplification technology: towards point of care diagnostics. PLoS Negl Trop Dis 6(6):e1572
88. Peeling RW, Mabey D (2010) Point-of-care tests for diagnosing infections in the developing world. Clin Microbiol Infect 16(8):1062–1069
89. Mitja O, Marks M, Bertran L, Kollie K, Argaw D, Fahal AH et al (2017) Integrated control and management of neglected tropical skin diseases. PLoS Negl Trop Dis 11(1):e0005136

Antimicrobial Treatment of *Mycobacterium ulcerans* Infection

Till Frederik Omansen, Tjip S. van der Werf, and Richard Odame Phillips

1 Historical Aspects

MacCallum and coworkers described Buruli ulcer (BU) as an infectious disease caused by *Mycobacterium ulcerans* in Victoria, Australia. They first considered the skin lesions in their patients to be caused by tuberculosis or leprosy, when they observed numerous acid-fast bacilli in the biopsy specimens [1]. The typical duration of illness was between 1 and 2 years; treatment was essentially surgical. With the advent of chemotherapy for tuberculosis [2–4], and later for leprosy, doctors made individual attempts to treat the lesions with anti-tuberculosis and anti-leprosy drugs. The anecdotal evidence suggested poor or no response to chemotherapy with rifampicin monotherapy [5], despite the fact that in vitro susceptibility of 33 strains of *M. ulcerans* was as good as for *M. tuberculosis* [6]. A randomized clinical trial by the British Medical Research Council in Buruli county (now called Nakasongola; Uganda) failed to show any benefit from clofazimine, a drug then first marketed for leprosy [7]. A small-sized trial with cotrimoxazole (18 participants; 12 evaluable) was inconclusive [8]. A small-sized randomized study in Côte d'Ivoire compared a

T. F. Omansen
Department of Tropical Medicine, Bernhard Nocht Institute for Tropical Medicine, Hamburg, Germany

I. Department of Medicine, University Medical Center Hamburg-Eppendorf, Hamburg, Germany

T. S. van der Werf (✉)
Department of Internal Medicine, Division of Infectious Diseases, and Pulmonary Diseases & Tuberculosis, University of Groningen, University Medical Centre Groningen, Groningen, The Netherlands
e-mail: t.s.van.der.werf@umcg.nl

R. O. Phillips
Department of Medicine, Kwame Nkrumah University of Science & Technology, Komfo Anokye Teaching Hospital, Kumasi Center for Collaborative Research, Kumasi, Ghana

G. Pluschke, K. Röltgen (eds.), *Buruli Ulcer*,
https://doi.org/10.1007/978-3-030-11114-4_11

combination of dapsone and rifampicin with placebo; the follow-up was limited; the ulcer size decreased slightly faster in the intervention group but the baseline characteristics of both groups differed, and the study did not allow to draw any firm conclusions about the effectiveness of these drugs [9]. By the turn of the millennium, the discrepancy between in vitro efficacy of rifampicin [6] or clarithromycin [10] and lack of clinical response prompted to stressing the need for well-designed and well-powered drug trials, but in the meantime, to also improve early detection and surgical treatment [11].

One important question that comes to mind when looking back at earlier studies that failed to show effectiveness of antimicrobial treatment for BU is how the clinical response was assessed. Even long before the chemical structure of the toxin mycolactone was discovered [12], it was realized that a toxin secreted by *M. ulcerans* was responsible for the extensive tissue necrosis [13, 14] as well as for the immune suppression observed and replicated in experimental animals, using a culture filtrate of *M. ulcerans* [15]. Radical surgical excision was to some extent effective but at the cost of tissue loss, and the need for plastic surgery, e.g., split-skin or full-thickness skin grafting. Reported recurrence rates following surgical treatment were variable; one follow-up study from Benin reported 6.1% recurrence in a subset of patients (66/150; 44%) followed up to 7 years after surgery [16]. In a case series of 346 patients operated in three centers in Côte d'Ivoire, the recurrence rate was 17.1% [17]. A large difference in recurrence rates was reported from two centers in Ghana; 21/45 (47%) of patients had a recurrence in one hospital, as compared to 6/33 (18%) in the other hospital [18]. In a case series from the Bas Congo, DR Congo, of 51 patients seen over 2 consecutive years, 14 (39%) had recurrent disease after previous surgery [19]. Lesions at a pre-ulcerative stage can be excised successfully with primary closure, however a 1-year recurrence rate of 16% was observed in a cohort of 50 patients in Ghana [20].

With increased understanding of the dominant role of the secreted toxin mycolactone in the pathogenesis of Buruli ulcer, and realizing that perhaps the best measure of efficacy of antimicrobial treatment might be complete healing without recurrence, the effectiveness of antimicrobial treatment might have been underestimated in the past. Indeed with killing of *M. ulcerans* in lesions, as exemplified by the beneficial effects of different classes of antimicrobials in vitro [6, 10, 21], mycolactone secretion might stop, but the impact on tissue damage including necrosis would conceivably take more time than in any other infectious disease. Likewise, there was increased understanding of the profound effects of mycolactone on the immune system, with little local [22] and systemic inflammation [15]. Yet another factor perhaps blurring the observation of response to antimicrobial therapy is a common paradoxical increase in lesion size, development of new lesions, and increased inflammation following antimicrobial therapy [23–25]. Apart from the in vitro studies [6, 10, 21], several studies in animal models had shown the efficacy of anti-mycobacterial drug combinations [26–28].

A formal proof-of-concept study to evaluate the potential of antimicrobial treatment without additional surgery was the landmark study conducted in Ghana, under the umbrella of the WHO Global Buruli Ulcer Initiative [29]. Patients having non-ulcerated lesions suspected to be *M. ulcerans* infection were randomly allocated to 0, 4, 8 or 12 weeks of antimicrobial treatment with a combination of streptomycin and rifampicin; later a 5th group receiving 2 weeks of treatment was added; the total number of study participants was 21. Lesions were subsequently surgically removed and submitted for culture, PCR and histopathology. As no bacteriological confirmation tests were carried out prior to the start of therapy, some of the lesions remained bacteriologically unconfirmed to be Buruli ulcer disease; none of the patients receiving at least 4 weeks of treatment had viable bacilli by culture of their excised lesion; none had recurrent disease 12 months after start of treatment. Although some questions remained unanswered, this study provided the first robust evidence that antimicrobial treatment alone was able to sterilize non-ulcerated lesions of *M. ulcerans* infection, with no recurrence or positive culture in patients treated for at least 4 weeks with the antimicrobial combination [29].

With the introduction of antimicrobial treatment, a more conservative, less aggressive approach of surgery in addition to antimicrobial treatment was subsequently advocated [30–33].

For over 10 years now, the WHO has recommended antimicrobial treatment as the primary treatment modality. Below, we systematically discuss the literature on antimicrobial susceptibility of *M. ulcerans* in vitro and in animal models; and the accumulated evidence for effectiveness as well as safety (i.e., adverse reactions) of antimicrobial treatment regimens emerging from clinical studies.

2 Antimicrobial Susceptibility of *M. ulcerans*

For the review of in vitro activity of antimicrobial agents against *M. ulcerans*, we used a systematic approach searching the literature in PubMed. Using '(Buruli OR ulcerans) AND (antibiotic OR antimicrobial OR in-vitro OR susceptible)', we retrieved 520 unique results that we further analyzed for duplicates and consistent reporting. We cross-searched references obtained from articles we analyzed.

The most widely used antimicrobials for *M. ulcerans* infection have been rifampin, streptomycin and clarithromycin. *M. ulcerans* is a phylogenetically close relative of *M. tuberculosis* and *M. leprae* and has evolved from a common *M. marinum*-like ancestor [34]. Subsequently, many attempts to treat *M. ulcerans* infection have been undertaken seeking to use and repurpose antibiotics active against *M. tuberculosis* and *M. leprae*. To estimate efficacy in vitro, the usual way to screen these agents is by determining the minimal inhibitory concentration (MIC) of inoculates of *M. ulcerans* in mycobacterial culture media. Next, inoculation of viable *M. ulcerans* in the mouse footpad is a well-established in vivo model to assess efficacy. The final test is eventually to test these agents in patients affected by Buruli ulcer disease. Such studies are extremely challenging, as the majority of potential study participants are patients (many being children) that live in underprivileged circumstances with limited access to health care, and generally low educational background [35–37].

2.1 Ansamycins/Rifamycins

Ansamycins or rifamycins act on mycobacteria by inhibiting RNA synthesis through interfering with bacterial RNA-polymerase. Rifampin is the rifamycin most widely used in the treatment of Buruli ulcer. The drug has a strong bactericidal effect and an MIC of approximately 0.5 μg/ml [38–40]. Rifampin is administered orally at a dose of 10 mg/kg to humans to treat Buruli ulcer and is the backbone-drug of most regimens evaluated to date. Other rifamycins that have been shown to kill *M. ulcerans* are the rifampin-analogues rifabutin [26] and rifapentine [41]. Rifapentine is highly active against *M. ulcerans* with an MIC of 0.125–0.5 μg/ml [41, 42]. Because of its longer half-life time compared to rifampin, an intermitted rifapentine-based regimen with only two or three times weekly administration of antibiotics has been tested in mice [41]. Such an intermittent regimen is suggested to allow outpatient management and to simplify clinical management. However intermittent regimens also bear the risk of confusion and low compliance and patients being subsequently lost during treatment. Even though antibiotic resistance is not a major concern in the slowly replicating organism *M. ulcerans*, rifampicin resistance has been detected in rare instances [43, 44]. In-vivo, resistant mutants were isolated after monotherapy with rifampin in mice, which should be avoided in the clinical setting. The resistance was conferred by mutations in the *rpo*B gene of *M. ulcerans* [45]. Albeit uncommon and of unknown clinical significance, researchers in Ghana found 5.7–11.4% of isolates from their study districts to be resistant to 40 μg/ml rifampin [40]. Yet another rifamycin tested in *M. ulcerans* is rifabutin. Rifabutin has an MIC of 0.1–0.4 μg/ml against *M. ulcerans* [39].

2.2 Aminoglycosides: Streptomycin

Streptomycin is an injectable aminoglycoside with a strong bactericidal effect on *M. ulcerans* and one of the first agents ever shown to be effective against this organism [46, 47]. It is a protein synthesis inhibitor that hinders binding of formyl-methionyl-tRNA to the bacterial ribosomal 30S subunit. In the mouse model, streptomycin remains one of the most active agents against *M. ulcerans* and is used as intramuscular injection of 15 mg/kg in humans [48]. In single drug in vitro and in vivo studies, streptomycin was more active than amikacin and linezolid, but less active than rifampicin and moxifloxacin [49]. In preclinical studies, streptomycin-containing regimens outperform all other commonly used drug combinations both in terms of colony-forming units remaining in infected mouse footpads and recurrence rate. However, the need to inject streptomycin is very impracticable in the rural African setting where Buruli ulcer is most common. It bears the risk of parenteral (needle-associated) infection and also threatens patient compliance due to fear of painful, daily injections, especially in children. Besides, streptomycin causes ototoxicity as well as nephrotoxicity as observed in patients followed after having been enrolled in the BURULICO trial [50].

2.3 Amikacin

Amikacin is another aminoglycoside that has been suggested for use in the treatment of patients with Buruli ulcer. It is highly active against *M. ulcerans* with an MIC of 0.5–1 μg/ml [21] and was shown to be equally efficient in the mouse model [26]. In a study testing a series of isolates, the MIC ranged between 0.25–1 (mean, 0.65) μg/ml [49]. Amikacin was as effective as streptomycin in the *M. ulcerans* mouse model [26].

2.4 Macrolides: Clarithromycin

Clarithromycin is a protein synthesis inhibitor of the macrolide family. It reversibly binds to the 23S rRNA on the 50s ribosomal subunit and subsequently prevents polypeptide synthesis. It acts mainly as a bacteriostatic drug in *M. ulcerans* therapy. The MIC of clarithromycin against *M. ulcerans* is approximately 0.12 μg/ml [10, 26, 51, 52]. Clarithromycin is well tolerated, it can be taken orally and it is widely available. It has been shown to be non-inferior when replacing streptomycin after 4 weeks in the 8 week rifampin-streptomycin regimen while treating small lesions [36]. Following preliminary results of a randomized controlled trial (ClinicalTrials.gov identifier: NCT01659437) comparing 8 weeks rifampicin plus streptomycin with 8 weeks rifampicin plus clarithromycin, the latter regimen is now recommended by WHO.[1]

2.5 Azithromycin

Azithromycin is a macrolide antimicrobial agent active against a variety of organisms. The mean MIC against a collection of different strains was 0.39 μg/ml [40]. However when 100 mg of azithromycin was administered to mice infected with *M. ulcerans*, only a modest bacteriostatic activity was observed [53].

2.6 Fluoroquinolones

Various fluoroquinolones such as ciprofloxacin, ofloxacin and moxifloxacin have been shown to be active against *M. ulcerans* with MICs ranging from 0.25 to 1 μg/ml

[1] http://www.who.int/neglected_diseases/events/WHO_BU_TAG_2017_report.pdf?ua=1.

[21]. Fluoroquinolones are bacterial DNA gyrase inhibitors. They impede DNA replication and transcription by hindering the DNA gyrase-catalyzed super-coiling of double-stranded DNA. In Australia, Buruli ulcer patients are frequently treated with rifampicin in combination with a fluoroquinolone, such as ciprofloxacin [54]. Such fluoroquinolones are administered orally and have favorable pharmacokinetic properties such as good tissue penetration. However, there is some concern about the use of fluoroquinolones in children; in Australia most Buruli ulcer patients are adults, most of them elderly, whereas in Africa where most of the cases are reported, the median age of patients is around 15 years. Fluoroquinolones may cause serious arthropathy in children and minors. However the evidence about increased toxicity in the young is incomplete. Although caution remains warranted, fluoroquinolones have increasingly been used in children without significantly increased adverse events or toxicity [55, 56]. The most potent fluoroquinolone against *M. ulcerans* is sparfloxacin, with an MIC at 0.25 μg/ml in most strains tested [21]. The MIC of ofloxacin in tested strains of *M. ulcerans* was 1.26 μg/ml [40]; in another study, ofloxacin was least active of all fluoroquinolones tested with an MIC of 2 μg/ml [21]. Moxifloxacin (MIC 0.14 μg/ml), sitafloxacin (MIC 0.125–0.5 μg/ml) and prulifloxacin (MIC 1–3 μM) are also effective [49, 57, 58]. Of note, fluoroquinolones have increasingly been used in Ghana, one of the countries highly burdened with Buruli ulcer; an alarming rate of fluoroquinolone-resistant *E. coli* was noticed [59].

2.7 Clofazimine

Clofazimine is an anti-leprosy drug that has emerged as a critically important sterilizing drug in the treatment of multi-drug resistant tuberculosis. Its effect on mycobacteria is delayed but strongly bactericidal [60]. The mechanism of action is multifactorial, complex and not fully understood. As a prodrug, it is reduced by type 2 NADH-quinone oxidoreductase resulting in the release of bactericidal quantities of reactive oxygen species. It is then believed to competitively inhibit menaquinone, a crucial electron acceptor in the mycobacterial respiration chain [61]. Clofazimine possesses also some poorly understood anti-inflammatory activity for which it is used in conditions like severe acne or discoid lupus erythematosus [62]. These anti-inflammatory properties might also be part of its mode of action in mycobacterial infections where tissue damage by inflammatory processes is a hallmark. The MIC for various strains of *M. ulcerans* was 2.19 μg/ml on average [40]. As mentioned above, in an early trial clofazimine monotherapy was inefficacious in human Buruli ulcer patients [7]. In mice, clofazimine in combination with rifampicin was very effective with relapse-free cure following a 6-weeks oral regimen [63]. Clofazimine causes yellow-orange skin-discoloration as well as gastrointestinal disturbances as adverse effects [64]. Newer clofazimine analogs with reduced accumulation and thus less risk for skin discoloration are under development [58].

2.8 Dapsone

Dapsone is an anti-leprosy drug that has widely been used. Activity against *M. ulcerans* is moderate—the MIC ranged between 2.0 and 4.0 μg/ml in one study [39] and the mean MIC was 0.94 μg/ml in another study [40].

2.9 Doxycycline

Doxycycline, a tetracycline, was inactive against *M. ulcerans* in one study [58].

2.10 Oxazolidinones

Linezolid showed only intermediate activity with an MIC ranging from 3 to 10 μM; as posizolid and sutezolid. This oxazolidinone is currently considered as a major component of second-line tuberculosis treatment [58, 65]. In another study with 29 isolates of *M. ulcerans* tested, the mean MIC was 0.73 μg/ml [49].

2.11 Avermectins

Avermectins are a class of anti-helminth and anti-parasite drugs that are thought to have intermediate efficacy on *M. tuberculosis* [66]. Subsequent testing showed the MIC for ivermectin and moxidectin ranging from 4 to 8 μg/ml [67]. In another study, the MIC for selamectin was 2–4 μg/ml and ivermectin and moxidectin showed no activity >32 μg/ml [68].

2.12 Trimethoprim and Epiroprim

Trimethoprim was not effective against *M. ulcerans*. However, epiroprim, another dihydrofolate reductase inhibitor, showed an MIC of 0.5–1.0 μg/ml [69].

3 Experimental Drugs

A series of experimental compounds, mainly originating from TB drug research pipelines were tested for use in *M. ulcerans* chemotherapy.

Diarylthiazoles and 1,3-diaryltriazenes are compounds with high potency against mycobacteria, including *M. ulcerans*. Different experimental 1,3-diaryltriazene analogues showed sub-micro molar inhibitory activity against *M. ulcerans 1615 lux* [70]. Similarly, diarylthiazoles, like fatostatin, were found to have a significant efficacy against *M. ulcerans* [58]. KRM-1648, a benzoxazinorifamycin with good efficacy in the TB mouse model was equally tested for use in *M. ulcerans* and showed good killing at an MIC of 12–25 μg/l [39]. The diarylquinoline bedaquiline, also a TB research drug had a mean MIC of 0.03 (0.015–0.12) μg/ml when tested against a set of 29 isolates [49].

GyrB is a common drug target in TB therapy and experimental GyrB inhibitors showed good to moderate activities (MIC 0.3–10 μM) against *M. ulcerans* [58]. In the same study, pyrrolamide and aminopyrazinamide exhibited activity with an MIC of 0.3–1.0 μM [58]. These investigators found one aminopyrazole, as well as pyrazolopyrimidine and one hydroxyquinolone to kill *M. ulcerans* with MICs from 0.3 to 3.0 μM [58].

There is high genetic similarity between *M. tuberculosis* and *M. ulcerans*. Yet some of the compounds that are active against the first, failed to prove efficacy in the latter. SQ641, PA-824 a promising nitroimidazopyran for tuberculosis, semisynthetic ketolides (HMR 3647 and HMR 3004), bisbenzaldehydes, as well as quinolinyl pyrimidines and phenothiazines showed only moderate or no antimicrobial effect against *M. ulcerans* (Table 1) [21, 39, 49, 51, 52, 57, 58, 67, 70].

Table 1 Summary of compounds tested against *M. ulcerans* in vitro

Compound	Class	Mic in-vitro
Rifampin	Rifamycin	0.1–0.81 μg/ml
Rifabutin	Rifamycin	0.1–0.4 μg/ml
Rifapentine	Rifamycin	0.125–0.5 μg/ml
Dapsone	Sulfone	0.94–4.0 mg/l
Doxycycline	Tetracycline	Inactive
Streptomycin	Aminoglycoside	0.33–1.10 μg/ml
Amikacin	Aminoglycoside	0.5–0.65 μg/ml
Azithromycin	Macrolide	0.39 μg/ml
Clarithromycin	Macrolide	0.125–1.25 μg/ml
Ofloxacin	Quinolone	1.26–2.0 μg/ml
Ciprofloxacin	Quinolone	1.15 μg/ml; 1–3 uM
Sparfloxacin	Quinolone	0.25 mg/l
Moxifloxacin	Quinolone	0.14 μg/ml
Sitafloxacin	Quinolone	0.125–0.5 μg/ml
Prulifloxacin	Quinolone	1–3 μM
GyrA_NTBI-analog	Topoisomerase II inhibitor	Inactive
GYRA-NTBI PubChem_15983305	Topoisomerase II inhibitor	Inactive
GyrB_Pyrollamide PubChem_25223515	Pyrrolamide	0.3–1.0 μM
GyrB_Aminopyrazinamide	Aminopyrazinamide	0.3–1.0 μM
Oxazolidinone PubChem_10251911	Oxazolidinone	0.3–1.0 μM
Linezolid	Oxazolidinone	3–10 μM; 0.73 μg/ml
Posizolid	Oxazolidinone	3–10 μM
PubChem_10251911	Oxazolidinone	0.3–1.0 μM
Clofazimine	Riminophenazine	2.19 μg/ml
Selamectin	Avermectin	2–4 μg/ml
Ivermectin	Avermectin	4–32 μg/ml
Moxidectin	Avermectin	4–32 μg/ml

Compound	Class	Mic in-vitro
HMR 3647	Ketolide	5–40 μg/ml
HMR 3004	Ketolide	5–40 μg/ml
1,3-diaryltriazenes		
Diarylthiazoles		
SQ641		8 μg/ml
KRM-1648	Benzoxazinorifamycin	0.012 and 0.025 μg/ml
R207910	Diarylquinoline	0.03 μg/ml
PA-824	Nitroimidazopyran	13.1 μg/ml
Aminopyrazoles		0.3–1.0 μM
Pyrazolopyrimidines		0.3–1.0 μM
hydroxyquinolones		1–3 μM
Quinolinyl pyrimidines		Inactive
Phenothiazines		Inactive
Diazene-1,2-dicarboxamides		5.67–7.25 μg/ml

4 Clinical Studies

As mentioned above, the shift to antimicrobial therapy was made following the landmark study by Etuaful et al. using 15 mg/kg body weight intramuscular streptomycin and 10 mg/kg body weight oral rifampin [29]. The earlier mentioned randomized placebo-controlled study from Côte d'Ivoire with 41 study participants testing a combination of rifampicin and dapsone lacked power and had too many dropouts with only 2 months follow-up [9].

During the past decade, case series and observational cohort studies of patients with BU were reported that we discuss here; case reports will not be discussed as BU patients may occasionally heal spontaneously [71, 72]. One observational cohort study in Benin reported on 224 patients, the majority with large (category III; >15 cm) lesions. 215 (96%) were categorized as treatment successes, and 9, including 1 death and 8 losses to follow-up, were treatment failures. Of the 215 successfully treated patients, 102 (47%) were treated exclusively with antibiotics and 113 (53%) were treated with antibiotics plus surgical excision and skin grafting; 73% of patients with lesions of >15 cm in diameter underwent surgery, whereas only 17% of patients with lesions of <5 cm had surgery [73]. Compliance with therapy was excellent; 208 of the 215 patients were actively retrieved 1 year after treatment and 3 (1.44%) of the 208 retrieved patients had recurrence of *M. ulcerans* disease, 2 among the 107 patients treated only with antibiotics and 1 among the 108 patients treated with antibiotics plus surgery. In a report from Australia, 40 patients received combined antimicrobial and surgical treatment. Failures occurred often but twice as often in the group that received surgery as the only treatment modality [74]. The Australian guidelines of 2006 (published in 2007) still proposed radical surgery including the removal of a rim of apparently healthy tissue for treatment [75]. Several other groups have reported on outcome of combined surgical and antimicrobial treatment—of 79 (61%) patients retrieved, 7 (9%) had a recurrence [76]. In a series of 92 patients treated in the Bas-Congo, DR Congo, patients received surgery and antimicrobial (streptomycin/rifampicin) treatment sometimes more than 8 weeks, with a high success rate (98.4% in PCR-confirmed patients) and low recurrence rate (1.1%).

With a substantial proportion of study participants receiving both surgery and antimicrobial therapy it has been difficult to tease out the potential of antimicrobial treatment alone. In larger lesions, the general assumption is that surgery as added therapy is essential to obtain wound closure; in smaller lesions, the hypothesis that antimicrobial treatment alone without extensive surgical debridement could heal BU with no recurrence, was tested in a controlled clinical trial including 151 study participants with small (<10 cm cross-sectional diameter) lesions, almost all being confirmed by PCR to have *M. ulcerans* infection. After 4 weeks of streptomycin/rifampicin combination therapy, patients were randomized to either continue for 4 more weeks or switch to oral treatment with rifampicin/clarithromycin combination therapy. A switch from streptomycin to clarithromycin after 4 weeks was non-inferior to 8 weeks streptomycin and rifampicin in these early lesions; recurrence or failure was very low—success rates were 96% in the group receiving 8 weeks of streptomycin/rifampicin, and 91% in the group that switched to oral treatment which was statistically non-inferior [36]. In an observational cohort of 160 PCR-confirmed patients with BU, treated with streptomycin/rifampicin for 8 weeks, 152 healed without surgery; of 158 patients seen 1 year later, no recurrences were noted [77]. In three centers in Ghana, 43 patients—16 of them having category II/III lesions—received the streptomycin/rifampicin combination regimen that was switched after 2 weeks to oral rifampicin/clarithromycin. Ninety-three percent had successful outcome, only one had surgery, and none had recurrence at follow-up [37]. A recent report reflects perhaps better what happens under routine service conditions [78]. Patients (n = 50) with confirmed BU in two centers in the Brong-Ahafo Region of Ghana were followed over time; a majority had first used traditional treatment; the patient population consisted predominantly of peasant farmers with no formal education, or children. Only 40 completed treatment and of those, only 28 healed; in the others the lesions reduced in size [78].

In summary, based on the above evidence from various studies and reports—notably, the potential to achieve cure without relapse, at much higher rates than previously with surgery alone, the role of antimicrobial treatment in the management of BU has become standard of care [79]. All drugs used for BU have significant side effects, especially in the elderly [80]; but aminoglycoside drugs—streptomycin and amikacin—are notoriously the most toxic compounds [81], and although most of the evidence supporting antimicrobial treatment for BU has been provided for the combination of rifampicin and streptomycin, there has been a search for an alternative, fully oral treatment for over a decade now. The first report came from Australia where the patient population is typically elderly and more vulnerable for the ototoxicity of aminoglycosides; the results of a small (n = 4) case series [82] was soon followed by larger series of 43 patients that had oral antimicrobials combined with surgery with only one patient that had a relapse [83]. Later, a series of 132 patients had oral therapy with less surgery than before and with excellent healing rates and only one relapse [84]. In Benin, all-oral treatment was given for 8 weeks to 30 patients with BU with a slightly higher daily dose of clarithromycin (12 mg/kg once daily, combined with rifampicin 10 mg/kg);

they all healed without relapse [35]. The WHO initiated a randomized clinical trial in 2012, that was recently completed (clinicaltrials.gov identifier: NCT01659437) comparing all-oral, 8-week rifampicin plus clarithromycin (in sustained release, once daily, 15 mg/kg) with standard streptomycin/rifampicin treatment; interim analysis showed non-inferiority of the all-oral group compared to the streptomycin/rifampicin group, and in the bi-annual meeting on BU in Geneva in the Spring of 2017, it was decided that sufficient evidence had been provided to change the treatment recommendation to all-oral treatment, using rifampicin and clarithromycin.[2] The Australian guidelines had already recommended all-oral antimicrobial treatment [85] earlier.

4.1 Secondary Infection

Wounds caused by *M. ulcerans* often harbor a multitude of secondary pathogens. Isolates of *S. aureus, Aeromonas hydrophila, Pseudomonas aeruginosa* and *Klebsiella pneumoniae* with a high degree of resistance to commonly used antibiotics were found in BU lesions in Ghana and Nigeria [86, 87]. *S. aureus* isolated from *M. ulcerans* lesions were found to harbor a large array of virulence factors. In Ghana, an association between the presence of these organisms and delayed healing was observed. The question whether these secondary bacterial invading pathogens are merely bystanders *of M. ulcerans* infection or actual contributors has not been resolved [88]. Apart from the question whether these secondary invaders inhibit and delay wound healing or not, rational use of antimicrobials other than the combination recommended by WHO, should be discouraged, in order to prevent further antimicrobial selection pressure and the further emergence of resistance among these secondary organisms [89]. The skin microbiome was significantly changed, however it is not clear if and how this contributes to pathology [90]. The option to explore topical instead of systemic antimicrobial agents, e.g., nitric oxide needs perhaps further attention [91]. For more detailed information the reader is referred to chapter “Secondary Infection of Buruli Ulcer Lesions” of this book.

4.2 HIV Co-infection

HIV co-infection complicates diagnosis and treatment of BU [92]. In 2015, a consensus statement was published by WHO; HIV testing is recommended for all patients with BU. For more detailed information the reader is referred to chapter “Management of BU-HIV Co-infection” of this book. Combined antiretroviral treatment might be postponed 4 weeks after starting streptomycin/rifampicin treatment for BU, because of drug-drug interactions and adverse drug effects, and the importance to use the most effective bactericidal drug combination available for *M. ulcerans* infection.[3]

---

[2] http://www.who.int/neglected_diseases/events/WHO_BU_TAG_2017_report.pdf?ua=1.

[3] http://apps.who.int/iris/bitstream/10665/154241/1/WHO_HTM_NTD_IDM_2015.01_eng.pdf?ua=1.

5 Conclusions; Areas of Uncertainty; and Future Directions

There is no doubt that antimicrobial therapy—preferably, all-oral rifampicin-based treatment is essential for cure without relapse. Not all questions have been resolved, however. Despite the overwhelming evidence that the combination of clarithromycin and rifampicin works well, with acceptable levels of toxicity especially in the relatively young patient population in West Africa, the dosage of these drugs has been different across different studies and regions. Drug-drug interactions of rifampicin and clarithromycin are complex—rifampicin induces the metabolism of clarithromycin into its inactive metabolite 14-OH clarithromycin; and it induces its own hepatic elimination as well. Clarithromycin inhibits the elimination of rifampicin [93]. Increased drug exposure to rifampicin is much safer than previously thought [94].

An important next step could be to improve the dosage by extensive pharmacokinetic modeling; a fixed dose combination drug might have considerable advantage for compliance, logistics and in order to prevent monotherapy; this approach has been shown to be highly successful in tuberculosis, malaria and HIV infection.

Further, duration of therapy has not been individualized; based on the initial observations in the small study by Etuaful et al. [29], where no viable *M. ulcerans* bacilli were recovered from lesions in patients treated for at least 4 weeks, an arbitrary safety margin of 4 more weeks was chosen. The question whether larger lesions need longer treatment, or smaller lesions perhaps less has not been addressed in well-designed studies. In a follow-up study among 56 patients in Ghana, who defaulted before the end of planned treatment, 92% of patients with category I lesions (<5 cm) were healed with treatment duration of 32 days or more [95], suggesting that smaller lesions indeed might do well with much less than standard 8-weeks treatment.

Next, the treatment of BU extends well beyond administration of antimicrobials alone. During the active phase of the disease adequate wound care is imperative to ensure healing and to prevent disabling scarring. Non-adhesive, absorbent dressing materials have been shown to improve the wound microenvironment and improve time to healing [96, 97], and might also play a role to prevent painful dressing changes. While *M. ulcerans* is generally thought to be a pain-free disease as the toxin mycolactone causes hypoalgesia [98–100], later on, as lesions start to heal and the mycolactone is washed out from the lesions and the system, patients report considerable pain especially during wound dressing changes and during physiotherapy needed to prevent and treat contractures and disability [101]. For patient comfort and compliance, a sound wound-care program is essential, with adequate pain management [102].

During and after the completion of antimicrobial treatment, many patients still experience restrictions through either physical disability caused by wound contractures or joint involvement of lesions; or psycho-social participation restrictions caused by social stigmatization [103, 104]. Physical therapy and societal inclusion and de-stigmatization are therefore important components of BU control programs

[81, 105]. Community-based approaches with early case finding are essential to prevent the large disease burden in individuals and communities [106, 107].

Future research thus should focus on evaluating shorter treatment for limited lesions while supplementing standard treatment with clofazimine or fluoroquinolones to provide a treatment tailored to the lesion size and estimated bacterial burden of individual patients.

Such attempts have so far been hampered by the fact that no good measurable surrogate parameter of treatment success, such as a blood marker, is available. However recent advances in the use of mass-spectrometry to carefully measure mycolactone in patient samples could aid in this regard.

Furthermore, there is dire need for the integration of control programs for tropical skin conditions such as BU, yaws, leprosy, etc. on a public health level [108]. Health officials should identify opportunities for systematic integration for the control of tropical neglected diseases depending on the prevalent diseases and available resources in any given setting.

Finally, BU treatment should be accompanied by surgery where needed, as well as good wound care, analgesia, physiotherapy and advocacy to reduce stigma.

References

1. MacCallum P, Tolhurst JC (1948) A new mycobacterial infection in man. J Pathol Bacteriol 60(1):93–122
2. Daniel TM (2006) The history of tuberculosis. Respir Med 100(11):1862–1870
3. Daniels M, Hill AB (1952) Chemotherapy of pulmonary tuberculosis in young adults; an analysis of the combined results of three Medical Research Council trials. Br Med J 1(4769):1162–1168
4. Schatz A, Waksman SA (2016) Effect of streptomycin and other antibiotic substances upon mycobacterium tuberculosis and related organisms. Proc Soc Exp Biol Med 57(2):244–248
5. Meyers WM, Shelly WM, Connor DH (1974) Heat treatment of Mycobacterium ulcerans infections without surgical excision. Am J Trop Med Hyg 23(5):924–929
6. Stanford JL, Phillips I (1972) Rifampicin in experimental Mycobacterium ulcerans infection. J Med Microbiol 5(1):39–45
7. Revill WD et al (1973) A controlled trial of the treatment of Mycobacterium ulcerans infection with clofazimine. Lancet 2(7834):873–877
8. Fehr H, Egger M, Senn I (1994) Cotrimoxazol in the treatment of Mycobacterium ulcerans infection (Buruli ulcer) in West Africa. Trop Dr 24(2):61–63
9. Espey DK et al (2002) A pilot study of treatment of Buruli ulcer with rifampin and dapsone. Int J Infect Dis 6(1):60–65
10. Portaels F et al (1998) In vitro susceptibility of Mycobacterium ulcerans to clarithromycin. Antimicrob Agents Chemother 42(8):2070–2073
11. van der Werf TS et al (1999) Mycobacterium ulcerans infection. Lancet 354(9183):1013–1018
12. George KM et al (1999) Mycolactone: a polyketide toxin from Mycobacterium ulcerans required for virulence. Science 283(5403):854–857
13. Hockmeyer WT et al (1978) Further characterization of Mycobacterium ulcerans toxin. Infect Immun 21(1):124–128
14. Read JK et al (1974) Cytotoxic activity of Mycobacterium ulcerans. Infect Immun 9(6):1114–1122

15. Pimsler M et al (1988) Immunosuppressive properties of the soluble toxin from Mycobacterium ulcerans. J Infect Dis 157(3):577–580
16. Debacker M et al (2005) Buruli ulcer recurrence, Benin. Emerg Infect Dis 11(4):584–589
17. Kanga JM et al (2003) Recurrence after surgical treatment of Buruli ulcer in Cote d'Ivoire. Bull Soc Pathol Exot 96(5):406–409
18. Teelken MA et al (2003) Buruli ulcer: differences in treatment outcome between two centres in Ghana. Acta Trop 88(1):51–56
19. Phanzu DM et al (2006) Mycobacterium ulcerans disease (Buruli ulcer) in a rural hospital in Bas-Congo, Democratic Republic of Congo, 2002-2004. Am J Trop Med Hyg 75(2):311–314
20. Amofah G, Asamoah S, Afram-Gyening C (1998) Effectiveness of excision of pre-ulcerative Buruli lesions in field situations in a rural district in Ghana. Trop Dr 28(2):81–83
21. Thangaraj HS et al (2000) In vitro activity of ciprofloxacin, sparfloxacin, ofloxacin, amikacin and rifampicin against Ghanaian isolates of Mycobacterium ulcerans. J Antimicrob Chemother 45(2):231–233
22. Guarner J et al (2003) Histopathologic features of Mycobacterium ulcerans infection. Emerg Infect Dis 9(6):651–656
23. Nienhuis WA et al (2012) Paradoxical responses after start of antimicrobial treatment in Mycobacterium ulcerans infection. Clin Infect Dis 54(4):519–526
24. O'Brien DP et al (2009) Paradoxical immune-mediated reactions to Mycobacterium ulcerans during antibiotic treatment: a result of treatment success, not failure. Med J Aust 191(10):564–566
25. Ruf MT et al (2011) Secondary Buruli ulcer skin lesions emerging several months after completion of chemotherapy: paradoxical reaction or evidence for immune protection? PLoS Negl Trop Dis 5(8):e1252
26. Dega H et al (2000) Activities of several antimicrobials against Mycobacterium ulcerans infection in mice. Antimicrob Agents Chemother 44(9):2367–2372
27. Dega H et al (2002) Bactericidal activity of rifampin-amikacin against Mycobacterium ulcerans in mice. Antimicrob Agents Chemother 46(10):3193–3196
28. Marsollier L, Prévot G et al (2003) Susceptibility of Mycobacterium ulcerans to a combination of amikacin/rifampicin. Int J Antimicrob Agents 22(6):562–566
29. Etuaful S et al (2005) Efficacy of the combination rifampin-streptomycin in preventing growth of Mycobacterium ulcerans in early lesions of Buruli ulcer in humans. Antimicrob Agents Chemother 49(8):3182–3186
30. Johnson PDR et al (2005) Buruli ulcer (M. ulcerans infection): new insights, new hope for disease control. PLoS Med 2(4):e108
31. Radford AJ (2009) The surgical management of lesions of ulcerans infections due to Mycobacterium ulcerans, revisited. Trans R Soc Trop Med Hyg 103(10):981–984
32. van der Werf TS et al (2005) Mycobacterium ulcerans disease. Bull World Health Organ 83(10):785–791
33. Wansbrough-Jones M, Phillips R (2006) Buruli ulcer: emerging from obscurity. Lancet 367(9525):1849–1858
34. Doig KD et al (2012) On the origin of Mycobacterium ulcerans, the causative agent of Buruli ulcer. BMC Genomics 13(1):258
35. Chauty A et al (2011) Oral treatment for Mycobacterium ulcerans infection: results from a pilot study in Benin. Clin Infect Dis 52(1):94–96
36. Nienhuis WA et al (2010) Antimicrobial treatment for early, limited Mycobacterium ulcerans infection: a randomised controlled trial. Lancet 375(9715):664–672
37. Phillips RO et al (2014) Clinical and bacteriological efficacy of rifampin-streptomycin combination for two weeks followed by rifampin and clarithromycin for six weeks for treatment of Mycobacterium ulcerans disease. Antimicrob Agents Chemother 58(2):1161–1166
38. Almeida D et al (2011) Activities of rifampin, Rifapentine and clarithromycin alone and in combination against Mycobacterium ulcerans disease in mice. PLoS Negl Trop Dis 5(1): e933

39. Dhople AM (2001) In vitro activity of KRM-1648, either singly or in combination with ofloxacin, against Mycobacterium ulcerans. Int J Antimicrob Agents 17(1):57–61
40. Owusu E et al (2017) In vitro susceptibility of Mycobacterium ulcerans isolates to selected antimicrobials. Can J Inf Dis Med Microbiol 2017(4):5180984–5180986
41. Chauffour A et al (2016) Sterilizing activity of fully oral intermittent regimens against Mycobacterium ulcerans infection in Mice. PLoS Negl Trop Dis 10(10):e0005066
42. Almeida DV et al (2013) Bactericidal activity does not predict sterilizing activity: the case of rifapentine in the murine model of Mycobacterium ulcerans disease. PLoS Negl Trop Dis 7(2):e2085
43. Beissner M et al (2010) A genotypic approach for detection, identification, and characterization of drug resistance in Mycobacterium ulcerans in clinical samples and isolates from Ghana. Am J Trop Med Hyg 83(5):1059–1065
44. Jansson M et al (2014) Comparison of two assays for molecular determination of rifampin resistance in clinical samples from patients with Buruli ulcer disease. J Clin Microbiol 52(4):1246–1249
45. Marsollier L, Honoré N et al (2003) Isolation of three Mycobacterium ulcerans strains resistant to rifampin after experimental chemotherapy of mice. Antimicrob Agents Chemother 47(4):1228–1232
46. Feldman WH, Karlson AG (1957) Mycobacterium ulcerans infections; response to chemotherapy in mice. Am Rev Tuberc 75(2):266–279
47. Pattyn SR, Royackers J (1965) Treatment of experimental infection with mycobacterium leprae in mice. Ann Soc Belg Med Trop 45:27–30
48. Lefrançois S et al (2007) Curing Mycobacterium ulcerans infection in mice with a combination of rifampin-streptomycin or rifampin-amikacin. Antimicrob Agents Chemother 51(2):645–650
49. Ji B et al (2006) In vitro and in vivo activities of rifampin, streptomycin, amikacin, moxifloxacin, R207910, linezolid, and PA-824 against Mycobacterium ulcerans. Antimicrob Agents Chemother 50(6):1921–1926
50. Klis S, Stienstra Y, Phillips RO, Abass KM, Tuah W, van der Werf TS (2014) Long term streptomycin toxicity in the treatment of Buruli ulcer: follow-up of participants in the BURULICO drug trial. PLoS Negl Trop Dis 8(3):e2739
51. Dubuisson T et al (2010) In vitro antimicrobial activities of capuramycin analogues against non-tuberculous mycobacteria. J Antimicrob Chemother 65(12):2590–2597
52. Rastogi N et al (2000) In vitro activities of the ketolides telithromycin (HMR 3647) and HMR 3004 compared to those of clarithromycin against slowly growing mycobacteria at pHs 6.8 and 7.4. Antimicrob Agents Chemother 44(10):2848–2852
53. Bentoucha A et al (2001) Activities of new macrolides and fluoroquinolones against Mycobacterium ulcerans infection in mice. Antimicrob Agents Chemother 45(11):3109–3112
54. O'Brien DP et al (2012) Successful outcomes with oral fluoroquinolones combined with rifampicin in the treatment of Mycobacterium ulcerans: an observational cohort study. PLoS Negl Trop Dis 6(1):e1473
55. Principi N, Esposito S (2015) Appropriate use of fluoroquinolones in children. Int J Antimicrob Agents 45(4):341–346
56. Treggiari MM et al (2011) Comparative efficacy and safety of 4 randomized regimens to treat early Pseudomonas aeruginosa infection in children with cystic fibrosis. Arch Pediatr Adolesc Med 165(9):847–856
57. Dhople AM, Namba K (2002) In vitro activity of sitafloxacin (DU-6859a) alone, or in combination with rifampicin, against Mycobacterium ulcerans. J Antimicrob Chemother 50(5):727–729
58. Scherr N, Pluschke G, Panda M (2016) Comparative study of activities of a diverse set of antimycobacterial agents against mycobacterium tuberculosis and Mycobacterium ulcerans. Antimicrob Agents Chemother 60(5):3132–3137
59. Namboodiri SS et al (2011) Quinolone resistance in Escherichia coli from Accra, Ghana. BMC Microbiol 11(1):44

60. Grosset JH et al (2013) Assessment of clofazimine activity in a second-line regimen for tuberculosis in mice. Am J Respir Crit Care Med 188(5):608–612
61. Lechartier B, Cole ST (2015) Mode of action of clofazimine and combination therapy with benzothiazinones against Mycobacterium tuberculosis. Antimicrob Agents Chemother 59(8):4457–4463
62. Arbiser JL, Moschella SL (1995) Clofazimine: a review of its medical uses and mechanisms of action. J Am Acad Dermatol 32(2 Pt 1):241–247
63. Converse PJ et al (2015) Efficacy of rifampin plus clofazimine in a Murine model of Mycobacterium ulcerans disease. PLoS Negl Trop Dis 9(6):e0003823
64. Hwang TJ et al (2014) Safety and availability of clofazimine in the treatment of multidrug and extensively drug-resistant tuberculosis: analysis of published guidance and meta-analysis of cohort studies. BMJ Open 4(1):e004143
65. Bolhuis MS et al (2015) Linezolid tolerability in multidrug-resistant tuberculosis: a retrospective study. Eur Respir J 46(4):1205–1207
66. Lim LE et al (2013) Anthelmintic avermectins kill Mycobacterium tuberculosis, including multidrug-resistant clinical strains. Antimicrob Agents Chemother 57(2):1040–1046
67. Omansen TF et al (2015) In-vitro activity of avermectins against Mycobacterium ulcerans. PLoS Negl Trop Dis 9(3):e0003549
68. Scherr N et al (2015) Selamectin is the avermectin with the best potential for Buruli ulcer treatment. PLoS Negl Trop Dis 9(8):e0003996
69. Dhople AM (2001) Antimicrobial activities of dihydrofolate reductase inhibitors, used singly or in combination with dapsone, against Mycobacterium ulcerans. J Antimicrob Chemother 47(1):93–96
70. Cappoen D et al (2014) Biological evaluation of diazene derivatives as anti-tubercular compounds. Eur J Med Chem 74:85–94
71. Beissner M et al (2012) Spontaneous clearance of a secondary Buruli ulcer lesion emerging ten months after completion of chemotherapy--a case report from Togo. PLoS Negl Trop Dis 6(7):e1747
72. Gordon CL et al (2011) Spontaneous clearance of Mycobacterium ulcerans in a case of Buruli ulcer. PLoS Negl Trop Dis 5(10):e1290
73. Chauty A et al (2007) Promising clinical efficacy of streptomycin-rifampin combination for treatment of buruli ulcer (Mycobacterium ulcerans disease). Antimicrob Agents Chemother 51(11):4029–4035
74. O'Brien DP et al (2007) Outcomes for Mycobacterium ulcerans infection with combined surgery and antibiotic therapy: findings from a South-Eastern Australian case series. Med J Aust 186(2):58–61
75. Johnson PDR et al (2007) Consensus recommendations for the diagnosis, treatment and control of Mycobacterium ulcerans infection (Bairnsdale or Buruli ulcer) in Victoria, Australia. Med J Aust 186:64–68
76. Schunk M et al (2009) Outcome of patients with buruli ulcer after surgical treatment with or without antimycobacterial treatment in Ghana. Am J Trop Med Hyg 81(1):75–81
77. Sarfo FS et al (2010) Clinical efficacy of combination of rifampin and streptomycin for treatment of Mycobacterium ulcerans disease. Antimicrob Agents Chemother 54(9):3678–3685
78. Iddrisah FN et al (2016) Outcome of Streptomycin-Rifampicin treatment of Buruli Ulcer in two Ghanaian districts. Pan Afr Med J 25(Suppl 1):13
79. Johnson PDR (2010) Should antibiotics be given for Buruli ulcer? Lancet 375(9715): 618–619
80. O'Brien DP et al (2017) Antibiotic complications during the treatment of Mycobacterium ulcerans disease in Australian patients. Intern Med J 47(9):1011–1019
81. Klis S, Ranchor A et al (2014) Good quality of life in former Buruli ulcer patients with small lesions: long-term follow-up of the BURULICO trial. PLoS Negl Trop Dis 8(7):e2964

82. Gordon CL et al (2010) All-oral antibiotic treatment for Buruli ulcer: a report of four patients. PLoS Negl Trop Dis 4(11):e770
83. Friedman ND et al (2013) Mycobacterium ulcerans disease: experience with primary oral medical therapy in an Australian cohort. PLoS Negl Trop Dis 7(7):e2315
84. Friedman ND et al (2016) Increasing experience with primary oral medical therapy for Mycobacterium ulcerans disease in an Australian cohort. Antimicrob Agents Chemother 60(5):2692–2695
85. O'Brien DP, Jenkin G et al (2014) Treatment and prevention of Mycobacterium ulcerans infection (Buruli ulcer) in Australia: guideline update. Med J Aust 200(5):267–270
86. Amissah NA et al (2015) Genetic diversity of Staphylococcus aureus in Buruli ulcer. PLoS Negl Trop Dis 9(2):e0003421
87. Yeboah-Manu D et al (2013) Secondary bacterial infections of buruli ulcer lesions before and after chemotherapy with streptomycin and rifampicin. PLoS Negl Trop Dis 7(5):e2191
88. Amissah NA et al (2017) Virulence potential of Staphylococcus aureus isolates from Buruli ulcer patients. Int J Med Microbiol 307(4-5):223–232
89. Barogui YT et al (2013) Towards rational use of antibiotics for suspected secondary infections in Buruli ulcer patients. PLoS Negl Trop Dis 7(1):e2010
90. Van Leuvenhaege C et al (2017) Bacterial diversity in Buruli ulcer skin lesions: Challenges in the clinical microbiome analysis of a skin disease. PLoS ONE 12(7):e0181994
91. Phillips R et al (2004) Pilot randomized double-blind trial of treatment of Mycobacterium ulcerans disease (Buruli ulcer) with topical nitrogen oxides. Antimicrob Agents Chemother 48(8):2866–2870
92. O'Brien DP, Ford N et al (2014) Management of BU-HIV co-infection. Tropical Med Int Health 19(9):1040–1047
93. Alffenaar JWC et al (2010) Pharmacokinetics of rifampin and clarithromycin in patients treated for Mycobacterium ulcerans infection. Antimicrob Agents Chemother 54(9):3878–3883
94. Boeree MJ et al (2015) A dose-ranging trial to optimize the dose of rifampin in the treatment of tuberculosis. Am J Respir Crit Care Med 191(9):1058–1065
95. Klis S et al (2016) Clinical outcomes of Ghanaian Buruli ulcer patients who defaulted from antimicrobial therapy. Tropical Med Int Health 21(9):1191–1196
96. Velding K et al (2016) The application of modern dressings to Buruli ulcers: results from a pilot implementation project in Ghana. Am J Trop Med Hyg 95(1):60–62
97. Velding K et al (2014) Wound care in Buruli ulcer disease in Ghana and Benin. Am J Trop Med Hyg 91(2):313–318
98. Anand U et al (2016) Mycolactone-mediated neurite degeneration and functional effects in cultured human and rat DRG neurons: mechanisms underlying hypoalgesia in Buruli ulcer. Mol Pain 12:1744806916654144
99. En J et al (2017) Mycolactone cytotoxicity in Schwann cells could explain nerve damage in Buruli ulcer. PLoS Negl Trop Dis 11(8):e0005834
100. En J et al (2008) Mycolactone is responsible for the painlessness of Mycobacterium ulcerans infection (Buruli ulcer) in a murine study. Infect Immun 76(5):2002–2007
101. de Zeeuw J et al (2015) Assessment and treatment of pain during treatment of Buruli ulcer. PLoS Negl Trop Dis 9(9):e0004076
102. Alferink M et al (2015) Pain associated with wound care treatment among Buruli ulcer patients from Ghana and Benin. S. C. Hausmann-Muela, ed. PLoS ONE 10(6):e0119926
103. Renzaho AMN et al (2007) Community-based study on knowledge, attitude and practice on the mode of transmission, prevention and treatment of the Buruli ulcer in Ga West District, Ghana. Tropical Med Int Health 12(3):445–458
104. Stienstra Y et al (2005) Factors associated with functional limitations and subsequent employment or schooling in Buruli ulcer patients. Tropical Med Int Health 10(12): 1251–1257

105. de Zeeuw J et al (2014) Psychometric properties of the participation scale among former Buruli ulcer patients in Ghana and Benin. PLoS Negl Trop Dis 8(11):e3254
106. Abass KM et al (2015) Buruli ulcer control in a highly endemic district in Ghana: role of community-based surveillance volunteers. Am J Trop Med Hyg 92(1):115–117
107. Barogui YT et al (2014) Contribution of the community health volunteers in the control of Buruli ulcer in Bénin. PLoS Negl Trop Dis 8(10):e3200
108. Mitjà O et al (2017) Integrated control and management of neglected tropical skin diseases. PLoS Negl Trop Dis 11(1):e0005136

Thermotherapy of Buruli Ulcer

Thomas Junghanss

For decades surgical excision has been the standard therapy for Buruli ulcer (BU) with several drawbacks, however [1]. Surgery is traumatizing with extensive soft tissue excisions required to prevent relapse. Long hospitalisation is expensive for both health services and patient families. Since 2004 an antibiotic regimen consisting of rifampicin and streptomycin for 8 weeks is recommended by WHO [2, 3]. Its efficacy has been demonstrated in various studies including a randomized controlled trial [4]. The parenteral application of streptomycin over 8 weeks is a substantial disadvantage in settings with limited resources and particularly in children who are mostly affected by BU in Africa. Ototoxicity has additionally been identified as a critical and prohibitive side effect [5]. In 2017 the WHO Technical Advisory Group on BU decided that the WHO recommendation for treatment should be changed to rifampicin and oral clarithromycin pending the publication of the full results of a recently completed clinical trial. On a national level, this combination has already been introduced in various countries.

The temperature sensitivity of *Mycobacterium ulcerans* has long been recognized [6–9]. *M. ulcerans* differs from most other pathogenic mycobacteria in that it grows best at 30–33 °C and not above 37 °C [8]. Meyers et al. treated eight patients from Zaire maintaining a temperature of approximately 40 °C in the ulcerated area for a mean duration of 68 days [8]. There was no evidence of local recurrence during follow-up periods of up to 22 months. Based on this impressive success rate, WHO guidelines listed the application of heat as a treatment option for BU [10]. However, the heat application devices initially employed were impractical in most endemic countries.

The phase change material (PCM) sodium acetate trihydrate, which is widely used in commercial pocket heat pads offers an ideal technical solution for

T. Junghanss (✉)
Section Clinical Tropical Medicine, Department of Infectious Diseases,
Heidelberg University Hospital, Heidelberg, Germany
e-mail: thomas.junghanss@urz.uni-heidelberg.de

G. Pluschke, K. Röltgen (eds.), *Buruli Ulcer*,
https://doi.org/10.1007/978-3-030-11114-4_12

thermotherapy of BU. The unique feature of PCM is its thermal energy storing capacity combined with an almost constant temperature during the liquid-solid phase transition. Sodium acetate trihydrate appears ideal with its supercooling behaviour, since, once completely molten, the material will stay liquid even when the temperature falls far below its melting point. It can thus be stored at room temperature without energy loss. With a starter, e.g. a piece of copper wire protected by a rubber tube placed within the PCM filled bag, the crystallization process is initiated when required. Once initiated, heat at constant temperature is emitted for around 6 h. Patients can be treated overnight. Further, the melting temperature of 58 °C allows sterile gauze to be placed between the wound and the PCM bag to protect the wound and to still maintain temperatures of above 39 °C at the skin surface without the risk of burning (Fig. 1).

In summary, sodium acetate trihydrate heat packs:

- Are free from significant side effects
- Are easy to apply, rechargeable in boiling water and can be used many times
- Are non-toxic and non-hazardous to the environment
- Are sterilized between rounds of application during the recharging process
- Exert gentle pressure and thus reduce peri-lesional edema
- Protect wounds from trauma
- Urge patients and medical staff to pay attention to the wound when reapplied daily

The heat application device was tested in healthy volunteers and a mathematical model was developed and validated to predict its thermal behaviour. The thermal model allowed the prediction of skin surface temperatures and an optimization of the amount of PCM with respect to discharge time [11].

In a next step commercially available heat packs were tested in a prospective observational single centre proof-of-principle trial in Ayos/Cameroon. "Six laboratory

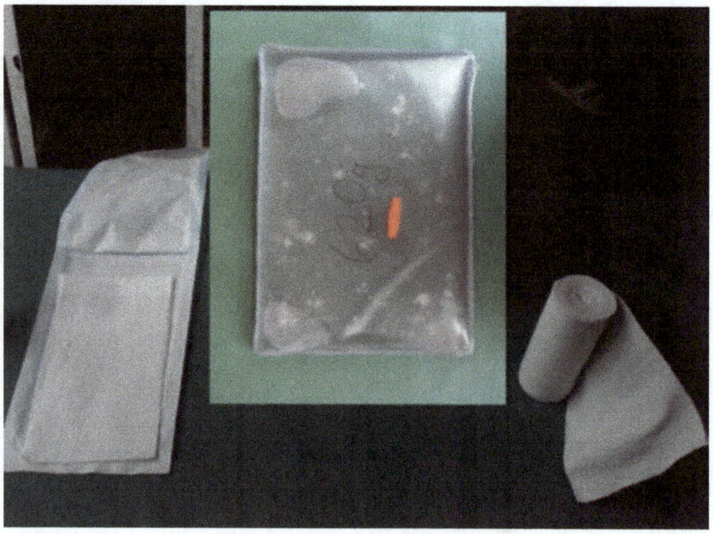

Fig. 1 Heat packs, sterile gauze and elastic bandage are the ingredients needed for thermotherapy

reconfirmed patients with ulcerative BU lesions received 28–31 (ulcers < 2 cm) or 50–55 (ulcers > 2 cm) days of thermotherapy with the PCM sodium acetate trihydrate as heat application system. All patients enrolled in the trial completed the heat treatment. Being completely mobile during the well-tolerated heat application, acceptability of the PCM bandages was very high. In patients with smaller ulcers, wounds healed completely without further intervention. Patients with large defects had skin grafting after successful heat treatment. Heat treatment was not associated with marked increases in local inflammation or the development of ectopic lymphoid tissue. One and a half years after completion of treatment, all patients were relapse-free" [12].

To confirm the findings of the pilot study a phase II open label single centre non-comparative clinical trial (ISRCTN 72102977) under GCP standards was carried out in Cameroon.

Laboratory confirmed BU patients received up to 8 weeks of heat treatment. The efficacy was assessed based on the endpoints 'absence of clinical BU specific features' or 'wound closure' within 6 months ("primary cure"), and 'absence of clinical recurrence within 24 month' ("definite cure"). Of 53 enrolled patients, 51 (96%) had ulcerative disease. Sixty two percent were classified as World Health Organization category II, 19% each as category I and III. The average lesion size was 45 cm^2. Within 6 months after completion of heat treatment, 92.4% (49 of 53, 95% confidence interval (CI), 81.8–98.0%) achieved cure of their primary lesion. At 24 months follow-up 83.7% (41 of 49, 95% CI, 70.3–92.7%) of patients with primary cure remained free of recurrence (see Fig. 2). Heat treatment was well tolerated and the only adverse effects were occasional mild local skin reactions [13].

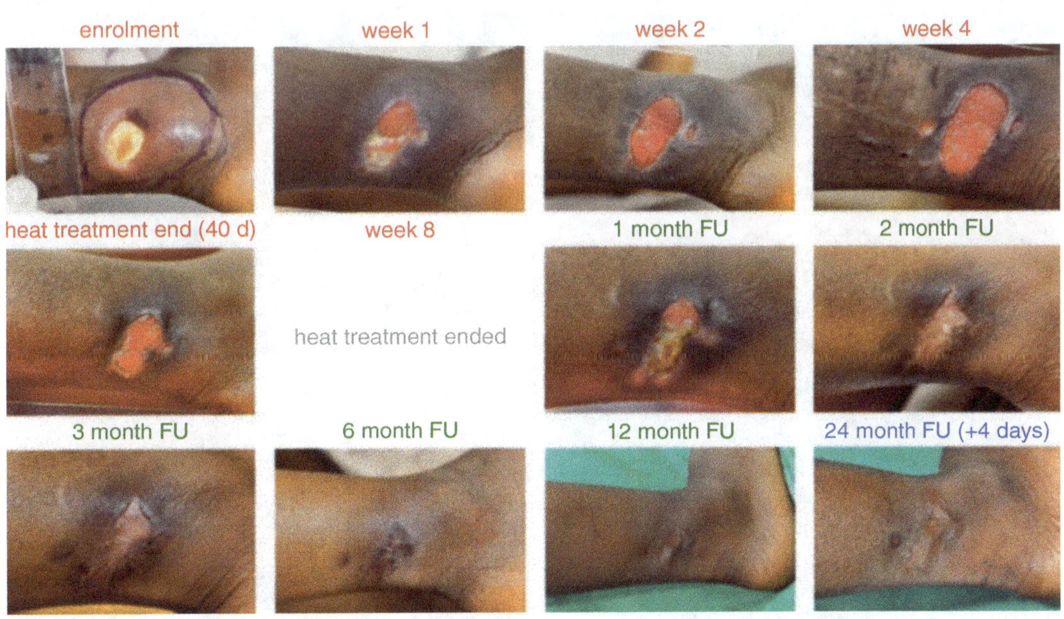

Fig. 2 Clinical evolution of patient 14 (laboratory confirmed ulcer of the right lateral ankle) exemplary for all patients. Images of all other patients are available as Supplemental Figures 1–65 of [13]. Pictures were taken within a range of ±2 days (heat treatment period), ±15 days (follow-up (FU) months 1–3), ±30 days (FU month 6) and ±60 days (FU months 12 + 24 months) from the designated time points (Source: [13])

Interestingly, there is evidence from other studies and our own observations that after controlling the bulk of the mycobacterial burden remaining viable *M. ulcerans* may be eliminated by the immune system without additional specific treatment. Equally, confirmed BU nodules distant to heat treated primary lesions healed (see Fig. 3). Protective immune responses may be triggered by thermotherapy and chemotherapy [4, 14, 15] but not surgery, where the mycobacterial antigens are largely removed [13].

Importantly, the skin lesions of ten patients, for which laboratory testing for *M. ulcerans* was negative, also healed under thermotherapy, indicating that heat might have a positive effect on the healing of wounds in general. This is encouraging for regions with limited access to laboratory confirmation and where quality assurance, in particular of PCR in reference labs, remains a problem [13]. With respect to reliability of PCR-based laboratory diagnosis it has to be noted that in a multicentre

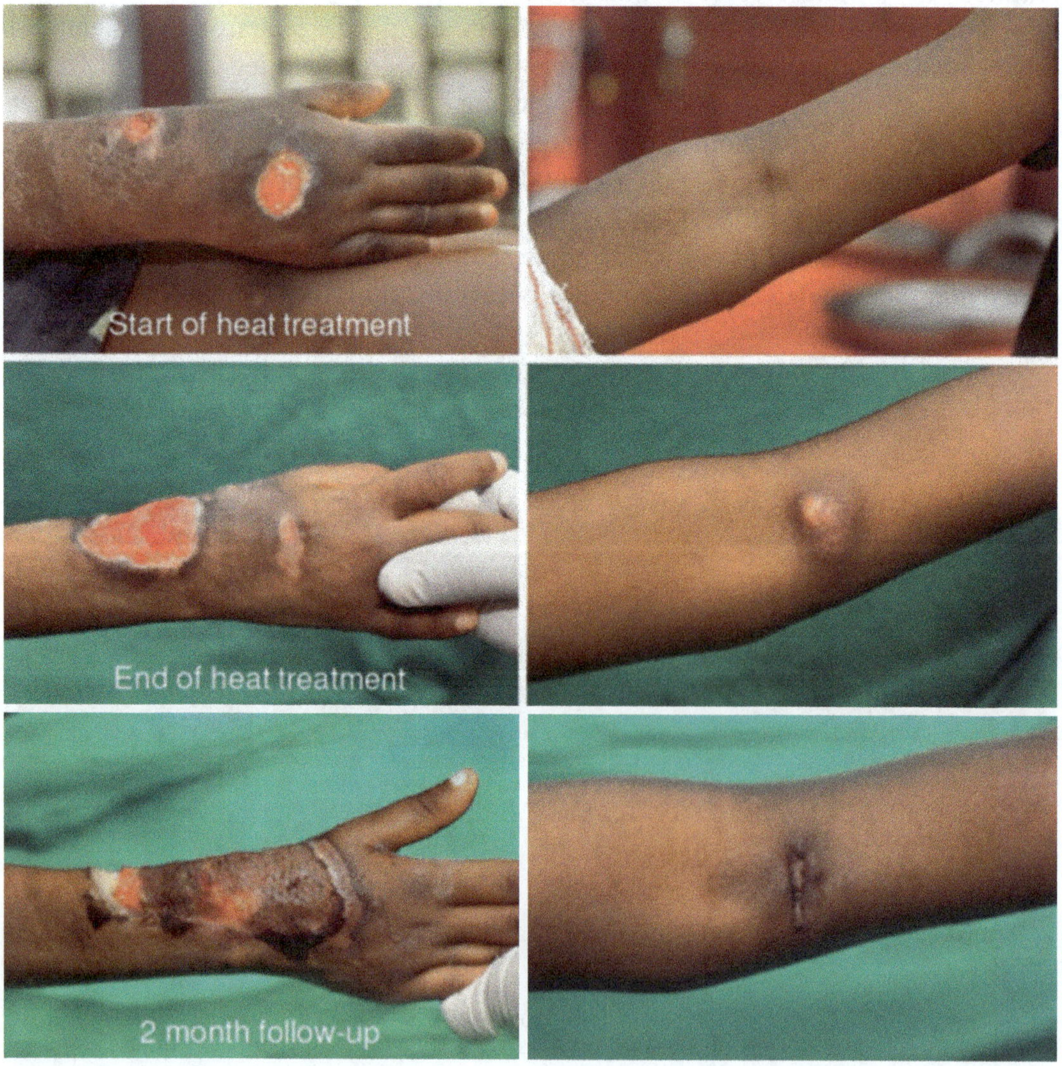

Fig. 3 Distant healing in a patient with a BU lesion at the right wrist and a laboratory confirmed BU nodule at the left elbow. Both of the lesions healed after heat was applied to the right wrist only (Source: [13])

external quality assessment for PCR detection of *M. ulcerans* only around one third of the participating laboratories had more than 90% concordant results [16].

Both, the proof-of-principle and the phase II trial showed that thermotherapy is well tolerated and effective. Daytime activities including schooling are not interrupted. In addition to its specific effect, heat treatment appears to promote wound healing by increasing blood circulation, reducing edema through gentle compression and protecting the wound. Renewal of heat packs is inevitably connected to inspection and care for the wound.

The thermotherapy-specific cost of heat packs and recharging for 10 min in boiling water are small. Nursing time mainly goes into wound management, which is not different from chemotherapy patients. Positioning of heat packs is straightforward and nurses are confident in heat pack application after few demonstrations.

In communities with limited resources treatment modalities which can be applied in early undifferentiated stages of diseases without causing harm and with the benefit of preventing advancement into severe complicated disease make a difference to peoples' lives. All wounds "count" and the patient's priority "I want to have my skin closed" (independent of the cause) is respected and promoted. This follows the broad understanding of universal health coverage [17]. The population impact lies in the rapid closure of a substantial proportion of wounds at the community level [18]. This avoids life threatening sequelae and disability and prevents the development of chronic wounds and the need for advanced treatment at the secondary or tertiary health services level. It saves costs both for the patient and the health care system. Patients with wounds prone to secondary bacterial infection benefit from treatment in the periphery because they are not exposed to nosocomial infections which are more frequent at higher levels of health services. This, again, saves people from sufferance and costs.

In summary, based on the results of the two studies with PCM heat packs, "thermotherapy can be considered an alternative to chemotherapy as primary treatment for BU for several reasons: it is highly effective, is easy to apply, cheap, well tolerated, free of relevant adverse effects, has nonspecific positive effects on wound healing and does not compromise wound healing in non-BU lesions in cases of misclassification. This is an undisputable advantage in settings where treatment decisions need to rely primarily on clinical diagnosis. Changing of heat packs urges health staff and patients to take notice of the wound and thus increases the probability of regular wound care. Heat therapy has potential as home remedy for BU lesions suspected to be BU, ideally, combined with general wound management" [13].

References

1. Junghanss T, Johnson RC, Pluschke G (2014) *Mycobacterium ulcerans* disease. In: Farrar J et al (eds) Manson's tropical diaseses. Elsevier Saunders, Philadelphia, pp 519–531
2. WHO, Asiedu K (2004) Provisional guidance on the role of specific antiobiotics in the management of *Mycobacterium ulcerans* disease (Buruli ulcer), vol 39. WHO/Department of Communicable Disease Prevention, Control and Eradication, Geneva
3. WHO, Asiedu K (2012) Treatment of *Mycobacterium ulcerans* disease (Buruli Ulcer). WHO/Department of Control of Neglected Tropical Diseases, Geneva, p 73

4. Nienhuis WA et al (2010) Antimicrobial treatment for early, limited *Mycobacterium ulcerans* infection: a randomised controlled trial. Lancet 375(9715):664–672
5. Klis S et al (2014) Long term streptomycin toxicity in the treatment of Buruli ulcer: follow-up of participants in the BURULICO drug trial. PLoS Negl Trop Dis 8(3):e2739
6. Glynn PJ (1972) The use of surgery and local temperature elevation in mycobacterium ulcerans infection. Aust N Z J Surg 41(4):312–317
7. Meleney FL, Johnson BA (1950) Supplementary report on the case of chronic ulceration of the foot due to a new pathogenic mycobacterium (MacCallum). Ann Soc Belg Med Trop (1920) 30(6):1499–1503
8. Meyers WM, Shelly WM, Connor DH (1974) Heat treatment of Mycobacterium ulcerans infections without surgical excision. Am J Trop Med Hyg 23(5):924–929
9. Reid IS (1967) *Mycobacterium ulcerans* infection: a report of 13 cases at the Port Moresby General Hospital, Papua. Med J Aust 1(9):427–431
10. WHO, Asiedu K, Scherpbier R, Raviglione M (2000) Buruli ulcer: *Mycobacterium ulcerans* infection. WHO/Department of Disease Control, Prevention and Eradication, Geneva, p 160
11. Braxmeier S et al (2009) Phase change material for thermotherapy of Buruli ulcer: modelling as an aid to implementation. J Med Eng Technol 33(7):559–566
12. Junghanss T et al (2009) Phase change material for thermotherapy of Buruli ulcer: a prospective observational single centre proof-of-principle trial. PLoS Negl Trop Dis 3(2):e380
13. Vogel M et al (2016) Local heat application for the treatment of Buruli ulcer: results of a phase II open label single center non comparative clinical trial. Clin Infect Dis 62(3):342–350
14. Beissner M et al (2012) Spontaneous clearance of a secondary Buruli ulcer lesion emerging ten months after completion of chemotherapy--a case report from Togo. PLoS Negl Trop Dis 6(7):e1747
15. Ruf MT et al (2011) Secondary Buruli ulcer skin lesions emerging several months after completion of chemotherapy: paradoxical reaction or evidence for immune protection? PLoS Negl Trop Dis 5(8):e1252
16. Eddyani M et al (2014) Multicenter external quality assessment program for PCR detection of *Mycobacterium ulcerans* in clinical and environmental specimens. PLoS One 9(2):e89407
17. Sakolsatayadorn P, Chan M (2017) Breaking down the barriers to universal health coverage. Bull World Health Organ 95(2):86
18. Addison NO et al (2017) Assessing and managing wounds of Buruli ulcer patients at the primary and secondary health care levels in Ghana. PLoS Negl Trop Dis 11(2):e0005331

Secondary Infection of Buruli Ulcer Lesions

Grace Semabia Kpeli and Dorothy Yeboah-Manu

1 Background

Proper wound care is increasingly becoming a very crucial component of the management of Buruli ulcer (BU). The cytopathic activity of the main virulent factor of the causative pathogen, *M. ulcerans,* leads to the formation of extensive necrotic ulcerative lesions, which are a good medium for the growth of other bacteria. Data from a number of clinical studies showed that in up to 80% of cases presenting with limited (category I and II) lesions, healing occurs within 6 month after onset of antibiotic treatment without the need for surgery [1–7]. However, wound healing is delayed in a proportion of affected patients, in particular in those reporting with large lesions [4, 8, 9]. In particular, after completion of antimycobacterial therapy good monitoring and wound care is important to avoid massive secondary bacterial infections [10], potentially affecting the healing potential of wounds and increasing the risk of more severe pathology and sepsis. The WHO recommends that secondary infection in BU should be suspected when a wound develops cellulitis or becomes painful [11]. Secondary infection in BU disease is not well characterized and recognized because it is assumed to be infrequent [11]. Thus, its occurrence has been documented only by few studies [10, 12–16]. Previously, it was speculated that mycolactone secreted by *M. ulcerans* during active disease may sterilize BU wounds and prevent secondary infection by other bacteria, since a number of macrolides have broad spectrum activity against many bacterial species including streptococci, pneumococci, staphylococci, enterococci, mycoplasma, mycobacteria, rickettsia, and chlamydia [17]. However, several studies [10, 14] have proven

G. S. Kpeli
University of Health and Allied Sciences, Ho, Ghana

D. Yeboah-Manu (✉)
Noguchi Memorial Institute for Medical Research, Accra, Ghana

University of Ghana, Accra, Ghana
e-mail: dyeboah-manu@noguchi.ug.edu.gh

G. Pluschke, K. Röltgen (eds.), *Buruli Ulcer*,
https://doi.org/10.1007/978-3-030-11114-4_13

that mycolactone does not prevent secondary bacterial infection of BU lesions. In these studies, the microbial flora of BU wounds was found to be very diverse and a broad range of bacteria species colonizing the lesions were identified. Furthermore, research by Scherr [18] using synthetic mycolactones also demonstrated the growth of the bacterial species *Streptococcus pneumoniae, Neisseria meningitidis* and *Escherichia coli,* as well as the yeast *Saccharomyces cerevisae* and the amoeba *Dictyostelium discoideum* in the presence of mycolactone. Thus, secondary infection in BU may be more common than formerly thought.

2 Bacterial Species Associated with Secondary Infections

2.1 Species Diversity

Extensive work on the microbial flora of BU wounds has been done by research groups in Ghana [10], Benin [14] and Nigeria [13]. Yeboah-Manu and Barogui studied the microbial flora of BU patients before, during and post antibiotic treatment while the study by Anyim included pre-treatment patients only. In the study by Yeboah-Manu et al., many different bacterial species were isolated, including *Staphylococcus aureus, Pseudomonas aeruginosa, Proteus mirabilis,* Coagulase negative Staphylococcus, *Chryseomonas luteola, Enterobacter cloacae, Klebsiella pneumonia, Escherichia coli, Streptococcus dysgalactia, Providencia stuartii, Staphylococcus haemolyticus, Morganella morganii, Streptococcus agalactia, Staphylococcus warneri, Proteus vulgaris* and other Gram negative bacteria. *S. aureus*, *P. aeruginosa* and *P. mirabilis* were frequently isolated pre-treatment. During treatment, *P. aeruginosa* and *P. mirabilis* were the dominating isolates, while *P. aeruginosa*, *P. mirabilis* and *S. aureus* were the most frequently isolated bacteria post antibiotic treatment. The study by Barogui isolated Group A streptococci, Group B or C streptococci, *S. aureus, S. epidermidis, Staphylococcus* sp., *P. aeruginosa* and Enterobacteriaceae with *S. aureus* and *P. aeruginosa* dominating pre and post treatment and *P. aeruginosa* being the most frequently isolated bacteria during treatment. The study by Anyim et al. isolated *S. aureus*, *A. hydrophila P. aeruginosa*, *K. pneumoniae*, *Enterobacter cloacae*, *Pseudomonas pseudomallei* and *Burkholderia cepacia*.

The species diversity isolated from the wounds represent the spectrum of microbial isolates commonly isolated from other wound types such as burn and infected diabetic foot ulcers. However, the main bacteria commonly cited as responsible for wound infection and healing delay are *S. aureus, P. aeruginosa* and β-hemolytic streptococci [19–21].

S. aureus and *P. aeruginosa* dominating among the isolates detected in the BU lesions, have been found to be associated with infection and healing delay in both chronic and acute wounds and are frequently implicated in healthcare-associated infections. The virulence of these bacterial species is based on their ability to produce a number of destructive enzymes and toxins. Through intrinsic and acquired mechanisms, they exhibit increased resistance to many antimicrobials. They also have the ability to form biofilms which increase their persistence and antibiotic tolerance [21–23]. Interestingly, in the two studies which assessed the species diversity at different time points [10, 14], *S. aureus* was not isolated from patient lesions during antibiotic treatment. Further research will be needed to explain the absence of this bacterial species during treatment and its probable "re-emergence" post treatment.

3 Bacterial Burden

The role of microorganisms in wound healing and infection is diverse and related not only to the type of microbial species colonizing a wound but also to the number of microorganisms or the wound's bioburden. Localized wound infection is a significant cause of impaired healing and wound chronicity [24]. The pathogenic effects of bacteria may be increased through the formation of biofilms and the release of toxic products [22]. Endotoxins released by Gram-negative bacteria in wounds lead to elevated levels of proinflammatory cytokines (IL-1 and tumour necrosis factor). In addition, factors including the release of free radicals, degradation of growth factors, production of metabolic products, consumption of local oxygen and interference with collagen formation may also result in a non-conducive wound environment. The effects of these mechanisms increase with high bacterial loads leading to impaired wound healing [20, 21].

Many studies have confirmed that, a wound's bioburden is an important predictor of wound infection and wound healing. In decubitus ulcers, wound healing was found to progress only when the microbial load of wound fluid was below 10^6 colony forming units (cfu) per ml of wound exudates (cfu/ml) or per gram of tissue (cfu/g) [25] with healing being inhibited above this value. The success of skin grafts has also been demonstrated to be associated with bacterial loads $<5 \times 10^4$ cfu/cm^2 [26]. These studies and others on diabetic foot ulcers and pressure ulcers led to the conclusion that the determination of bacterial loads could be useful in the prediction of wound healing and infection [27–32]. Thus traditionally, bacterial levels above 10^6 cfu/ml are used as predictors of wound infection and related wound healing delay. An exception to this is, where a wound is colonized by β-haemolytic streptococci, which has been found to cause disease even at levels $<10^5$ cfu/g of tissue.

A recent study [33] assessing the bacterial load in the lesions of BU patients longitudinally to gain insight into the evolution of the bacterial load showed that levels of bacteria are not constant throughout the period of antibiotic treatment of BU patients. High bacterial loads were reported before and after treatment compared to decreased loads during treatment. A study by Gardner [34] reported a decrease in bacterial load in subjects on systemic antibiotics compared to those not on antibiotics. This could account for the decreased loads reported in BU patients during treatment with the 8-week regimen of streptomycin and rifampicin (SR8), which have broad spectrum activity. This result was also in agreement with clinical signs observed during the treatment period such as fewer lesions presenting with slough, necrosis and high wound exudate production. The increase in bacterial burden in the BU lesions was found to be highest after the antibiotic treatment phase. The impact of an increased bacterial load on wound healing cannot be underestimated and the presence of bacteria in wounds even in the absence of obvious clinical signs can inhibit the normal wound healing process. BU wound management guidelines should therefore consider and include strategies for preventing secondary infection.

4 Diagnosing Secondary Infection in BU

Generally, for a wound to be considered as infected, bacterial multiplication should increase strong enough to induce immune reactions from the host. Clinically, such reactions will present with signs and symptoms such as pain, swelling, erythema (redness), increased temperature, malodour and discoloured granulation tissue etc. Diagnosing wound infection is challenging and ideally, a holistic analysis of lesions employing all available diagnostic methods is favourable [21].

Diagnosis should consider both the clinical presentation of the wound and the results of microbiological investigations through quantitative and qualitative investigations involving direct microscopy and cultures. In diagnosing BU secondary infection, where possible, histopathological analysis of suspected secondarily infected lesions gives an added advantage to the quality of results obtained. In the absence of specimen for histopathological analysis, bacterial loads above 10^5 cfu/g or cfu/ml are the accepted gold standard in diagnosing localized infection [32, 35, 36] as research has shown that the bacterial burden of a wound has an inverse relationship with wound healing and wound healing is likely to progress only when bacterial counts are below 10^6 cfu/ml [25, 27–32].

Yeboah-Manu [10] combined clinical observations, histopathological analysis, qualitative and quantitative microbiological methods to identify secondary infection in BU wounds. Results revealed correlations between clinical signs and microbiological and histopathological features (Fig. 1). Lesions of 28 patients with clinical indications of infection after SR8 treatment were analyzed and 75% of these lesions yielded quantitative cfu counts $>10^6$ cfu/ml with an average value of 1.2×10^9 cfu/ml, clearly above the levels representing the lower limits of infection by quantitative microbiology (10^5 cfu/g or cfu/ml). Among these patients, clinical signs highly predictive for infection were pain and yellow discharge. Histopathology also confirmed the presence of infecting bacteria in 75% of the lesions microbiologically classified as infected.

5 Drug Susceptibility Patterns of Bacterial Isolates from BU Lesions

Countries with a high burden of infectious diseases rely on antibiotics as an important part of their health-care. There are currently no guidelines for the management of secondarily infected BU lesions; however, it is common practice among health

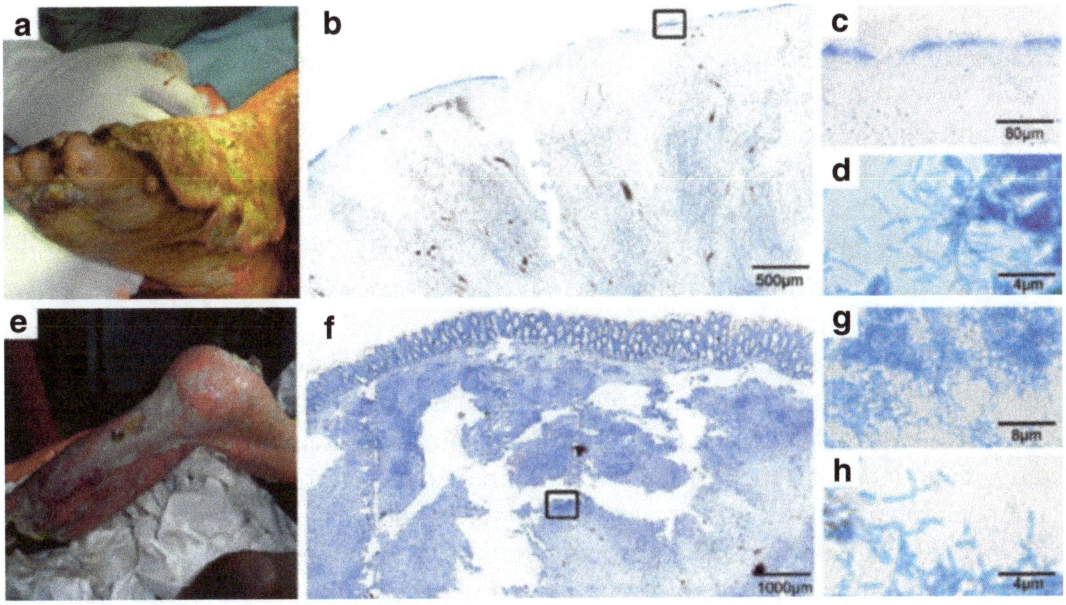

Fig. 1 Histopathological analysis of tissue from two patients excised 6 and 8 weeks respectively after SR8 treatment respectively. Histological sections were stained with Ziehl-Neelsen (acid fast bacteria) and methylene blue (DNA, secondary infection). (**a**) clinical presentation of a patient presenting with a large lesion on the right foot. (**b**) overview over excised tissue specimen (open ulcer surface) revealing the presence of an infection (blue band, box). (**c/d**) higher magnification confirming the presence of densely packed rods. (**e**) clinical presentation of a patient presenting with a large lesion covering the left leg. (**f**) overview over excised tissue specimen revealing an epidermal hyperplasia as well as a strong edema. (**g/h**): secondary infection of the dermal and subcutaneous tissue with rod shaped bacteria [10]

personnel to prescribe additional antibiotics for BU patients suspected of having a secondary infection. Studies of the antibiotic susceptibility patterns of bacterial isolates from BU wounds has revealed a high level of resistance to the commonly prescribed first-line drugs against infectious diseases including the BU treatment drugs streptomycin and rifampicin [10, 14, 37].

The high level of streptomycin resistance reported in Ghana is not surprising as it is a widely used antibiotic in animal husbandry for treatment and disease prevention. It was also until recently used as a first line treatment drug for tuberculosis in Ghana. Thus years of streptomycin use in both humans and animals appears to have resulted in selection pressure leading to high resistance rates. Lower levels of resistance were reported for amikacin, gentamicin and tobramycin (also aminoglycosides) as well as for the carbapenem imipenem [10]. Amikacin, gentamicin and tobramycin are expensive injectable drugs only prescribed for serious and life-threatening infections. These factors discourage the abuse of these antibiotics and contribute to the low resistance levels reported. Carbapenems are the last choice of antibiotics used in treatment of infections due to extended spectrum β-lactamase (ESBL)-producing bacteria and are not widely used or easily accessible. Meropenem is the only carbapenem approved for use in Ghana and has so far not recorded any resistance though it has been on the market since 2002.

Drug resistant bacteria including multi-drug resistant (MDR) *P. aeruginosa*, Methicillin resistant *S. aureus* (MRSA) and Extended Spectrum β-lactamase producers were also detected in BU lesions [10, 14, 37–39]. Of increased concern is the high frequency of MRSA with prevalence rates of 33%, 38%, 13% and 22.8% reported [10, 14, 37, 38].

The resistome of whole genome sequenced *S. aureus* isolates from BU patients was investigated by Kpeli [37] and Amissah [12]. Kpeli's study identified antibiotic resistance genes coding for resistance to β-lactams (*blaZ*), chloramphenicol (*cat* and *catpC221*), trimethoprim (*dfrG*), methicillin (*mecA*), quinolone (*norA*), streptomycin (*str*) and tetracycline (*tetK, tetL* and *tetM*). The study also further investigated the *rpoB* gene of rifampicin resistant strains and identified two known amino acid substitutions H481N and I527M implicated in rifampicin resistance. Amissah's study identified resistance to penicillin characterized by the presence of various blaZ operons; resistance to chloramphenicol (*catA* and *fexB*), tetracycline (*tetK, tetL* and *tetM*), trimethoprim (*drfG*), streptomycin (*str*), and rifampicin (*rpoB* mutation encoding an amino acid substitution that changed Asp471 into Gly). Six methicillin resistant isolates were also identified with five encoding the *mecA* gene and one identified as a borderline oxacillin resistant strain.

Based on the results from various studies on the antimicrobial susceptibilities of bacteria isolated from BU wounds, it is clear that the use of antibiotics needs to be minimized. The definition of secondary infection is subjective in many cases and clinicians are unsure of how to manage wounds, especially if they show no clinical evidence of infection. Specific guidelines on the management of BU wounds secondarily infected by other bacterial species are urgently needed [14]. Guidelines on whether to use antibiotics for treating these wounds, the choice of antimicrobial agents to prescribe, the duration of treatment and whether topical antimicrobials should be prescribed and prioritized over systemic antimicrobials need to be clarified.

6 Molecular Epidemiology

6.1 Sources of Infection

All wounds healing by secondary intention are prone to microbial contamination. Wounds can be infected from the environment, surrounding skin and also from endogenous sources such as the nasal mucosa, gastrointestinal tract and the genitourinary tract. Sources of wound contamination in the health-center can be from health-care workers (HCW), other patients and the inanimate environment. Most of the bacterial pathogens recovered from the lesions of BU patients are known nosocomial pathogens. A study by Kpeli et al. [37] to identify the possible routes of infection of BU lesions identified three modes of infection of BU lesions in two health centers in Ghana; two health-care facility related sources, through a HCW and the environment, and a self-infection (Fig. 2). *S. aureus* was the main bacterial pathogen identified in many of the samples analyzed and these isolates were selected for further studies. The study employed molecular (*S. aureus* protein A, (*spa*) typing) and whole genome sequencing (WGS) methods to identify transmission events

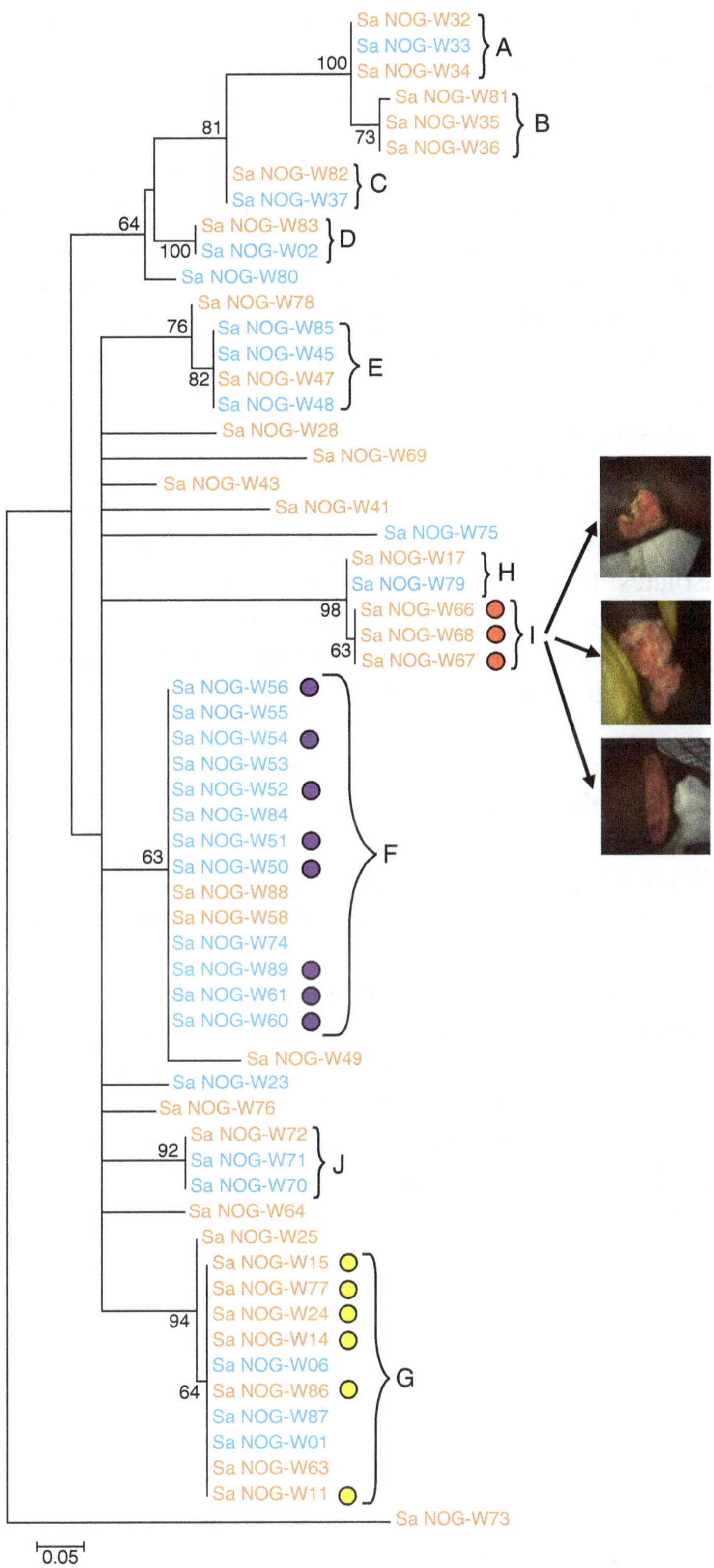

Fig. 2 *Spa* phylogeny showing clusters and relationships between isolates: Maximum likelihood phylogeny tree based on *spa* gene typing. The tree was rooted in the midpoint. Numbers in nodes indicate support values in the form of proportions of bootstrap pseudoreplicates. Branches with support values higher than 55% are collapsed. A–J = the clusters identified. The green colored strains are from Health center A, and the blue ones from Health center B. The yellow coloured circles represent MRSA and the violet circles show isolates from a HCW, patients and equipment in health center B. The red coloured strains in Cluster I were from three lesions of the same patient which had the same *spa* (t2500) and ST type (ST 3248) [37]

ongoing within the health centers. Phylogenetic analysis of *spa* types found in *S. aureus* isolates from samples collected from the health center, from HCW's and patients were analyzed. Different clusters were identified and comprised strains isolated in both health centers (clusters A, C, D, H and J), strains found only in Facility

A (clusters B, G, and I) and strains isolated only in Facility B (clusters E and F) (Fig. 2). Isolates within cluster F included isolates from samples taken from the hand of a health care worker and from patients dressed consecutively by this worker on the same day, isolates taken on another day from a forceps used to dress patients after it had been used on four different patients consecutively and two isolates from a patient who attended the health facility for treatment at two different time points (weeks 2 and 8) during treatment (Fig. 2). Isolates in cluster I were from three BU lesions of a patient. These isolates had the same *spa* type (t2500), and sequence type (ST 32489). They had SNP differences of 29 bp, 51 bp and 34 bp between them, and these small numbers of SNPs suggests that they could have been spread from a common source. Similar clusters were also predicted by the WGS analysis [37].

Using a WGS-based-gene-by-gene typing approach, Amissah et al. [12] also identified transmission events between different patients at a health center. Results from this study showed that transmission events could have occurred during overlapping visits of the patients to the health center. Further analysis of the *S. aureus* isolates detected from the wounds showed that they belonged to lineages which have also been reported by other studies from the Ghanaian health-care setting strongly suggesting that these lineages could be associated with transmission within our health centers [12, 40].

7 Predisposing Socio-Economic Factors

Several factors could predispose BU patients to wound infection. Firstly, because of the painlessness of early lesions, most affected people do not report early to the formal health centers for treatment. They resort to home management of the wounds until they have deteriorated and lesions become clinically and microbiologically infected [33]. Secondly, BU is seen in many endemic areas as a spiritual disease. An unhealing wound is thought to be caused by spiritual forces, charms, witches, ancestral spirits or even the gods of the land [41]. These beliefs drive the affected to seek for help from traditional healers and witch doctors as their first point of call and not biomedical health facilities [41–44]. These witch doctors and traditional healers also apply various kinds of concoctions to manage the lesions and these concoctions may not have been prepared under hygienic conditions and may serve as conduits for the introduction of contaminants into the wounds.

Cases are only reported to the biomedical health centers after all efforts have been exhausted by these various approaches and no improvement has been seen. Therefore frequently large wounds are presented which may have been treated over long periods of time with various concoctions usually not prepared under aseptic conditions.

The socio-cultural beliefs and perceptions of the patients also affect their response to treatment even at the formal health centers. Some categories of people such as pregnant women, 'promiscuous people', breastfeeding mothers are considered unqualified to manage wounds in a community. Our research has shown that when patients encounter any of these categories of people in the formal health center, they resort to re-dressing their wounds after they have been dressed by these people and thus end up introducing contaminants into their wounds [41].

Economic hardships have also been found to predispose BU patients to wound infection. In Ghana, treatment of the disease is free; however, affected patients have to bear the cost of wound management. This places a huge financial burden on patients and their families, which they are sometimes unable to bear. Thus instead of using fresh bandages anytime the wounds are dressed, patients resort to recycling of old bandages. Some of these bandages are not washed properly after use and analysis of some of the washed bandages by Yeboah-Manu et al. [10] showed that potential pathogens similar to those isolated from the wounds can be isolated from the recycled washed bandages.

8 Prevention of Secondary Infection

In order to prevent secondary infection, the cycle of events that lead to the contamination of wounds by bacteria must be broken. This will entail dealing with the sources of potential pathogens in order to minimize their spread or transmissibility. In BU wound management, observing aseptic techniques at every stage of wound management is critical to preventing super-infection by other bacteria. Thus education of both HCWs and patients is important and periodic monitoring to ensure compliance with laid down guidelines will also go a long way in reducing the occurrence of secondary infection in BU disease [45]. Decentralized treatment may reduce the risk of nosocomial wound infection compared to in-patient facilities [8]. Guidelines for the prevention and management of wound infection were released in 2010 by the WHO. These guidelines which can be accessed at www.who.int/gpsc/SSI-outline.pdf?ua=1 provide core principles for the appropriate prevention and management of infected wounds as well as protocols to guide the management of the wound site, tetanus prone wounds and antibiotic prophylaxis and treatment. Separate guidelines for the prevention of surgical site infection (www.who.int/hac/techguide/tools/guidelines_prevention_and_management_wound_infection.pdf) have also been released by the WHO and these can also be used to strengthen infection prevention and control programmes in the health centers and applied in cases where infected patients require surgical interventions.

9 Wound Care

The median time to healing in early, limited BU lesions has been found to be 18 weeks [1]. For large lesions, this can be more than 2 years. Therefore, after antibiotic therapy, wound care will form the core of BU wound management until wound closure is achieved and the procedure should be adequate to ensure wound healing by decreasing time to healing, pain and morbidity [46–48].

Basic principles of wound management seek to treat or manage relevant systemic conditions, maintain a moist wound environment [49], protect the wound from trauma, promote a clean wound base, prevent infection, and control edema and lymphoedema. Basic wound care depends on the type of wound being managed; however fundamental wound management practices involve cleansing the wound, debriding and applying appropriate dressing materials. The correct type of dressing is essential in maintaining a moist wound environment and decreasing rates of infection [45]. Traditional gauze dressings have been shown to have increased infection rates compared to moisture retaining dressings [47, 49], therefore selection of appropriate dressing is important. Personal observations show that different clinicians use varying dressing materials for BU patients to reduce edema, however none of the dressings have been evaluated in a controlled trial. Other practices that ensure timely wound healing are irrigation with physiological saline solution during cleansing, pain management and working under aseptic conditions. The WHO has two guidelines on wound management that can be applied to BU [50, 51]. However, a study by Velding [45] accessing the wound care practices for BU in Ghana and Benin concluded that they differed from the recommended WHO guidelines [45]. It is important that health centers adhere to the available guidelines on wound care to ensure timely healing of wounds. Though some of the differences in wound management techniques were related to the economic conditions of the health centers, Velding recommends some simple and low cost solutions which can be implemented in the health centers to enhance the standard of wound care [45]. As such, health care workers need to be trained on the appropriate guidelines for wound management and a retraining and monitoring machinery should be put in place to ensure compliance with laid down guidelines.

Many lines of evidence indicate that good wound care practices can decrease the length of hospital stay and also increase the number of wounds healing without surgical intervention. Wound care is expensive and the burden usually falls on the affected patients and their families who are sometimes ill-equipped to handle the financial costs over a long period of time. Good wound care will therefore reduce this burden on the affected individuals by ensuring they do not stay for prolonged periods in the health centers.

References

1. Nienhuis WA, Stienstra Y, Thompson WA, Awuah PC, Abass KM, Tuah W, Awua-Boateng NY, Ampadu EO, Siegmund V, Schouten JP, Adjei O, Bretzel G, van der Werf TS (2010) Antimicrobial treatment for early, limited *Mycobacterium ulcerans* infection: a randomised controlled trial. Lancet 375(9715):664–672. https://doi.org/10.1016/S0140-6736(09)61962-0
2. Beissner M, Arens N, Wiedemann F, Piten E, Kobara B, Bauer M, Herbinger KH, Badziklou K, Banla Kere A, Loscher T, Nitschke J, Bretzel G (2015) Treatment outcome of patients with Buruli ulcer disease in Togo. PLoS Negl Trop Dis 9(10):e0004170. https://doi.org/10.1371/journal.pntd.0004170
3. Chauty A, Ardant MF, Adeye A, Euverte H, Guedenon A, Johnson C, Aubry J, Nuermberger E, Grosset J (2007) Promising clinical efficacy of streptomycin-rifampin combination for treatment of buruli ulcer (*Mycobacterium ulcerans* disease). Antimicrob Agents Chemother 51(11):4029–4035. https://doi.org/10.1128/AAC.00175-07

4. Kibadi K, Boelaert M, Fraga AG, Kayinua M, Longatto-Filho A, Minuku JB, Mputu-Yamba JB, Muyembe-Tamfum JJ, Pedrosa J, Roux JJ, Meyers WM, Portaels F (2010) Response to treatment in a prospective cohort of patients with large ulcerated lesions suspected to be Buruli ulcer (*Mycobacterium ulcerans* disease). PLoS Negl Trop Dis 4(7):e736. https://doi.org/10.1371/journal.pntd.0000736
5. Phillips RO, Sarfo FS, Abass MK, Abotsi J, Wilson T, Forson M, Amoako YA, Thompson W, Asiedu K, Wansbrough-Jones M (2014) Clinical and bacteriological efficacy of rifampin-streptomycin combination for two weeks followed by rifampin and clarithromycin for six weeks for treatment of *Mycobacterium ulcerans* disease. Antimicrob Agents Chemother 58(2):1161–1166
6. Sarfo FS, Phillips R, Asiedu K, Ampadu E, Bobi N, Adentwe E, Lartey A, Tetteh I, Wansbrough-Jones M (2010) Clinical efficacy of combination of rifampin and streptomycin for treatment of *Mycobacterium ulcerans* disease. Antimicrob Agents Chemother 54(9):3678–3685. https://doi.org/10.1128/AAC.00299-10
7. Vincent QB, Ardant MF, Adeye A, Goundote A, Saint-Andre JP, Cottin J, Kempf M, Agossadou D, Johnson C, Abel L, Marsollier L, Chauty A, Alcais A (2014) Clinical epidemiology of laboratory-confirmed Buruli ulcer in Benin: a cohort study. Lancet Glob Health 2(7):e422–e430. https://doi.org/10.1016/S2214-109X(14)70223-2
8. Addison NO, Pfau S, Koka E, Aboagye SY, Kpeli G, Pluschke G, Yeboah-Manu D, Junghanss T (2017) Assessing and managing wounds of Buruli ulcer patients at the primary and secondary health care levels in Ghana. PLoS Negl Trop Dis 11(2):e0005331. https://doi.org/10.1371/journal.pntd.0005331
9. Andreoli A, Ruf MT, Sopoh GE, Schmid P, Pluschke G (2014) Immunohistochemical monitoring of wound healing in antibiotic treated Buruli ulcer patients. PLoS Negl Trop Dis 8(4):e2809. https://doi.org/10.1371/journal.pntd.0002809
10. Yeboah-Manu D, Kpeli GS, Ruf MT, Asan-Ampah K, Quenin-Fosu K, Owusu-Mireku E, Paintsil A, Lamptey I, Anku B, Kwakye-Maclean C, Newman M, Pluschke G (2013) Secondary bacterial infections of buruli ulcer lesions before and after chemotherapy with streptomycin and rifampicin. PLoS Negl Trop Dis 7(5):e2191. https://doi.org/10.1371/journal.pntd.0002191
11. WHO (2016) Buruli ulcer-diagnosis of *Mycobacterium ulcerans* disease. WHO, Geneva
12. Amissah NA, Chlebowicz MA, Ablordey A, Sabat AJ, Tetteh CS, Prah I, van der Werf TS, Friedrich AW, van Dijl JM, Rossen JW, Stienstra Y (2015a) Molecular characterization of *Staphylococcus aureus* isolates transmitted between patients with Buruli ulcer. PLoS Negl Trop Dis 9(9):e0004049. https://doi.org/10.1371/journal.pntd.0004049
13. Anyim MC, Meka AO, Chukwu JN, Nwafor CC, Oshi DC, Madichie NO, Ekeke N, Alphonsus C, Mbah O, Nwaekpe C, Njoku M, Fakiyesi D, Ulodiaku V, Ejiofor I, Bisiriyu AH, Ukwaja KN (2016) Secondary bacterial isolates from previously untreated Buruli ulcer lesions and their antibiotic susceptibility patterns in Southern Nigeria. Rev Soc Bras Med Trop 49(6):746–751. https://doi.org/10.1590/0037-8682-0404-2016
14. Barogui YT, Klis S, Bankole HS, Sopoh GE, Mamo S, Baba-Moussa L, Manson WL, Johnson RC, van der Werf TS, Stienstra Y (2013) Towards rational use of antibiotics for suspected secondary infections in Buruli ulcer patients. PLoS Negl Trop Dis 7(1):e2010. https://doi.org/10.1371/journal.pntd.0002010
15. Phanzu DM, Bafende EA, Dunda BK, Imposo DB, Kibadi AK, Nsiangana SZ, Singa JN, Meyers WM, Suykerbuyk P, Portaels F (2006) Mycobacterium ulcerans disease (Buruli ulcer) in a rural hospital in Bas-Congo, Democratic Republic of Congo, 2002-2004. Am J Trop Med Hyg 75(2):311–314
16. van der Werf TS, van der Graaf WT, Tappero JW, Asiedu K (1999) Mycobacterium ulcerans infection. Lancet 354(9183):1013–1018. https://doi.org/10.1016/S0140-6736(99)01156-3
17. Katz L, Donadio S (1993) Polyketide synthesis: prospects for hybrid antibiotics. Annu Rev Microbiol 47:875–912. https://doi.org/10.1146/annurev.mi.47.100193.004303
18. Scherr N, Gersbach P, Dangy JP, Bomio C, Li J, Altmann KH, Pluschke G (2013) Structure-activity relationship studies on the macrolide exotoxin mycolactone of *Mycobacterium ulcerans*. PLoS Negl Trop Dis 7(3):e2143. https://doi.org/10.1371/journal.pntd.0002143

19. Bowler PG, Duerden BI, Armstrong DG (2001) Wound microbiology and associated approaches to wound management. Clin Microbiol Rev 14(2):244–269. https://doi.org/10.1128/CMR.14.2.244-269.2001
20. Guo S, Dipietro LA (2010) Factors affecting wound healing. J Dent Res 89(3):219–229. https://doi.org/10.1177/0022034509359125
21. Percival SL, Dowd SE (2010) The microbiology of wounds. CFC Press, Boca Raton, FL
22. Davis SC, Ricotti C, Cazzaniga A, Welsh E, Eaglstein WH, Mertz PM (2008) Microscopic and physiologic evidence for biofilm-associated wound colonization in vivo. Wound Repair Regen 16(1):23–29. https://doi.org/10.1111/j.1524-475X.2007.00303.x
23. Percival SL, McCarty SM, Lipsky B (2015) Biofilms and wounds: an overview of the evidence. Adv Wound Care (New Rochelle) 4(7):373–381. https://doi.org/10.1089/wound.2014.0557
24. Tarnuzzer RW, Schultz GS (1996) Biochemical analysis of acute and chronic wound environments. Wound Repair Regen 4(3):321–325. https://doi.org/10.1046/j.1524-475X.1996.40307.x
25. Bendy RH Jr, Nuccio PA, Wolfe E, Collins B, Tamburro C, Glass W, Martin CM (1964) Relationship of quantitative wound bacterial counts to healing of decubiti: effect of topical gentamicin. Antimicrob Agents Chemother (Bethesda) 10:147–155
26. Majewski W, Cybulski Z, Napierala M, Pukacki F, Staniszewski R, Pietkiewicz K, Zapalski S (1995) The value of quantitative bacteriological investigations in the monitoring of treatment of ischaemic ulcerations of lower legs. Int Angiol 14(4):381–384
27. Daltrey DC, Rhodes B, Chattwood JG (1981) Investigation into the microbial flora of healing and non-healing decubitus ulcers. J Clin Pathol 34(7):701–705
28. Edwards R, Harding KG (2004) Bacteria and wound healing. Curr Opin Infect Dis 17(2):91–96. https://doi.org/10.1097/00001432-200404000-00004
29. Krizek TJ, Robson MC, Kho E (1967) Bacterial growth and skin graft survival. Surg Forum 18:518–519
30. Lookingbill DP, Miller SH, Knowles RC (1978) Bacteriology of chronic leg ulcers. Arch Dermatol 114(12):1765–1768
31. Pruitt BA Jr, McManus AT, Kim SH, Goodwin CW (1998) Burn wound infections: current status. World J Surg 22(2):135–145
32. Robson MC, Heggers JP (1969) Bacterial quantification of open wounds. Mil Med 134(1):19–24
33. Kpeli G, Owusu-Mireku E, Hauser J, Pluschke G, Yeboah-Manu D (2017b) Longitudinal assessment of the bacterial burden of buruli ulcer wounds during treatment. BBRJ 1(1):65–70. https://doi.org/10.4103/bbrj.bbrj_37_17
34. Gardner SE, Hillis SL, Frantz RA (2009) Clinical signs of infection in diabetic foot ulcers with high microbial load. Biol Res Nurs 11(2):119–128. https://doi.org/10.1177/1099800408326169
35. Gardner SE, Frantz RA, Doebbeling BN (2001) The validity of the clinical signs and symptoms used to identify localized chronic wound infection. Wound Repair Regen 9(3):178–186
36. Schraibman IG (1990) The significance of beta-haemolytic streptococci in chronic leg ulcers. Ann R Coll Surg Engl 72(2):123–124
37. Kpeli G, Darko Otchere I, Lamelas A, Buultjens AL, Bulach D, Baines SL, Seemann T, Giulieri S, Nakobu Z, Aboagye SY, Owusu-Mireku E, Pluschke G, Stinear TP, Yeboah-Manu D (2016) Possible healthcare-associated transmission as a cause of secondary infection and population structure of *Staphylococcus aureus* isolates from two wound treatment centres in Ghana. New Microbes New Infect 13:92–101. https://doi.org/10.1016/j.nmni.2016.07.001
38. Amissah NA, Glasner C, Ablordey A, Tetteh CS, Kotey NK, Prah I, van der Werf TS, Rossen JW, van Dijl JM, Stienstra Y (2015b) Genetic diversity of *Staphylococcus aureus* in Buruli ulcer. PLoS Negl Trop Dis 9(2):e0003421. https://doi.org/10.1371/journal.pntd.0003421
39. Kpeli G, Buultjens AH, Giulieri S, Owusu-Mireku E, Aboagye SY, Baines SL, Seemann T, Bulach D, Goncalves da Silva A, Monk IR, Howden BP, Pluschke G, Yeboah-Manu D, Stinear T (2017a) Genomic analysis of ST88 community-acquired methicillin resistant *Staphylococcus aureus* in Ghana. PeerJ 5:e3047. https://doi.org/10.7717/peerj.3047
40. Egyir B, Guardabassi L, Sorum M, Nielsen SS, Kolekang A, Frimpong E, Addo KK, Newman MJ, Larsen AR (2014) Molecular epidemiology and antimicrobial susceptibility of clinical

Staphylococcus aureus from healthcare institutions in Ghana. PLoS One 9(2):e89716. https://doi.org/10.1371/journal.pone.0089716
41. Koka E, Yeboah-Manu D, Okyere D, Adongo PB, Ahorlu CK (2016) Cultural understanding of wounds, Buruli ulcers and their management at the Obom sub-district of the Ga south municipality of the Greater Accra region of Ghana. PLoS Negl Trop Dis 10(7):e0004825. https://doi.org/10.1371/journal.pntd.0004825
42. Asiedu K, Etuaful S (1998) Socioeconomic implications of Buruli ulcer in Ghana: a three-year review. Am J Trop Med Hyg 59(6):1015–1022
43. Aujoulat I, Johnson C, Zinsou C, Guedenon A, Portaels F (2003) Psychosocial aspects of health seeking behaviours of patients with Buruli ulcer in southern Benin. Tropical Med Int Health 8(8):750–759
44. Renzaho AM, Woods PV, Ackumey MM, Harvey SK, Kotin J (2007) Community-based study on knowledge, attitude and practice on the mode of transmission, prevention and treatment of the Buruli ulcer in Ga West District, Ghana. Tropical Med Int Health 12(3):445–458. https://doi.org/10.1111/j.1365-3156.2006.01795.x
45. Velding K, Klis SA, Abass KM, Tuah W, Stienstra Y, van der Werf T (2014) Wound care in Buruli ulcer disease in Ghana and Benin. Am J Trop Med Hyg 91(2):313–318. https://doi.org/10.4269/ajtmh.13-0255
46. Beldon P (2004) Comparison of four different dressings on donor site wounds. Br J Nurs 13(6 Suppl):S38–S45. https://doi.org/10.12968/bjon.2004.13.Sup1.12541
47. Brolmann FE, Eskes AM, Goslings JC, Niessen FB, de Bree R, Vahl AC, Pierik EG, Vermeulen H, Ubbink DT (2013) Randomized clinical trial of donor-site wound dressings after split-skin grafting. Br J Surg 100(5):619–627. https://doi.org/10.1002/bjs.9045
48. Chang KW, Alsagoff S, Ong KT, Sim PH (1998) Pressure ulcers--randomised controlled trial comparing hydrocolloid and saline gauze dressings. Med J Malaysia 53(4):428–431
49. Singh A, Halder S, Menon GR, Chumber S, Misra MC, Sharma LK, Srivastava A (2004) Meta-analysis of randomized controlled trials on hydrocolloid occlusive dressing versus conventional gauze dressing in the healing of chronic wounds. Asian J Surg 27(4):326–332. https://doi.org/10.1016/S1015-9584(09)60061-0
50. Lehman L, Simonet V, Saunderson P, Agbenorku P (2006) Buruli ulcer: prevention of disability (POD). World Health Organization, Geneva
51. Macdonald JM, Geyer MJ (2010) Wound and lymphoedema management. World Health Organization, Geneva

Management of BU-HIV Co-infection

Daniel P. O'Brien, Vanessa Christinet, and Nathan Ford

1 Epidemiology

In many Buruli ulcer (BU) endemic countries worldwide there is also a high HIV prevalence. This is especially the case in Africa with adult HIV prevalence rates between 1 and 5% in BU endemic countries. Evidence from Benin, Cameroon, Ghana and Gabon suggest that HIV may increase the risk of BU [1–5]. For example, in the *Médecins Sans Frontières* programme in Akonolinga, Cameroon, the prevalence of HIV was approximately 3–6 times higher in BU treated patients compared to the regional estimated HIV prevalence (37% vs 8% in women; 20% vs 5% in men; and 4% vs 0.7% in children) [2]. Likewise in Benin, in one region patients with BU were eight times more likely to have HIV infection than those without BU (2.6% vs 0.3%) [3], and in another region adults with BU were five times more likely to have HIV compared to the regional estimated HIV prevalence (5.0% vs 1.1%) [4]. In Ghana HIV prevalence was 4–5 times higher in BU patients than those without BU (5% vs 0.9% in one study and 8.2% vs 2.5% in another study) [1, 6]. Therefore, there is a significant potential for the two infections to overlap in the same individual and HIV needs to be considered in all BU patients in settings with high background HIV prevalence.

D. P. O'Brien (✉)
Department of Infectious Diseases, Barwon Health, Geelong, VIC, Australia

Department of Medicine and Infectious Diseases, Royal Melbourne Hospital, University of Melbourne, Melbourne, VIC, Australia
e-mail: Danielo@BarwonHealth.org.au

V. Christinet
Centre International de Recherches, d'Enseignements et de Soins en milieu tropical (CIRES), Akonolinga, Cameroon

N. Ford
HIV Department, World Health Organization, Geneva, Switzerland

G. Pluschke, K. Röltgen (eds.), *Buruli Ulcer*,
https://doi.org/10.1007/978-3-030-11114-4_14

2 Clinical Effects of BU-HIV Co-Infection

HIV appears to affect the clinical presentation and severity of BU disease with a reported increased incidence of multiple, larger and ulcerated BU lesions in HIV-infected individuals (Fig. 1) [1, 2, 4, 5]. In Cameroon BU-HIV co-infected patients have a median lesion size of 5.5 cm (equivalent to WHO category 2) and up to 24% have multiple BU lesions (equivalent to WHO category 3) [2]. This results in increased patient morbidity as well as more time and resources required to treat lesions, and increased rates of long-term disability. Additionally it appears that the presence and severity of BU may reflect the level of underlying immune suppression in an HIV-infected person. In Akonolinga, Cameroon, 80% of patients with category 2 or 3 BU lesions had a CD4 count ≤500 cells/mm^3 compared to 55% of those with category 1 lesions, and the main lesion size was significantly increased with decreasing CD4 cell counts [2].

BU-HIV co-infected patients often present with severe immunosuppression and thus in urgent need of antiretroviral therapy (ART). In Akonolinga, 22% of patients diagnosed with BU were classified as having advanced HIV disease (defined as CD4 counts ≤200 cells/mm^3) and 48% had CD4 counts between 200 and 500 cells/mm^3 [2]. In this programme, mortality in BU-HIV co-infected patients treated for BU without ART was significantly higher than for HIV non-infected BU patients (11% vs 1%, $p < 0.001$). This is despite median CD4 cell counts at baseline among the 8 deceased HIV-infected patients being only moderately reduced at 229 cells/mm^3, levels usually not associated with such a high mortality rate. Death occurred early with a median time to death post BU diagnosis of 41.5 days and none had received ART. This suggests that a delay in ART initiation until after the completion of the recommended 8 weeks BU antibiotic treatment may adversely affect mortality rates [2]. In Ghana,

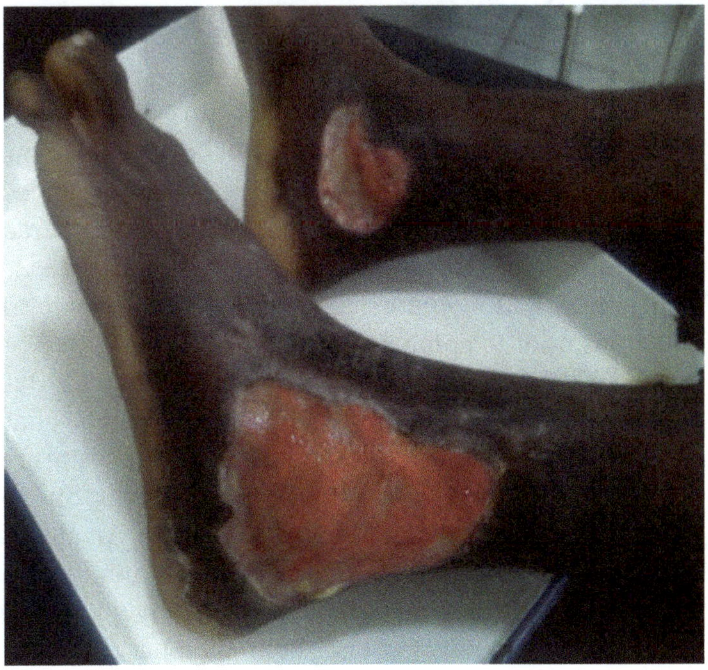

Fig. 1 Bilateral severe BU lesions on the lower limbs of an HIV co-infected patient

during the management of BU lesions, two of four patients (50%) who did not start ART died compared with none of three (0%) who commenced ART [1]. In Gabon, all nine (100%) patients with known HIV infection died during the treatment of BU, although data regarding their ART status was not reported [5]. There are also several case reports describing deaths in BU-HIV patients not commenced on ART [7–9].

3 Diagnosis

It is not known whether co-infection with HIV affects the sensitivity or specificity of BU diagnostic tests, but it appears unlikely. Therefore recommended methods of BU diagnosis are the same as for HIV negative patients. This includes mainly IS*2404* PCR testing of specimens, and microscopic detection of acid fast bacilli in stained smears from wound specimens [10]. As BU patients appear to represent a sentinel group with a higher prevalence of HIV, and knowing that HIV status will affect treatment options and may influence mortality and BU outcomes, it is recommended that all BU patients presenting in settings with a high prevalence of HIV have their HIV status determined by offering quality provider-initiated HIV testing and counseling [11].

4 Management of BU-HIV Co-Infection

The management of BU-HIV co-infection presents a complex and challenging situation for clinicians involved in patient care. Effective management requires attention to the treatment of each infection, but also consideration of the potential interactions between the treatments.

4.1 BU Treatment

If the patient is not already on ART, BU treatment should be commenced first, with combination antibiotic treatment. This is recommended prior to commencing ART for HIV to minimize pill burden and avoid drug interactions and side effects in the early stages of BU treatment, to allow the time needed for patient preparation for ART, and to follow the usual principle of HIV care to treat and stabilize any co-infections prior to commencing ART [12].

In 2017 the WHO Technical Advisory Group on BU decided that the WHO recommendation for BU treatment in Africa should be changed to rifampicin 10 mg/kg daily up to a maximum of 600 mg/day plus clarithromycin 7.5 mg/kg twice daily (up to a maximum of 1000 mg daily). However, due to potential drug interactions this combination should be used with caution. An alternative is rifampicin 10 mg/kg daily up to a maximum of 600 mg/day plus moxifloxacin 400 mg daily. The use of the combination of rifampicin and streptomycin is no longer recommended due to unacceptable toxicity levels such as renal dysfunction and ototoxicity, drug interactions with tenofovir, as well as the negative effects on adherence of daily painful injections. There is no evidence that the duration of BU antibiotic treatment needs to be prolonged beyond the standard recommended 8-week course for BU-HIV co-infected patients [13].

4.2 HIV Treatment

As currently recommended for all HIV-infected individuals, a CD4 cell count should be determined for all BU-HIV positive patients to assess the level of HIV-associated immune suppression. If the CD4 cell count is equal to or <350 cells/mm^3 then prophylactic cotrimoxazole (960 mg tablet daily) should be commenced immediately to reduce mortality, morbidity and HIV disease progression [14]. If a CD4 cell count is not available and the patients have advanced HIV disease (WHO clinical stages 3 or 4) they should likewise receive cotrimoxazole prophylaxis. In areas with high prevalence of malaria and/or bacterial infections, cotrimoxazole should be commenced regardless of CD4 cell count in all BU-HIV-infected patients and continued for life. Patients with advanced HIV disease are at risk of a range of other opportunistic infections including cryptococcal meningitis, and it is recommended to screen with a cryptococcal antigen test if the CD4 cell count is <200 cells/mm^3, and assess for danger signs of severe illness [14].

If patients are already receiving ART then this should be continued. All patients with active BU disease who are known or diagnosed as HIV positive but not on ART should initiate ART. This aims to reduce HIV associated mortality and morbidity, and HIV transmission [15]. Furthermore, as the immune system plays an important role in curing BU disease and in healing lesions, optimization of immunity with ART may be important to combat BU disease and potentially improve treatment outcomes (healing times, cure rates, long-term disability and recurrence rates). It has been found that in BU HIV coinfected patients, a CD4 cell count >500 cells/mm^3 was associated with a reduction in the time needed to heal BU lesions by more than 50% compared to those with a CD4 count ≤500 cells/mm^3 [2]. This suggests that healing times are significantly more prolonged among immune suppressed HIV positive individuals.

ART should begin as soon as possible after the start of BU treatment, preferably within 8 weeks, and as a priority in those with a CD4 cell count <350 cells/mm^3 or WHO stage 3 or 4 disease. Patients with advanced HIV are at immediate risk of further life-threatening opportunistic infections and delay in ART initiation may result in significant HIV-associated morbidity and mortality. As described earlier, this risk appears to be significant in BU-HIV co-infected patients, perhaps related to an increased risk of bacterial sepsis from secondarily infected BU lesions. Therefore, early reconstitution of immunity with ART in advanced BU disease may be important. Furthermore, this recommendation to start early also takes into account the fact that in routine programmes there may be delays in ART initiation whilst patients wait for assessment, training and availability of ART after completing their BU treatment. Also, as patients may receive BU treatment at a significant distance from ART centers, they may be lost to HIV care if ART initiation is delayed [16].

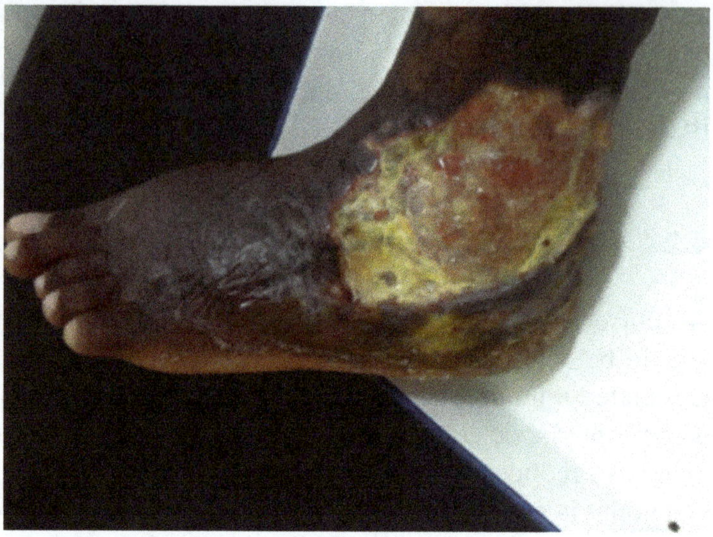

Fig. 2 A severe paradoxical reaction associated with a severe BU lesion on the left ankle of an HIV co-infected patient 2 weeks after commencing antiretroviral treatment

It is possible that early initiation of ART will lead to an increased incidence of paradoxical reactions associated with immune reconstitution when combining BU antibiotic treatment and ART [1, 17, 18]. This may lead to significant tissue damage and undesired consequences, especially if lesions are in sensitive areas (e.g. the face) (Fig. 2). Thus care needs to be taken when initiating ART. If severe paradoxical reactions occur ART should be continued and the use of prednisolone to minimize the severity of the effects should be considered [17, 19].

4.3 HIV and BU Treatment Interactions

There are a number of important issues regarding the use of antiretroviral drugs in patients receiving antibiotic treatment for BU. Firstly, as recommended for TB/HIV patients on ART using rifampicin containing regimens, the non-nucleoside reverse transcriptase inhibitor (NNRTI) component of the ART regimen should be efavirenz [20]. If this option is not available or appropriate then nevirapine can be used, but the lead-in dose of nevirapine should be omitted in the presence of rifampicin at the start of treatment. Additionally, caution should be exercised in the use of nevirapine particularly in patients with high or unknown CD4 cell-counts at initiation due to a potential increased risk of hypersensitivity and Stevens Johnson's syndrome [21]. Close monitoring during the initial weeks of therapy is recommended when nevirapine is initiated in these patients. Another alternative in adults and non-pregnant women is the integrase inhibitor dolutegravir which is preferred over efavirenz as a first-line antiretroviral in high-income settings and is being introduced in low- and middle-income settings. However, due to an interaction with rifampicin, it is currently recommended that the dose of dolutegravir should be doubled to 100 mg daily.

There are concerns about significantly reduced levels of protease inhibitor (PI) medications and increased toxicity when they are used with rifampicin and therefore they are ideally avoided during BU antibiotic treatment. If the patient is already receiving a PI-based regimen, and they are NNRTI-naïve and not infected with the

HIV-2 virus, the PI-based regimen should be changed to an NNRTI-based regimen using efavirenz. If they are not NNRTI-naïve or are infected with HIV-2 virus, then the recommended PI regimen to use is lopinavir/ritonavir (LPV/RTV) at either double dose (800 mg/200 mg twice daily) or standard LPV dose with increased dose of RTV (400 mg/400 mg twice daily). However this combination in higher doses is frequently associated with high levels of toxicity and requires close clinical and laboratory monitoring. Again an alternative, if available, is for the integrase inhibitor dolutegravir to replace the PI during the 8-week BU treatment.

Efavirenz can reduce clarithromycin levels by up to 39% [22] which likely further compounds the known significant reduction of clarithromycin levels when co-administered with rifampicin [23]. Although the clinical consequences of these interactions are unknown, it could potentially lead to reduced effectiveness of the rifampicin/clarithromycin regimen for BU treatment, with secondary treatment failure and drug resistance. Increased toxicity is also reported when the two drugs are combined, with 46% of patients reported to develop a rash [24]. Therefore this combination should be used with caution. The alternative that avoids this interaction is rifampicin 10 mg/kg daily up to a maximum of 600 mg/day plus moxifloxacin 400 mg daily. This regimen has the added benefit of reducing the risk of rifampicin resistant TB if BU treatment is used in patients also infected with undetected drug sensitive active tuberculosis [25].

4.4 Children

All children should be commenced on ART as soon as possible within 8 weeks of the start of BU treatment. Efavirenz is not approved for clinical use in children <3 years of age. Therefore in this age-group if initiating ART whilst on BU treatment with rifampicin, nevirapine should be used instead of efavirenz at a dose of 200 mg/m^2. An alternative is to use a triple NRTI ART regimen (AZT/3TC/ABC or AZT/3TC/TDF). If already on a PI-based ART regimen when commencing BU treatment with rifampicin, LPV/RTV can be continued but the dose of ritonavir should be increased to achieve a 1:1 ratio with LPV. Alternative options include either replacing the LPV/RTV with nevirapine at a dose of 200 mg/m^2 or using a triple NRTI regimen.

4.5 Tuberculosis

All patients should be actively screened for tuberculosis (TB) before commencing BU treatment and before starting ART [26]. As most BU-HIV co-infected patients live in highly endemic areas for TB, there is a significant risk of TB co-infection. As HIV-infected patients have a higher risk of TB reactivation, especially when severely immunosuppressed, there is a risk of co-existent active TB disease. Therefore it is important to exclude active TB disease prior to commencing BU treatment, as BU treatment regimens are not adequate to treat active TB, which may result in TB-related mortality and morbidity and the development of drug resistant TB. For those with symptoms suggestive of TB, this includes the use of Xpert MTB/RIF on sputum, and if CD4 count $\leq$100 cells/mm^3 or the patient is seriously ill (at any CD4 cell count) the use of LF-LAM on urine. In those without symptoms of TB, TB preventative treatment should be started according to local guidelines [14].

4.6 Service Provision

BU patients found to be HIV-positive should be referred to clinicians trained in clinical management of HIV infection. Ideally, management should be integrated within the BU treatment centres to facilitate timely initiation of ART and avoid loss of patients to follow up which may occur during an external referral process for HIV care. If HIV management capacity in BU treatment centres is not possible, then referral to the nearest HIV treatment centre for care is recommended. Good cooperation between the BU and HIV treatment programmes at local, regional and national levels should be implemented to ensure the highest standard of care for BU-HIV co-infected patients. Approaches to support adherence to drug treatments for BU and HIV should be integrated, and programmes should implement a monitoring and reporting system to monitor and evaluate the outcomes of BU-HIV interventions.

References

1. Tuffour J, Bonney K, Owusu-Mireku E, Paintsil A, Pereko J, Yeboah-Manu D et al (2015) Challenges associated with management of Buruli ulcer/human immunodeficiency virus coinfection in a treatment center in Ghana: a case series study. Am J Trop Med Hyg 93(2):216–223
2. Christinet V, Comte E, Ciaffi L, Odermatt P, Serafini M, Antierens A, Rossel L, Nomo AB, Nkemenang P, Tsoungui A, Delhumeau C, Alexandra C (2014) Impact of human immunodeficiency virus on the severity of Buruli ulcer disease: results of a retrospective study in Cameroon. Open Forum Infect Dis 1(1):ofu021
3. Johnson RC, Nackers F, Glynn JR, de Biurrun BE, Zinsou C, Aguiar J et al (2008) Association of HIV infection and *Mycobacterium ulcerans* disease in Benin. AIDS 22(7):901–903
4. Vincent QB, Ardant MF, Marsollier L, Chauty A, Alcais A, Franco-Beninese Buruli Research Group (2014) HIV infection and Buruli ulcer in Africa. Lancet Infect Dis 14(9):796–797
5. Bayonne Manou LS, Portaels F, Eddyani M, Book AU, Vandelannoote K, De Jong BC (2013) *Mycobacterium ulcerans* disease (Buruli ulcer) in Gabon: 2005-2011. Me'decine et Sante' Tropicales 23:450–457
6. Raghunathan PL, Whitney EA, Asamoa K, Stienstra Y, Taylor TH Jr, Amofah GK et al (2005) Risk factors for Buruli ulcer disease (*Mycobacterium ulcerans* infection): results from a case-control study in Ghana. Clin Infect Dis 40(10):1445–1453
7. Kibadi K, Colebunders R, Muyembe-Tamfum JJ, Meyers WM, Portaels F (2010) Buruli ulcer lesions in HIV-positive patient. Emerg Infect Dis 16(4):738–739
8. Johnson RC, Ifebe D, Hans-Moevi A, Kestens L, Houessou R, Guedenon A et al (2002) Disseminated *Mycobacterium ulcerans* disease in an HIV-positive patient: a case study. AIDS 16(12):1704–1705
9. Bafende AE, Lukanu NP, Numbi AN (2002) Buruli ulcer in an AIDS patient. S Afr Med J 92(6):437
10. World Health Organisation (2014) Laboratory diagnosis of Buruli ulcer disease: a manual for health-care providers. WHO, Geneva

11. World Health Organization (2014) Management of Buruli ulcer-HIV coinfection: technical update. WHO, Geneva
12. O'Brien DP, Ford N, Vitoria M, Christinet V, Comte E, Calmy A et al (2014) Management of BU-HIV co-infection. Tropical Med Int Health 19(9):1040–1047
13. World Health Organisation (2012) Treatment of *Mycobacterium ulcerans* disease (Buruli ulcer): guidance for health workers. WHO, Geneva
14. World Health Organization (2017) Guideline for managing advanced HIV disease and the timing for initiating antiretroviral therapy. WHO, Geneva
15. World Health Organization (2015) Guideline on when to start antiretroviral therapy and on preexposure prophylaxis for HIV. WHO, Geneva
16. Rosen S, Fox MP (2011) Retention in HIV care between testing and treatment in sub-Saharan Africa: a systematic review. PLoS Med 8(7):e1001056
17. Wanda F, Nkemenang P, Ehounou G, Tchaton M, Comte E, Toutous Trellu L et al (2014) Clinical features and management of a severe paradoxical reaction associated with combined treatment of Buruli ulcer and HIV co-infection. BMC Infect Dis 14(1):423
18. Komenan K, Elidje EJ, Ildevert GP, Yao KI, Kanga K, Kouame KA et al (2013) Multifocal Buruli ulcer associated with secondary infection in HIV positive patient. Case Rep Med 2013:348628
19. O'Brien DP, Jenkin G, Buntine J, Steffen CM, McDonald A, Horne S et al (2014) Treatment and prevention of *Mycobacterium ulcerans* infection (Buruli ulcer) in Australia: guideline update. Med J Aust 200(5):267–270
20. World Health Organization (2013) Consolidated guidelines on the use of antiretroviral drugs for treating and preventing HIV infection: recommendations for a public health approach. WHO, Geneva
21. Shubber Z, Calmy A, Andrieux-Meyer I, Vitoria M, Renaud-Thery F, Shaffer N et al (2013) Adverse events associated with nevirapine and efavirenz-based first-line antiretroviral therapy: a systematic review and meta-analysis. AIDS 27(9):1403–1412
22. Kuper JI, D'Aprile M (2000) Drug-drug interactions of clinical significance in the treatment of patients with Mycobacterium avium complex disease. Clin Pharmacokinet 39(3):203–214
23. Alffenaar JW, Nienhuis WA, de Velde F, Zuur AT, Wessels AM, Almeida D et al (2010) Pharmacokinetics of rifampin and clarithromycin in patients treated for *Mycobacterium ulcerans* infection. Antimicrob Agents Chemother 54(9):3878–3883
24. Bristol-Myers-Squibb (2010) Sustiva product information. Bristol-Myers-Squibb, Princetown
25. O'Brien DP, Comte E, Ford N, Christinet V, du Cros P (2013) Moxifloxacin for Buruli ulcer/HIV coinfected patients: kill two birds with one stone? AIDS 27(14):2177–2179
26. World Health Organisation (2011) Guidelines for intensified tuberculosis case-finding and isoniazid preventive therapy for people living with HIV in resourceconstrained settings. WHO, Geneva

Social Science Contributions to BU Focused Health Service Research in West-Africa

Mark Nichter

1 Introduction

Health social science (HSS) research has contributed in significant ways to our understanding of how rural communities in West Africa perceive and respond to Buruli ulcer (BU), and the challenges facing those involved with BU health service delivery. These challenges range from the need to mount more effective community outreach education programs leading to earlier BU detection to the need to identify and then address predisposing, enabling, and service delivery related factors [1] that result in treatment delay, drop out, and non-adherence. In this chapter, we briefly highlight HSS research that has broadened our knowledge of community response to BU and then consider HSS-inspired BU interventions. Most of the sources we draw upon involve research carried out in Benin, Cameroon, and Ghana. To date, few HSS studies have been conducted in other endemic countries such as Cote d'Ivoire, The Democratic Republic of the Congo, Nigeria and Uganda.[1] We include in our overview a selection of studies conducted by both social scientists and medical researchers investigating social and cultural factors that influence health care decision-making.

Two things are important to bear in mind at the onset of this review. First, there is considerable cultural variability both within and between countries. It is therefore best not to overgeneralize the findings of one study to an entire country. While we do identify general patterns that emerge in the literature related to overarching themes, these patterns do not necessarily apply to all groups. Second, different studies have employed different research methodologies. These range from in-depth ethnographic research employing participant observation and detailed case studies

[1] Two recent studies conducted in Cote' D'ivoire by Konan et al. [2] and Uganda by Pearson [3] are worthy of mention.

M. Nichter (✉)
School of Anthropology, University of Arizona, Tucson, AZ, USA

G. Pluschke, K. Röltgen (eds.), *Buruli Ulcer*,
https://doi.org/10.1007/978-3-030-11114-4_15

to structured and semi-structured interviews, surveys, and focus groups. The findings of one type of study may differ significantly from another study using different methods even when conducted in the same region. Each method has its own unique contribution. This is one reason social scientists often employ multiple methods and triangulate data to get at different aspects of any social phenomenon. The importance of doing so is demonstrated in a study of BU conducted by Grietens et al. [4], who speak to the value of establishing trust with research participants as a means of getting at sensitive issues related to health care seeking before attempting to conduct a survey.

2 Part One

We begin with research on factors that influence health care seeking beginning with predisposing factors. More attention is given to this topic than enabling and service related factors because they provide the most insight into how BU is perceived and experienced culturally.

2.1 Predisposing Factors: Cultural Perceptions of BU Causality, Social Stigma, and Preference for Traditional Healing

2.1.1 Cultural Perceptions of BU Causality

Cultural perceptions of illness causality are best assessed in terms of multiple levels of causality that encompass predisposing, efficient, instrumental, and ultimate causes of illness [5, 6]. Predisposing factors range from social transgressions to vulnerable traits and states. Efficient causes of illness involve agents whose actions reflect intention such as spirits or witchcraft. Instrumental causes of illness, such as insects, worms, or germs, have no intention. Ultimate causes of illness are associated with such things as God's judgement or some notion of fate. Multiple ideas about the cause of an illness often coexist and it is common for efficient causes (agents) held responsible for causing serious or unusual illnesses to be viewed as working through or in concert with instrumental causes. This is particularly the case when the characteristics of an illness associate it with malevolent intent. In the case of BU, the unusual physical characteristics of BU—ulceration, long term suffering without death, and the negative social and economic hardship associated with the disease—lead many to consider it unnatural even when biomedical treatment is accepted.

Community members often maintain "what if" subjunctive reasoning [7, 8] when considering possible causes of an illness and may draw attention to one type or level of causality at a particular time for any of a variety of reasons. For example, drawing attention to mystical causes of a case of BU may entail rationalization for a course of action or involve the moral identity of the afflicted, a decision maker, the household, or even an entire community when illness etiology indexes cultural affiliation or difference. Disease attribution may also be a means of "othering" a population. For example in Cote d'Ivoire, a country in which both physical and mystical causes of BU are thought to exist, a perceived rise in the number of BU cases has been attributed to an influx of migrant workers from other West African countries. These populations are thought to somehow harbor the causative agent of the disease. Notably, the importance of identifying an efficient cause of a "sent" illness (like witchcraft or sorcery) is generally not to seek retribution from a specific person or group. It is rather to give an unusual event some sense of coherence. What follows is a brief overview of how BU is perceived by members of the three index countries along with observations on the extent to which illness perceptions influence health care seeking behavior.

Benin

Mulder et al. [9] interviewed 153 former/active BU patients (107 treated at hospital, 46 by healers) in Southern Benin about their perceptions of BU causality using a semi structured questionnaire. Fifty-eight percent of former patients reported attributing their illness to witchcraft, often for reasons associated with jealousy. However, only 13% reported that their perception of causality was an important factor in health care decision making when directly asked if such was the case.

Aujoulat et al. [10] carried out a more in-depth community based study of BU causality in the same region. Community members exposed to BU education as well as several health workers maintained the perception that BU-like symptoms could either be signs of a natural disease or an induced disease (*sasa*) caused by sorcery, witchcraft, or a curse only treatable by healers. Community members generally relied on diagnosis by treatment. The underlying cause of one's illness was thought to become apparent based on the aliment's response to different types of treatment and the trajectory of the illness over time. Notably, in this study, women were far more likely to ascribe BU-like symptoms to natural causes in part because children are often affected. Children are not seen as embroiled in problematic social relations associated with jealousy or any social infraction that might lead to witchcraft or sorcery.

One of the most common natural causes of BU recognized by women was worms. Informants described worms as a possible instrumental cause of BU linked to predisposing factors such as body weakness, fatigue, or environmental factors. However, they did not discount the idea that worm activity might be related to an efficient cause interfering with wound healing. The researchers also explored perceptions of contagion. They found that contagion was commonly related to the foul smell of a necrotizing BU-related wound. Notably, while some community members and health workers feared that BU might be contagious, most women who cared for a child with BU did not believe this to be the case because they had been in close proximity to an afflicted child and not gotten sick. Healers contributed to the perception that BU was contagious by insisting that a patient should remain in isolation during treatment. Notably, a major factor underlying why children having BU were kept out of school by some parents was fear others might consider the child contagious.

A third study by Kpadonou et al. [11] further documented differences of opinion about causality among a sample of 244 former BU patients treated at Allada Hospital and residing within a 100 km radius of the hospital. This study found that while a majority of adults and teens attributed BU to witchcraft, a majority of children perceived BU to have natural causes. The authors attributed this difference to children's exposure to health education messages while being treated in hospital. Among teens and elders exposed to the same messages, cultural beliefs about efficient causes of disease remained salient. Regardless of what cause BU was attributed to, 90% of all informants stated they did not feel guilt associated with having the disease.

A fourth study by Boyer [12] carried out in Benin's Ouémé River region investigated stories and rumors circulating about BU and its treatment at a local Catholic mission hospital renown for BU lesion excision, and the social relations of BU causal attribution in the context of health care seeking. Boyer's in-depth ethnography of BU included case studies of household decision making related to BU treatment. He found that in some instances mystical attribution of BU was employed by a male decision maker to justify treatment of a family member by traditional healers instead of treatment at a hospital for pragmatic reasons. In this region of Benin, decentralized care did not exist while Boyer was conducting his research, and going to a hospital resulted in a significant loss of revenue and/or labor on the part of the afflicted and/or a patient caretaker required by the hospital. Focusing on mystical causes of the illness and the need for traditional treatment upheld the moral identity of some decision makers opting for traditional treatment. Notably, Boyer documented cases where there was ambivalence about this decision on the part of other family members. Boyer draws attention to the fact that explanatory models of BU involving mystical causes can sometimes serve as post hoc rationalizations of action or inaction as well as being determinants of health care seeking. He further points to the folly of describing "beliefs" about BU as durable unquestioned ideas fixed in the mind of the afflicted instead of seeing them as ad hoc explanatory models subject to revision given the rich semantic illness network [13] of the Beninese. Ideas about BU, he argues, are hardly fixed and subject to the circulation of stories and rumors about the ailment and its treatment. All of this leads Boyer to caution against simplistically blaming "beliefs" for BU treatment delay and assuming that once beliefs (read superstitions) are addressed through education a preference for hospital care will follow.

A fifth study by Kennell [14, 15] conducted among the Aja ethnic group draws attention to why some ethnic groups wait before treating an ulcer to determine its cause and decide on a course of action, a process the Aja refer to as *zohwiʒi*. Among the Aja, there is a belief that skin eruptions/wounds may be a sign of *Sakpata*, the *vodun* deity of the earth long associated with small pox and measles. If deemed a sign of the spirit's presence, it is believed the wound should not be treated by hospital medicine for fear that the spirit might be angered and the patient face dire consequences. A watch and wait policy is deemed prudent to see if the wound erupts as a common abscess or is seen as unusual. If seen as a sign of Sakpata, the afflicted becomes an initiate in a spirit cult. Notably, while hospital medicines may not be used during the zohwiʒi process, herbal medicines may be applied to hasten the eruption process.

Cameroon

Several ethnographic studies have looked at perceptions of BU causality in Cameroon. A detailed study of BU by Grietens et al. [4] in Cameroon highlights the importance of perceptions of multiple or "double causality." This study found that in Southern Cameroon BU (referred to as *atom*) is commonly attributed to mystical causes associated with either some type of social infraction or witchcraft. Social infractions include such things as theft or trespassing (knowingly or unknowingly) on someone else's agricultural plot protected by a powerful fetish. Guard fetishes are thought to have the power to inflict particular kinds of illnesses, including *atom*. Sorcery may also cause *atom*. Sorcery operates through antisocial malignant forces present in an invisible world (*evu*). These forces may be controlled and used by either a sorcerer or a healer. They may also be unconsciously present within some people and only become active at night when they are asleep. The characteristics of BU (a wound that does not heal, becomes more severe, and inflicts immense suffering) match the imagery of how people think sorcery operates. A common perception is that in order for a sorcerer to gain power to inflict suffering and in order for a healer to gain power to engage in "night battles" with malevolent forces, sacrifices must be made to strengthen the *evu* under their control. One way to do so is through the mystical eating of flesh, even the flesh of family members. Slow progressing BU wounds appearing on the limbs of children as well as adults are thought to be evidence of such flesh eating.

Atom is also thought to have a natural origin related to insect bites, especially the bites of horseflies that fester and become more severe. Some community members exposed to outreach education programs have come to think of these wounds as being infected with microbes. Others perceive these insects to be sent by sorcerers to inflict harm and negatively influence disease progression. The possibility of "double causality" is shared by health staff who maintain an open mind when patients express a desire to seek the assistance of traditional healers while being treated at hospital. This is especially the case when the healing process is not going well. A point emphasized by Grietens et al. is that although perceptions of "double causality" are strong in the region, they are not necessarily the major reason for treatment delay at hospitals and health posts.

A study conducted by Awah et al. [16] in the Bankim region of Cameroon found perceptions of BU causality that are similar, yet distinct from those described in the Southern region of the country. BU like symptoms are seen as possibly having a natural cause, being the sign of a powerful spirit (*mbouati*), or being caused by witchcraft. Natural causes of the disease are associated with changes in ecology resulting from the building of a large dam in the region, and the introduction of rice cultivation. While residents recognize that the illness has become more common following these events, the instrumental cause of the disease is not understood. *Mbouati* is a spirit affliction that gives special powers to the afflicted, powers that some of the afflicted may not want to relinquish. Informants across several ethnic groups inhabiting the area voiced the opinion that both *mbouati* and BU co-existed, and that only traditional healers could determine the difference between the two as well as afflictions caused by witchcraft. Only the most powerful of healers, it is thought, can transform a mystical ailment into a chronic physical ulcer (*nbong*) amenable to successful treatment by herbs or hospital medications. Furthermore, when a wound does not heal when treated it is conjectured that *mbouati* may be toying with healers or health staff and leading them to believe their treatment is successful in early stages only to undermine their efforts latter.

Local observation that BU-related diseases have increased as a result of ecological and social change is a subject investigated by Giles-Vernick et al. [17] in the central region of Cameroon. Adopting an ethno historical perspective and the use of multiple methodologies ranging from in-depth interviews to group discussions, these researchers investigated the circumstances that have led to a widely held perception that the number of cases of *atom* have steadily increased during the last two to four decades. An increase in the number of *atom* cases is associated with ruptured social relations that took place at a time of heightened environmental degradation and ecological decline as well as rising economic hardship, competition, and jealousy. *Atom* is seen as a sign and symptom of diminished *eding*, a term that indexes both love and social cohesion leading to fruitful and productive ecological extraction by community members. The researchers note that social, environmental, and ecological disruption are perceived to have provoked and expanded the prevalence of *atom* through both natural and mystical causes. This study calls our attention to the necessity of studying NTDs in a broader "One world one health" context [18] attentive to both political ecology and those factors leading to structural vulnerability. In the case of BU, this entails the need to document how changes in the environment and shifts in social relations associated with such factors as structural adjustment policies and changing cropping patterns, contribute to disease susceptibility and increased exposure to the pathogen responsible for BU.

Ghana

The greatest number of social science studies of BU have been carried out in Ghana. A few studies of causality may be highlighted to illustrate the range of observations documented. A large community-based study of local perceptions of BU was carried out by Renzaho et al. [19] in Ga district. Ten percent of the district was surveyed and seven focus groups conducted. The survey found that 53% of informants had no idea of the cause of BU, 16% thought it might be caused by drinking non-potable water, 8% by poor hygiene and unclean surroundings, and 6% by swimming or wading in ponds. Only 5% of participants reported that BU might be caused by witchcraft. During focus groups, however, witchcraft surfaced as a common possible cause of BU and a major source of concern. The survey also suggested that children were far more likely to think BU was contagious than adults.

Ackumey et al. [20, 21] investigated perceptions of BU among those afflicted with BU in both the pre-ulcer and ulcer state in West and South Ga municipalities near Accra. They employed an explanatory model interview guide adapted for BU research. The patient-based data generated in this study was quite different from the community-based survey data generated by the aforementioned study. In the pre-ulcer state, respondents commonly reported behavior-based BU causality, especially swimming in ponds and rivers. However, informants who did not engage in such behaviors questioned how this could be the cause of the illness. Still others questioned how this could be the major cause of BU when so many people who had contact with the same water sources did not become ill. Half of all informants believed their condition might be related to drinking unclean water, a general (not BU-specific) health message repeatedly heard from health workers. Many other informants attributed their ulcers to small wounds and insect bites that became aggravated as a result of their scratching them. About a third of all informants stated they had no idea of the cause of their condition. For those with large ulcers, witchcraft was the most common cause reported followed by swimming in rivers, weakness of blood, and drinking unclean water. Weak blood was reported to be a common cause of BU among children.

Notably, the authors found that 40% of informants with pre-ulcers and 50% of those with ulcers said their condition could not have been prevented because witchcraft cannot be stopped. Nevertheless, approximately half of all informants stated that avoiding bathing in rivers or ponds prevented the disease. Many noted, however, that this was hardly feasible. The researchers concluded on the basis of their interviews that regardless of perception of causality, those afflicted with BU were willing to seek medical care if it was accessible. The afflicted maintained a pragmatic attitude and were open to trying any treatment that might help them recover from their condition.

Another study of 86 BU patients in Ga district by Owusu and Adamba [22] reported data similar to Ackumey with a few notable differences. In this study, only 15% of informants attributed the cause of their disease to "enemies"(witchcraft), while 19% associated the disease with water use, walking through swampy areas while working in the fields (because BU commonly affects the lower limbs), and children swimming in the river. Thirteen percent thought the disease might come from insect bites, but informants could not specify which insect might be causing the disease. Of those who reported insect bite as a possible cause, many maintained dual casualty. They found it hard to believe that an insect bite could lead to such a serious illness unless witchcraft was involved. Of those who imagined dual diagnosis, it was deemed prudent to use herbal medicine before hospital medicine to deal with witches first.

A more recent study by Koka [23] in Ga district found that while 78% of informants ($N = 300$) thought BU had natural causes, 61% thought BU could be caused by supernatural agents as well. Notably, there was a statistically significant gender difference in perceptions of causality. Sixty nine percent of women reported natural causes of the disease while only 36% of men did so ($p = 0.006$). Education also proved to be a factor influencing perceptions of causality. Far more respondents with at least primary education (46%) were likely to report that Buruli ulcer can be caused by both natural and supernatural causes when compared to those with no education (29%) ($p < 0.001$). Moreover, those with at least primary education were far more likely to report only natural causes than those with no education.

Stienstra et al. [24] conducted a study of BU in three regions of Ghana (Agogo, Denkyira, and Amansie districts) and interviewed BU patients at hospitals as well as community members. More than half of all respondents associated poor personal hygiene with the disease, but many others thought this was not possible. They observed that people with good hygiene suffered from the disease and that others with very poor hygiene living in the same village did not experience the disease. Nearly half of all respondents also identified the environment as a possible risk factor for BU, especially walking through swampy areas and being bitten by insects that lead one to scratch and open the skin to diseases. Only a few people mentioned drinking water or eating contaminated food as possible causes of BU. Twenty-three percent of all respondents feared BU might be contagious in some way (touch, clothing, flies, feeding, etc.) and a few thought the disease might be sexually transmitted. Fifty-nine percent of respondents thought the disease might be caused by witchcraft, 47% by a curse, and a few mentioned God's will as an ultimate cause. In the latter case, it was thought that God removed his protection against diseases because of some sin that one had committed. Notably, even though ancestors play an important role in Ghanaian life, they were not mentioned as a cause of BU.

Shifts in environmental landscapes have also been connected to a rise in BU cases in Ghana. Tschakert et al. [25] found that community members in an area of Ghana exposed to artisanal gold mining and illegal logging connected a rise in BU cases to ecological destruction, flooding, and increased bodies of stagnant and/or contaminated water. How these conditions led to increases in the disease was open to considerable conjecture. Notably, community members identified having to cross disturbed spaces to engage in daily activities with increased risk of experiencing BU.

2.1.2 Stigma and Social Risk

It has been widely reported in all three index countries that BU is stigmatized. Stigma is a concept that covers a broad range of social responses to a condition or group that is negatively valued. Applied to an illness like BU, it encompasses both felt stigma and enacted stigma. Felt stigma entails a sense of shame, embarrassment, inferiority, and fear of enacted stigma. Enacted stigma entails social exclusion, discrimination, and isolation. Stienstra et al. [24] conducted what has become a landmark study of stigma related to BU attentive to both types of stigma. These researchers employed an eleven-question, pretested instrument to explore social exclusion in various areas of everyday life when one suffered from BU. They administered the instrument to subsamples of both infected and non-infected community members. From their study of three regions of Ghana, they concluded that stigma for BU was pervasive. Half of all afflicted and non-afflicted informants, regardless of educational background, reported that those afflicted with BU were avoided by others, avoidance being slightly higher for men than women. Social avoidance was primarily reported for non-relatives and was particularly high in endemic areas.

Many of the afflicted reported feeling ashamed and stated that they tried to hide their wounds from others. Others, however, were concerned that their illness might be contagious. They displayed their wounds to let others know not to come near. Fear of transmitting BU was related to both natural causes and witchcraft. However, when witchcraft was associated with BU, stigma scores rose appreciably. The study found that not only was BU feared as contagious, but it was also thought to lead to infertility because the disease was perceived to render the blood of the afflicted dirty. Having the disease made it more difficult for both the afflicted and family members to marry.

Notably, the study did not find that stigma scores correlated with any pattern of health-care seeking or treatment delay. However, it did find that stigma affected school attendance. Other studies in Ghana have likewise reported that stigma and fear of contamination is particularly strong among school children, leading the afflicted to drop out of school. Koka [23] found that in West Ga district it was not just students who feared that BU might be contagious, but also teachers. As a result, teachers did not intervene when students shunned those having the symptoms of BU.

In Benin, Mulder et al. [9] employed the stigma questionnaire developed by Stienstra et al. and found that about half of all current BU patients interviewed experienced some degree of stigma. Patients treated at health posts and by traditional healers did not report statistically significant differences in levels of stigma. Stigma appeared to decrease post treatment with less than a third of former patients reporting stigma.[2] Stigma was largely associated with the smell emanating from the wound. Patients often engaged in self-stigmatization, isolating themselves from others out of embarrassment. The actions of traditional healers also reinforced stigma. Many traditional healers demanded isolation while treating BU patients for months and in some cases years. Anthropologists working on a Stop Buruli research project in Benin documented another factor leading to self-isolation: a perception of vulnerability. They found that some patients with open wounds felt vulnerable to witchcraft and other malevolent forces. For this reason, they hid their wounds and when compelled to move through public spaces, would not display bandages.

Some social scientists have cautioned against overemphasizing the role of stigma as a factor causing BU treatment delay. Ribera et al. [27] found that social isolation was common among BU patients in Cameroon, but not a primary factor influencing health care seeking. Paying too much attention to stigma, they argue, deflects attention away from enabling and health service related factors causing delay. They also argue that while BU is commonly associated with social transgressions and witchcraft in central Cameroon, stigma is not the main cause for patients abandoned at the hospitals. Their findings suggest that social isolation and abandonment are often part of a household coping strategy, an attempt to avoid plunging the household into a spiral of impoverishment.

Social risk has been identified as a factor influencing health seeking behavior for BU. Social risk refers to risk to one's reputation that negatively affects present or future social relationships. When a woman is compelled to travel to a clinic for decentralized care or to remain in a hospital for long term BU care she may become the subject of rumors suggesting she is engaging in illicit sexual relations when travelling outside her community. Social risk associated with seeking health care was identified in each of the three index countries, but poorly documented in all but

[2] Some studies suggested some types of ongoing social exclusion. A study by de Zeeuw et al. [21] carried out in Benin and Ghana documented persisting social and work participation restrictions among former patients with BU. These restrictions remained long after treatment was completed and wounds had healed.

Benin [28, 29]. Concern about social risk is one reason women travel to health posts with children or a chaperon, and that children serve as patient care takers at hospitals. Notably, Amoussouhoui et al. [30] found that when outreach education about treatment and decentralized care was made available near one's village in Benin, fear of social risk was reduced.

2.1.3 Traditional Healers

It has been widely reported that traditional healers are often a first resort for wound care in all three index countries [10, 16, 24, 31, 32], and that healers often treat BU related wounds for extended periods of time. It is beyond the scope of this chapter to provide a detailed account of the practices of traditional healers. These practices range from exorcism, the placating of spirits, and prayer to herbal treatment, and in many cases provision of both mystical and physical treatment. Researchers like Grietens et al. [4] and Johnson et al. [33] have aptly described some of the more common types of therapy offered by various kinds of healers in all three countries.[3] Traditional healers often engage in some form of divination to determine the etiology of a wound and then decide whether they should treat the wound or refer the afflicted to another healer who specializes in another type of treatment. If a social infraction is suspected on the part of the afflicted, then a confession may be required and regret communicated in some way. Purification may be required in the form of ritual washing with magically infused water or the blood of a sacrificial animal. Stop Buruli Consortium researchers from all three index countries found that when herbal medicines are employed, medicines are commonly applied to dry a wound. Drying is seen as necessary for the healing process [34]. In some cases, burning or cutting a patient to remove bad blood may be undertaken. Interdictions may also be required in the form of prohibitions, which may involve various kinds of food, contact with others during treatment, and/or suspension of sexual relations on the part of the patient. When witchcraft or a spirit's presence is suspected, herbs and mystical protection offered by a traditional healer are commonly thought necessary even if biomedical treatment is eventually sought.

There has been heated discussion within the international Buruli research community about whether traditional healers should play a role in BU outreach and whether collaboration with health staff is feasible. Two factors that emerge from the literature as key to collaboration are an appreciation for dual illness causality, and mutual respect. Stated more succinctly, collaboration depends on whether the expertise of healers is acknowledged when it comes to dealing with psychosocial and mystical aspects of patients' BU experience as the "work of culture" [35]. The "work of culture" refers to the process whereby distressful states, perceived risk and motives, negative affects, and sensations are transformed into publicly accepted sets of meanings and symbols that can be manipulated or dealt with in some culturally salient manner. On the part of traditional healers, there has to be recognition of the expertise of clinical staff in treating BU wounds as a systemic health problem

[3] Also see Pearson for a detailed study of why herbalists are preferred for treatment of ulcers, boils, and BU in Uganda.

requiring timely referral and treatment adherence. In short, collaboration requires task sharing.

Few rigorous studies have been carried out to date to test the conditions under which collaborative relations between hospital staff and healers might be established toward the end of managing BU. One exception is a pilot project in Bankim, Cameroon [16]. The project is briefly summarized below in a section on social science inspired interventions. The project was successful in establishing long-term collaborative relations between traditional healers, health staff, and community volunteers. Findings of this study as well as observations made in Obom, Ghana suggest that in regions where healer groups are supported by chiefs and well functioning, it is far easier to establish collaborative agreements and lines of referral than where such groups do not exist. In the case of Bankim, healer groups signed a contract specifying that they would treat suspected cases of BU for 10 days or less, and not treat a patient's skin. Health staff in return granted traditional healers privileges to visit patients in health posts and offer spiritual protection and psychosocial support.

Three lessons learned in the Bankim case study may be highlighted. First, before introducing training courses for healers in general, it was found important to work with a select group of healers to establish best practices. These healers then served as role models for other healers. Second, it was necessary to insure that collaboration was a win-win proposition. Credit for healing a patient was shared between traditional healers and health staff, so that the reputation of both increased. Third, offering incentives to traditional healers to refer BU cases to hospital was found to be a culturally sensitive issue that needed to be handled carefully. In Bankim, traditional healer collaboration was not established by offering cash payments to healers for patient referrals. While an honorarium enabling a healer to bring a patient to a clinic was greatly appreciated, this had to be done in such a way that there was not a perception that healers gained from the sale of sick bodies to the hospital. This would render them morally suspect. One reason for this is that traditional healers are thought capable of engaging in witchcraft to generate more cases. Healers valued the symbolic capital they gained from collaborating with health staff more than the honorarium they received. This was documented in an impact evaluation conducted after funds for honorariums were no longer available. What traditional healers lost in terms of direct and indirect payments for treating BU cases was eclipsed by a rise in their status. The respect they received from health staff at outreach functions and in the hospital increased their status and was highly valued.

In Benin, Johnson et al. [33] supported collaboration with traditional healers in principle, but latter, some of the same authors urged caution when considering healer participation in BU programs based on their experience offering ad hoc training to a few independent healers. Following the training, a few of these healers represented themselves as BU treatment experts and kept instead of referring cases [36]. A more recent study in Benin, however, identified conditions under which healers were willing to refer cases and collaborate with health staff and community volunteers. In a region of the country where decentralized care was recently introduced for BU and other chronic ulcers, the popularity of clinic-based treatment soared due to the demonstration effect of good wound management and culturally sensitive outreach education [30]. Traditional healers responded by wanting to be associated with the community based BU outreach efforts being promoted. As in the case of Bankim, traditional healers did not see collaboration as a loss of status when invited to be part of outreach activities and offered respect.

Concern about cost and labor loss have been identified as major factors leading to the consultation of traditional healers who live near the afflicted, and are often relatives [37]. It is important to recognize the accessibility of healers as an important factor influencing health seeking behavior, but it would be simplistic to conclude that accessibility is the only reason traditional healers are consulted at different stages of health problems. As noted by many medical anthropologists (cf. [38, 39]), traditional healers are consulted before or in conjunction with biomedical care for myriad reasons having social and cultural salience. Some social scientists have argued that patient delay as a result of visiting traditional healers will cease to be a "problem" once adequate BU care is provided by the government [40] or educating the population [41] will lead them to give up superstitions. Based on a review of the BU literature to date, it would be premature if not erroneous to conclude that traditional healing remains popular only because of a lack of affordable biomedical care.

At the same time, it is important to recognize that healers are often chosen over poorly staffed and resourced health facilities. Hausermann [42] presents a compelling case study of health care seeking for BU in a border region of Ghana-Cote d'Ivoire to make this case as well as to call attention to the impact of poorly implemented BU health policy. Hausermann documents the disjunction between official policy narratives about BU and the lived experiences of people in endemic regions. Following the work of medical anthropologist Stacey Langwick [43] in Tanzania, her analysis shows that nurses refer patients to healers in part due to the inadequacics of hcalth posts under their charge. She raises an issue long debated in international/global health. Is government support for the involvement of traditional healers (and community health volunteers) really about providing more holistic patient-centered care or deflecting attention from the shortcomings of the health system and those responsible for developing and implementing policy?

Enabling Factors

Studies of health care seeking in all three index countries have reported that enabling factors are a major reason for delays in seeking treatment at hospitals and health posts [9, 10, 16, 20, 24, 41, 44, 45]. Enabling factors include the seasonal availability and cost of transportation, direct and indirect treatment costs, and opportunity costs in the form of labor loss, especially during peak agricultural seasons. Despite free antibiotic treatment for BU, these costs can be catastrophic [44–46].[4] Another enabling factor is the identification of appropriate substitute caretakers for children

[4] Chukwuet al. [38] points out that the high costs of pre-diagnosis treatment for BU in Ghana can also put households in a precarious situation.

left at home when a mother needs to travel to receive treatment for herself or a child [28]. For this reason it is important to look at the household as a unit of analysis [45, 48] and to employ a household production of health [49, 50] lens when examining BU health care decisions and patient abandonment.[5] Another useful lens employed by researcher Ines Agbo [28] is that of gender. A consideration of gender relations leads us to an appreciation of the impact BU treatment has on social networks beyond the immediate household of the afflicted given that assistance is commonly requested from extended kin during times of need. Agbo et al. draw attention to the ripple effect of BU beyond households as well as the impact that requests for assistance on social relations over time [29].

Interventions offering patient support have been piloted in each of the three index countries (Benin: Ouinhi, Cameroon: Bankim, Ghana: Obom and Agogo). In each case, when free transportation to health posts and/or free food is offered to patients, clinic popularity increased and treatment delay and treatment drop-out decreased [16, 31, 40, 51].

2.2 Service Level Factors Affecting Health Care Seeking and Treatment Adherence

Several service related factors have been identified as leading to treatment delay, drop out, and perceptions of quality of care. Factors identified by social scientists beyond cost, waiting time, and lack of resources at health posts include poor staff-patient communication about the wound healing process, patient concern about prolonged hospital visits with little feedback on likely duration, and fear of amputation and skin grafts [52]. Other concerns reported related to bandaging and pain management. In the case of bandaging, concerns were raised not only about the procedure itself, but the person doing the bandaging. If the intention of the person bandaging toward the patient was not positive, there was concern the healing process would be delayed. And in Ghana concern was raised if a pregnant nurse did a patient's bandaging as her condition was thought to impede healing [34]. In the case of pain, a common perception was that pain was not appreciated or responded to by clinic staff, leading patients to often seek medicines for pain from the market or traditional healers. The most serious concern reported to researchers was lack of trust in clinic staff by patients in some clinical settings [19]. The positive impact of establishing trust through culturally sensitive education outreach, proactive CHWs, and psychosocial as well as material support for patients has been demonstrated at NGO and mission-supported hospitals treating BU cases such as the Agogo Presbyterian Hospital and the Global Evangelical Mission Hospital serving the Ashanti region of

[5] The household production of health conceptual framework analyzes practices in which household members engage to maintain, protect, and promote health, as well as how household members respond to health related issues. The HHPH model situates health and illness within the larger environment of activities occurring in households and thus provides a valuable complementary component to the study of health care seeking.

Ghana [51, 53], and the clinic and community-based interventions described below. An issue not adequately investigated is what kind of incentives might keep community stakeholders and clinic staff in different settings motivated [30, 36], and what factors might strengthen as well as weaken collaborative relationships. Social scientists tend to think of incentives in terms of different types of convertible capital (economic, cultural, social, symbolic; [54, 55]). In need of consideration are the ramifications of offering different types of capital as incentives in various ways in different cultural settings, and the impact of these incentives on both the identity of stake holders and teamwork.

3 Part Two: Social Science Inspired Interventions

Social scientists have worked closely with their public health/medical colleagues in developing, testing, and implementing health service interventions to raise the consciousness of community members about BU in an effort to detect and treat cases earlier, reduce treatment no-show and drop-out, and increase treatment adherence. Interventions have been mounted in the areas of outreach education, patient support, community stakeholder collaboration, the formation of BU communities of practice, and the transforming of hospitals for long-term patients into therapeutic communities. In this review I briefly highlight seven interventions that have gone beyond offering food and transport to outpatients as forms of support. All seven interventions were sponsored by the Stop Buruli Consortium. For more information on the Stop Buruli Initiative the reader is referred to chapter "Transdisciplinary Research and Action to Stop Buruli Ulcer: A case Study from Philanthropy" of this book.

3.1 Outreach Education

An innovative form of outreach education was introduced in Benin, Cameroon, and Ghana based on a review of research documenting what community members did and did not know about BU, misconceptions about BU and its treatment, awareness of the availability of free treatment, and concerns about treatment. This intervention served as a foundation for building a new generation of community-based BU interventions, and the cornerstone for building BU communities of practice (described below).

The outreach program adopted a question–answer format and iterative process. In brief, the program employed an image-rich PowerPoint presentation on BU delivered by local teams equipped with portable generators and sound systems, laptop computers, and LCD projectors. Outreach meetings were interactive, not passive, and questions were invited from community members. Social scientists recorded and investigated how best to respond to questions posed in a way that was at the same time scientific and understandable to local audiences. Messages and visuals were tested and changed as needed. The following table summarizes the ten major sections of the outreach program (Table 1). Notably, specific sets of messages were designed to inform and educate the community about BU, reassure community members about the quality of care available at hospitals and health posts, offer hope

Table 1 Format of Stop Buruli Consortium outreach education

Ten sections of outreach education program	Key messages conveyed	Issues downplayed or emphasized
Signs and symptoms of BU, how to recognize the disease, and the need to treat it early	• Visuals of physical signs of BU in different stages • Visual and tactile cues suggesting that a lesion, abscess ulcer, or edema may be BU • Progression of disease if not treated	Category I and II BU depicted, but not category III as this evoked great fear
High risk environments and modes of transmission	• High risk environments where one is more likely to be exposed to Mycobacterium ulcerans • Focus was on addressing incorrect ideas about BU transmission and contagion	Less time and attention allotted to risk environments and possible modes of transmission as the science is inconclusive and behavior change related to exposure to water sources difficult given the local reality
What clinic staff do to determine if the affliction is BU or some other disease	Why health staff take swabs, what they look for under the microscope, why medicine for BU is specific and not the same as medications used for other ulcers	Step-by-step explanation of what staff is actually doing along with pictures to offset fears and rumors about what they are doing as a means to increase trust
Effective and ineffective treatments for BU	• Why 56 days of pills and injections (or more) are needed • Why herbal medicine for this disease does not lead to a cure even if a wound is dried	• Agricultural analogies used to convey the idea that medication is taken beyond treatment for the visible wound, as a means to get at the roots and seeds of BU as a systemic infection in the body • Pictures used to show inappropriate treatment, how drying wound is not curing, and effectiveness of medication after herbal medicine has failed to treat the wound
• Traditional healers and rapid referral to hospitals and health posts • Emphasis on rapid referral	Positive messages about exemplar healers who recognize signs of BU and rapidly refer patients to clinic after spiritual protection is offered	• No message disrespecting local practices as superstitious • Respect for traditional healers' role in offering spiritual protection for those for whom this is a concern
Quality of care at the clinic	Quality of care offered by staff: pictures of what care in the clinic looks like, approachable staff, hygienic conditions, empathetic caretakers, etc.	To offset fear and evoke confidence
Care available free of charge or subsidized for BU patients at clinics	Types of support available ranging from free medications, to food, transport, and housing	

(continued)

Table 1 (continued)

Ten sections of outreach education program	Key messages conveyed	Issues downplayed or emphasized
Before and after pictures of BU related wounds successfully treated	• Pictures of BU treatment, and the healing process at different stages • Depict the healing of ulcers on different parts of the body	Pictures depict children and male and female patients of different ages so members of the audience can personally relate
The presentation ends on a note of hope	Testimonials of patients who have been cured speak of their experiences and to the quality of care they have received at the clinic	Open microphone: some testimonials are planned and others are spontaneous
Questions from the audience	On any topic related to information presented or any other issue related to BU	• Open microphone empowers people to speak • Questions are recorded and responses to questions assessed as part of iterative process of ongoing research

of a cure, display stakeholder collaboration, and inform community members about the kind of support available during treatment.

In each of the three countries, large-scale community outreach education meetings were held in the evenings. Community health workers were responsible for organizing meetings and inviting chiefs, local healers, and former patients to attend. This increased their status in the community, enhanced lines of communication with health staff, and motivated them to become more actively involved in BU outreach. In contexts where multiple languages were spoken, CHWs were trained to act as translators from French or English into local dialects. Following community programs, BU screening took place, CHWs accompanied suspected cases to hospital and health posts, and followed up on confirmed cases. The outreach program was piloted in over 200 communities in the three countries and reached an estimated 75,000 people [56]. Evaluations carried out in each country confirmed that the program was effective in identifying significant numbers of BU cases and raising community confidence in clinic based care [16, 30].

3.2 Introducing Decentralized BU Care in Ouinhi, Benin

Although decentralized treatment of category I and II BU cases has been well established in several regions of Benin, it has not been introduced into the Ouinhi region. In this endemic region, centralized hospital care and surgery was the norm. Despite outreach efforts by a very proactive Catholic mission hospital, few early category I BU cases were brought to the hospital due in large part to fear of surgery. In the intervention, two local nurses were trained in the decentralized treatment of BU along with other types of chronic wounds. Outreach education programs were conducted and a message delivered that outpatient care was now available at local health posts for BU, and that staff had been trained to treat chronic wounds and early stages of BU without surgery. The community responded well to the intervention [30]. Self-referral following outreach programs increased significantly and many category I cases were treated at the clinic. Notably, over 70% of all BU cases brought to the clinic and confirmed as BU could be treated by decentralized care with over a 90% success rate. Community volunteer groups were established by social scientists to ensure that patients finished a complete course of medications and followed wound care advice. A lesson learned was the importance of treating all chronic wounds in the community as a means of establishing broad based clinic credibility. This project inspired research currently underway to develop a pragmatic approach to clinic and community based wound care management in Benin and Cote d'Ivoire.

3.3 Establishing a BU Community of Practice: Cameroon

In Cameroon, previous efforts to identify BU through the mobilization of CHWs yielded poor results. A majority of BU cases were first treated by traditional practitioners and few cases of category I BU were seen at hospital or health stations. Social scientists established a successful BU community of practice (BUCOP). A community of practice (COP) is an assemblage of stakeholders committed to a common objective, a common basic understanding of a focal problem, and mutual respect for what each stakeholder contributes to a process of problem-solving. In the case of BU, this entails health staff, CHWs, and traditional healers sharing a common understanding of the signs of BU, collaboration in encouraging the afflicted to seek and continue BU treatment, open lines of communication between stakeholders, and mutual respect for what each contributes to the process of healing that includes, but extends beyond the management of BU as a disease.

The success of the BUCOP has been measured in terms of numbers of suspected BU cases referred and confirmed, decline in patient treatment drop-out, and sustained collaboration among stakeholders both during and following the 3-year pilot project. The pilot project implemented the aforementioned BU outreach education program, increased levels of patient assistance beyond food and transport to include setting up halfway houses where patients could reside during treatment if they came from distant places, and established close working relations with traditional practitioners groups. Traditional healers not only referred cases of suspected BU to health posts; they also participated in outreach programs, and at the invitation of health staff provided psychosocial support for hospital patients. Nearly all of the 50 traditional healers participating in the BUCOP adhered to a collaboration contract, and after 2 years were presented with identification badges by district health authorities based upon their performance.

The success of the pilot project is far beyond what researchers had thought possible to achieve [16]. During the project over 90% of suspected and confirmed cases of BU seen at the Bankim hospital were referred by community stakeholders. Many category I BU cases were identified and treated successfully, and ethnic groups like the Fulani that previously had not sought treatment at the hospital began doing so. Notably, following 3 years of mass outreach education programs, the large majority of BU cases seen in Bankim Hospital were self-referred. Community members commonly checked with CHWs and healers if they suspected a case of BU before bringing the afflicted to a clinic. The status of healers, CHWs and the hospital all improved.

3.4 Yaws Cases Identified as a Result of BU Outreach in Cameroon: A Case for Integrated Skin Neglected Tropical Disease (NTD) Programs

One unanticipated outcome of BU outreach in Bankim, Cameroon was the identification of cases of yaws. Up until a household survey identified 29 cases of yaws, a year before the intervention [57], yaws was thought to be rare, if not eliminated. It had not been seen at hospital or health posts in Bankim district for years. Mass BU outreach programs attracted community members with other chronic skin diseases. When yaws cases were recognized by health staff during wound screening, schools in the community were also screened. Over 850 cases of yaws were identified and treated [58]. This illustrates the effectiveness of mass outreach programs as well as the potential of integrated NTD programs. Notably, the demonstration effect of yaws treatment requiring only a single dose of azithromycin or penicillin, increased community confidence in clinic treatment for BU.

3.5 BU Children's Support Group Ghana

In Ghana a children's program was developed to accompany the community-based outreach program. The necessity for this intervention became clear when it was discovered that children having BU were stigmatized by classmates and often dropped out of school. Teachers did not intervene nor follow up on such children. Stop Buruli researcher Eric Koka realized that in addition to providing education about BU to students and teachers, the negative perception of BU needed to be transformed. An intervention was introduced that had three parts. First, students with category I–II BU were offered free transportation by motorcycle to a clinic, where they were treated and then returned to school. Motor cycle taxi rides are seen by children as an exciting adventure. Once at the clinic children were also given a nutritious snack, which they relished. A second component was establishing a peer mentorship program. Children who had been successfully treated mentored new patients and helped them deal with the fear of daily injections and informed them about the healing process, as well. Children found having a peer mentor reassuring. Third, an BU patients club was formed. T-shirts identifying patients and ex-patients were proudly worn and members were asked to be on the lookout for new cases as well as to give testimonials on the healing process at school.

Eleven former patients' clubs were formed between 2011 and 2014 involving 132 school children. Most of the club members were between the ages of 6 and 18 years with a median age of 12 years. A more positive and less stigmatized identity created by club activities contributed to more children willing to come for treatment, and more parents willing to have their children screened and treated. In the 3 years prior to the intervention (2007–2010) 18 of the 32 BU patients attending Obom health centre were school children. During the 3 years of the intervention (2011–2014) the number of BU cases increased to 273 of which 145 were children. Notably, out of the 145 school children receiving treatment, 138 were category one cases, and 42 (30%) were referred by members of the children's BU club [59].

3.6 Mhealth as a Tool in Monitoring BU Healing in Ghana

Social scientist Mercy Ackumey teamed up with physician Nana Kotey to experiment with using mobile phones to monitor wound and scar care in Ghana. Oral therapy has been on the horizon of BU care for some time and in 2017 the WHO Technical Advisory Group on BU provisionally recommended treatment with oral clarithromycin and rifampicin pending the full results of a clinical trial. Anticipating this therapeutic option, the researchers deemed it imperative to test how wound care might be monitored by mobile phone if patients were not visiting health posts daily for BU treatment. Mobile phones are widely available in the region of Ghana where the project took place. The team created a check list of possible signs of a wound that was not healing well. Using the check list, short, semi-structured phone calls were made to patients and questions asked about the status of their wound. If danger signs were indicated, the patient was asked to visit the clinic. Patients found the calls reassuring and the intervention enhanced the reputation of the clinic. The pilot project identified several issues that will need to be addressed in the future should the approach go to scale [60].

3.7 Transforming a BU Hospital into a Therapeutic Community for Inpatients

Reducing social distance between hospital staff and patients and establishing clear lines of communication is a major challenge when providing in-patient care for people afflicted by BU and other chronic ulcers. Research on hospitals as therapeutic communities is virtually non-existent in Africa. An ethnography of Allada reference hospital in Benin responsible for treating BU and other chronic ulcers identified several sources of psychosocial distress and communication patterns compromising quality of care. Based on this research, an intervention was mounted to transform the hospital into a higher functioning therapeutic community. Question-answer education sessions were introduced to provide patients the opportunity to inquire about their illness and its treatment and trajectory, weekly open-forums were established to give patients and hospital staff a chance to air grievances, patient representatives met with hospital staff to resolve problems related to daily living in a non-confrontational manner, and psychosocial support for individual patients was provided through drop-in counseling sessions with social scientists in residence.

Patients reported positive changes in the quality of their care and interactions with care providers, care providers reported that the problem-solving process instituted was productive, and hospital administrators actively supported efforts to improve social relations and lines of communication [52]. Former patients were seen as a community outreach resource as both expert patients and ambassadors of the National BU control program were able to speak firsthand about the disease and the quality of care being offered. One systemic problem identified in both this and the Bankim project was community response to linear disease focused programs. Preferential treatment for BU patients in the form of subsidized treatment supported by the government and international funding agencies was deeply resented by patients being treated for other chronic wounds and who had to pay for their treatment out of their pocket. As far as patients and community members could see, all chronic wounds appear to warrant the same level of clinical attention, and special care for BU could not be defended on the basis of being contagious and therefore a community wide risk like TB.

4 Conclusion

Social science has contributed in important ways to health service research related to BU and to the development and testing of innovative community and clinic based interventions. This brief overview of research illustrates the importance of using mixed methods to investigate predisposing, enabling, and service related factors influencing health care seeking, treatment adherence, and subjective perceptions of quality of care. It also draws attention to the importance of community outreach and the active participation of community stakeholders in BU programs. This entails giving community health volunteers, former patients, and traditional healers a proactive and not just a passive role in outreach and patient support, learning how to establish respectful partnerships and effective lines of communication, and assessing how to keep all stakeholders motivated. Lessons learned from the BU interventions highlighted in this chapter have broad relevance to other NTD and (re)emerging disease preparedness programs. They also illustrate why integrated skin NTD and wound care programs are needed and ways in which linear programs, privileging the care of one kind of chronic wound over another, undermine community trust.

References

1. Kroeger A (1983) Anthropological and socio-medical health care research in developing countries. Soc Sci Med 17(3):147–161
2. Konan DO, Mosi L, Fokou G, Dassi C, Narh CA, Quaye C, Saric J, Abe NN, Bonfoh B (2018) Buruli ulcer in southern Côte D'ivoire: dynamic schemes of perception and interpretation of modes of transmission. J Biosoc Sci:1–14

3. Pearson G (2018) Understanding perceptions on 'Buruli' in northwestern Uganda: a biosocial investigation. PLoS Negl Trop Dis 12(7):e0006689
4. Grietens KP, Toomer E, Boock AU, Hausmann-Muela S, Peeters H, Kanobana K, Gryseels C, Ribera JM (2012) What role do traditional beliefs play in treatment seeking and delay for Buruli ulcer disease?–insights from a mixed methods study in Cameroon. PLoS One 7(5):e36954
5. Glick LB (1967) Medicine as an ethnographic category: the gimi of the New Guinea highlands. Ethnology 6(1):31–56
6. Nichter M (1987) Kyasanur forest disease: an ethnography of a disease of development. Med Anthropol Q 1(4):406–423
7. Good BJ, Good MJ, Togan I, Ilbars Z, Güvener A, Gelişen I (1994) In the subjenctive mode: epilepsy narratives in Turkey. Soc Sci Med 38(6):835–842
8. Whyte SR (2005) Uncertain undertakings: practicing health care in the subjunctive mood. In: Managing uncertainty. Museum Tusculanum, Copenhagen, pp 245–264
9. Mulder AA, Boerma RP, Barogui Y, Zinsou C, Johnson RC, Gbovi J, van der Werf TS, Stienstra Y (2008) Healthcare seeking behaviour for Buruli ulcer in Benin: a model to capture therapy choice of patients and healthy community members. Trans R Soc Trop Med Hyg 102(9):912–920
10. Aujoulat I, Johnson C, Zinsou C, Guédénon A, Portaels F (2003) Psychosocial aspects of health seeking behaviours of patients with Buruli ulcer in southern Benin. Trop Med Int Health 8(8):750–759
11. Kpadonou TG, Alagnidé E, Azanmasso H, Fiossi-Kpadonou E, Moevi AH, Niama D, Houngbédji G (2013) Psychosocioprofessional and familial becoming of formers Buruli ulcer patients in Benin. Ann Phys Rehabil Med 56(7–8):515–526
12. Boyer MN (2017) Questioning assumptions about decision-making in West African households: examples from longitudinal studies in Benin and Mali. PhD Dissertation, University of Arizona
13. Good BJ (1977) The heart of what's the matter the semantics of illness in Iran. Cult Med Psychiatry 1(1):25–58
14. Kennell J (2011) The senses and suffering: medical knowledge, spirit possession, and vaccination programs in Aja. PhD dissertation, Southern Methodist University
15. Sargent C, Kennell JL (2017) Elusive paths, fluid care. In: Olsen WC, Sargent C (eds) African medical pluralism. Indiana University Press, Indianapolis, p 227
16. Awah PK, Boock AU, Mou F, Koin JT, Anye EM, Noumen D, Nichter M, Stop Buruli Consortium (2018) Developing a Buruli ulcer community of practice in Bankim, Cameroon: a model for Buruli ulcer outreach in Africa. PLoS Negl Trop Dis. 12(3):e0006238
17. Giles-Vernick T, Owona-Ntsama J, Landier J, Eyangoh S (2015) The puzzle of Buruli ulcer transmission, ethno-ecological history and the end of "love" in the Akonolinga district, Cameroon. Soc Sci Med 129:20–27
18. Leach M, Scoones I (2013) The social and political lives of zoonotic disease models: narratives, science and policy. Soc Sci Med 88:10–17
19. Renzaho A, Woods PV, Ackumey MM, Harvey SK, Kotin J (2007) Community-based study on knowledge, attitude and practice on the mode of transmission, prevention and treatment of the Buruli ulcer in Ga West District, Ghana. Tropical Med Int Health 12(3):445–458
20. Ackumey MM, Gyapong M, Pappoe M, Maclean CK, Weiss MG (2012) Illness meanings and experiences for pre-ulcer and ulcer conditions of Buruli ulcer in the Ga-West and Ga-South Municipalities of Ghana. BMC Public Health 12(1):264
21. Ackumey MM, Gyapong M, Pappoe M, Weiss MG (2011) Help-seeking for pre-ulcer and ulcer conditions of Mycobacterium ulcerans disease (Buruli ulcer) in Ghana. Am J Trop Med Hyg 85(6):1106–1113
22. Owusu AY, Adamba C (2012) Household perceptions, treatment-seeking behaviour and health outcomes for Buruli ulcer disease in a peri-urban district in Ghana. Adv Appl Sociol 2(3):179–186. https://doi.org/10.4236/aasoci.2012.23024

23. Koka E (2018) Community knowledge and perceptions about Buruli ulcers in Obom Sub-District of the Ga south municipality in the Greater Accra region of Ghana. Adv Appl Sociol 8(09):621
24. Stienstra Y, van der Graaf WT, Asamoa K, van der Werf TS (2002) Beliefs and attitudes toward Buruli ulcer in Ghana. Am J Trop Med Hyg 67(2):207–213
25. Tschakert P, Ricciardi V, Smithwick E, Machado M, Ferring D, Hausermann H, Bug L (2016) Situated knowledge of pathogenic landscapes in Ghana: understanding the emergence of Buruli ulcer through qualitative analysis. Soc Sci Med 150:160–171
26. de Zeeuw J, Omansen TF, Douwstra M, Barogui YT, Agossadou C, Sopoh GE, Phillips RO, Johnson C, Abass KM, Saunderson P, Dijkstra PU (2014) Persisting social participation restrictions among former Buruli ulcer patients in Ghana and Benin. PLoS Negl Trop Dis 8(11):e3303
27. Ribera JM, Grietens KP, Toomer E, Hausmann-Muela S (2009) A word of caution against the stigma trend in neglected tropical disease research and control. PLoS Negl Trop Dis 3(10):e445
28. Agbo I (2015) The gendered impact of Buruli ulcer on the household production of health: why decentralization favors women. Tropical Med Int Health 20:298
29. Agbo I, Johnson C, Sopoh G, Nichter M (2019) The gendered impact of Buruli ulcer on the household production of health and social beyond: why decentralization favors women. Plos (forthcoming)
30. Amoussouhoui AS, Sopoh GE, Wadagni AC, Johnson RC, Aoulou P, Agbo IE, Houezo JG, Boyer M, Nichter M (2018) Implementation of a decentralized community-based treatment program to improve the management of Buruli ulcer in the Ouinhi district of Benin, West Africa. PLoS Negl Trop Dis 12(3):e0006291
31. Ackumey MM, Kwakye-Maclean C, Ampadu EO, de Savigny D, Weiss MG (2011) Health services for Buruli ulcer control: lessons from a field study in Ghana. PLoS Negl Trop Dis 5(6):e1187
32. Webb BJ, Hauck FR, Houp E, Portaels F (2009) Buruli ulcer in West Africa: strategies for early detection and treatment in the antibiotic era. East Afr J Public Health 6:144–147
33. Johnson RC, Makoutode M, Hougnihin R, Guédénon A, Ifebe D, Boko M, Portaels F (2004) Le traitement traditionnel de l'ulcere de Buruli au Benin. Med Trop 64(2):145–150
34. Koka E, Yeboah-Manu D, Okyere D, Adongo PB, Ahorlu CK (2016) Cultural understanding of wounds, Buruli ulcers and their management at the Obom sub-district of the Ga south municipality of the Greater Accra region of Ghana. PLoS Negl Trop Dis 10(7):e0004825
35. Nichter M (2008) Coming to our senses: appreciating the sensorial in medical anthropology. Transcult Psychiatry 45(2):163–197
36. Barogui YT, Sopoh GE, Johnson RC, de Zeeuw J, Dossou AD, Houezo JG, Chauty A, Aguiar J, Agossadou D, Edorh PA, Asiedu K (2014) Contribution of the community health volunteers in the control of Buruli ulcer in Benin. PLoS Negl Trop Dis 8(10):e3200
37. Ackumey MM, Gyapong M, Pappoe M, Maclean CK, Weiss MG (2012) Socio-cultural determinants of timely and delayed treatment of Buruli ulcer: implications for disease control. Infect Dis Poverty 1(1):6
38. Labhardt ND, Aboa SM, Manga E, Bensing JM, Langewitz W (2010) Bridging the gap: how traditional healers interact with their patients. A comparative study in Cameroon. Trop Med Int Health 15(9):1099–1108
39. Nichter M, Quintero G (1996) Pluralistic therapy systems: why do they co-exist and how are they used. In: Levinson D, Ember M (eds) Encyclopedia of cultural anthropology. Henry Holt, New York, pp 1–17
40. Ahorlu CK, Koka E, Yeboah-Manu D, Lamptey I, Ampadu E (2013) Enhancing Buruli ulcer control in Ghana through social interventions: a case study from the Obom sub-district. BMC Public Health 13(1):59
41. Owusu-Sekyere E (2013) The buruli ulcer morbidity in the amansie West District of Ghana: a myth or a reality? J Pub Health Epidemiol 5(10):402–409
42. Hausermann HE (2015) 'I could not be idle any longer': buruli ulcer treatment assemblages in rural Ghana. Environ Plan A 47(10):2204–2220

43. Langwick SA (2008) Articulate (d) bodies: traditional medicine in a Tanzanian hospital. Am Ethnol 35(3):428–439
44. Asiedu K, Etuaful S (1998) Socioeconomic implications of Buruli ulcer in Ghana: a three-year review. Am J Trop Med Hyg 59(6):1015–1022
45. Grietens KP, Boock AU, Peeters H, Hausmann-Muela S, Toomer E, Ribera JM (2008) "It is me who endures but my family that suffers": social isolation as a consequence of the household cost burden of Buruli ulcer free of charge hospital treatment. PLoS Negl Trop Dis 2(10):e321
46. Amoakoh HB, Aikins M (2013) Household cost of out-patient treatment of Buruli ulcer in Ghana: a case study of Obom in Ga south municipality. BMC Health Serv Res 13(1):507
47. Chukwu JN, Meka AO, Nwafor CC, Oshi DC, Madichie NO, Ekeke N, Anyim MC, Chukwuka A, Obinna M, Adegbesan J, Njoku M (2017) Financial burden of health care for Buruli ulcer patients in Nigeria: the patients' perspective. Int Health 9(1):36–43
48. Goudge J, Govender V (2000) A review of experience concerning household ability to cope with the resource demands of ill health and health care utilisation. EQUINET
49. Berman P, Kendall C, Bhattacharyya K (1994) The household production of health: integrating social science perspectives on micro-level health determinants. Soc Sci Med 38(2):205–215
50. Nichter M, Kendall C (1991) Beyond child survival: anthropology and international health in the 1990s. Med Anthropol Q 5(3):195–203
51. Agbenorku P, Agbenorku M, Amankwa A, Tuuli L, Saunderson P (2011) Factors enhancing the control of Buruli ulcer in the Bomfa communities, Ghana. Trans R Soc Trop Med Hyg 105(8):459–465
52. Amoussouhoui AS, Johnson RC, Sopoh GE, Agbo IE, Aoulou P, Houezo JG, Tingbe-Azalou A, Boyer M, Nichter M (2016) Steps toward creating a therapeutic community for inpatients suffering from chronic ulcers: lessons from Allada Buruli Ulcer Treatment Hospital in Benin. PLoS Negl Trop Dis 10(7):e0004602
53. Abass KM, Van Der Werf TS, Phillips RO, Sarfo FS, Abotsi J, Mireku SO, Thompson WN, Asiedu K, Stienstra Y, Klis SA (2015) Buruli ulcer control in a highly endemic district in Ghana: role of community-based surveillance volunteers. Am J Trop Med Hyg 92(1):115–117
54. Bourdieu P (1986) The forms of capital. In: Richardson JG (ed) Handbook of theory and research for the sociology of education. Greenwood Press, New York, pp 241–258
55. Magrath P, Nichter M (2012) Paying for performance and the social relations of health care provision: an anthropological perspective. Soc Sci Med 75(10):1778–1785
56. Nichter M, Amoussouhoui A, Ferdinand M, Koka E, Kum AP, Mbah E, Tohnain K, Boyer M (2015) Buruli ulcer outreach education: an exemplar for community based tropical disease interventions. Oral presentation, World Health Organization Biannual Buruli Conference
57. Bratschi MW, Bolz M, Minyem JC, Grize L, Wantong FG, Kerber S, Tabah EN, Ruf MT, Mou F, Noumen D, Boock AU (2013) Geographic distribution, age pattern and sites of lesions in a cohort of Buruli ulcer patients from the Mapé Basin of Cameroon. PLoS Negl Trop Dis 7(6):e2252
58. Boock AU, Awah PK, Mou F, Nichter M (2017) Yaws resurgence in Bankim, Cameroon: the relative effectiveness of different means of detection in rural communities. PLoS Negl Trop Dis 11(5):e0005557
59. Koka E, Okyere D, Aboagye S, Ahorlu CK, Yeboah-Manu D, Nichter M. Mobilising youth in support of Buruli ulcer outreach in Southern Ghana: lessons learned from former patients groups. Unpublished manuscript
60. Ackumey M, Kotey NK, Nichter M (2015) Using cell phones to improve compliance to treatment and monitor wound care in the Nsawam-Adoagyiri and Akwapem-South Districts of Ghana. Oral presentation at World Health organization Biannual Buruli Conference, March 24, 2015

Transdisciplinary Research and Action to Stop Buruli Ulcer: A case Study from Philanthropy

Susanna Hausmann-Muela and Ann-Marie Sevcsik

1 Introduction

Over the past decades, advances in Buruli ulcer (BU) research and control have been substantially shaped by philanthropic and civil society engagement. It has been a long journey for advocates to attract public attention and funding for BU. In 1998, the World Health Organization (WHO) established the Global Buruli Ulcer Initiative, supported by the Nippon Foundation. A first international conference on BU, organized by WHO, led to the 1999 Yamoussoukro Declaration, which brought greater public awareness to the disease, for the first time [1].

Intensified research and control activities started thereafter. In 2004, the World Health Assembly adopted a resolution on BU, which called upon the international community, organizations and bodies of the United Nations system, donors, non-governmental organizations, foundations and research institutions to cooperate directly with disease-endemic countries in order to strengthen control and research activities; and to develop partnerships and to foster collaboration with organizations and programs involved in health-system development, in order to ensure that effective interventions can reach all those in need [2].

Despite increased attention and the coordinated efforts by the WHO-Global Buruli Ulcer Initiative, research and control activities remained highly underfunded, with advances limited. In 2007, the UBS Optimus Foundation (UBS-OF) identified BU as a disease where a relatively small investment had the potential to make a big difference. With the goal to catalyse a transdisciplinary consortium bundling BU

Susanna Hausmann-Muela was responsible for the UBS Optimus Foundation led Stop Buruli Initiative from 2007–2011, Ann-Marie Sevcsik from 2012–2018.

S. Hausmann-Muela (✉)
Fondation Botnar, Basel, Switzerland
e-mail: shausmann@fondationbotnar.org

A.-M. Sevcsik
UBS Optimus Foundation, Zürich, Switzerland

G. Pluschke, K. Röltgen (eds.), *Buruli Ulcer*,
https://doi.org/10.1007/978-3-030-11114-4_16

research into a concerted effort, the UBS-OF engaged researchers from different disciplines and countries, with a prominent role of endemic countries and good links to national BU control programs.

Philanthropic engagement aspires for innovation and impact. This chapter takes a critical look back at the whys, hows, and major results of the Stop Buruli Initiative initiated by the UBS-OF and partially co-funded by the Medicor Foundation. It discusses the long-term and comprehensive approach of the UBS-OF in research and control of BU, and the lessons learned for achieving the aspiration of innovation and impact.

2 The Beginnings: Six Reasons Why

2.1 Reason #1: Lack of Funding Incentives

BU is one of the Neglected Tropical Diseases (NTDs) listed under the WHO classification system. These disease conditions disproportionately affect populations living in tropical and sub-tropical areas and people living in poverty.

Research and development efforts for BU were marginal. Like for many other NTDs, industry has no incentive to invest in products such as diagnostics, medicines, vaccines and other control tools defined by global health needs or public priorities and targeting populations and governments with low purchasing power. According to G-Finder data, total investment in BU research by the private sector between 2007 and 2016 was merely around USD 300,000.

Furthermore, philanthropic and public research funding for BU has been far lower than for other tropical diseases. Between 2007 and 2016, BU research funding amounted to as little as 38 million USD in total. Slightly more than one third (14 million USD) originated from philanthropic sources, with the remaining (24 million USD) from the public sector, which majorly invested in basic or vaccine research. The biggest single commitment was the 2010–2013, EU-supported project to develop a vaccine protecting against BU (BuruliVac).

The lack of interest and funding for BU also affects the academic world. BU remains one of the least studied diseases because this is not where researchers can make a great academic career. Little attention means small academic impact, difficulties to publish in high-ranking journals and to acquire research funding. For a foundation with relatively small grant-making resources, this 'niche' situation was a compelling argument to invest in BU.

2.2 Reason #2: A Disease of Social Justice

The second compelling argument was based on social justice and a desire for global health equity. The neglect of the disease goes along with the neglect of the affected populations. BU affects the most vulnerable and voiceless people. Children living in remote and poverty-stricken areas of Africa are suffering the most.

Findings from epidemiological studies indicated that the BU burden in Africa was underestimated [3] and solid world-wide prevalence data were not available. BU was reported in 33 countries, but not all endemic countries regularly reported case numbers to the WHO and in some countries, endemic regions still have to be mapped. Assessment of the global disease burden was impeded by the remoteness and poverty of the most affected populations and countries. The Global Burden of disease study 2016, the most comprehensive worldwide observational epidemiological study, did not include estimates on BU burden.

Organizations working on BU were confronted with a spiral of neglect: lack of epidemiological data veiled the evidence of the real dimensions of the disease. This, in turn, compromised attention and support, which hindered the collection of more robust and comprehensive evidence. This, characteristic for most neglected diseases, is a matter of social justice, exacerbating the shadowy existence of the victims of the disease.

2.3 Reason #3: A Disease with Devastating Consequences

The poorest populations, lacking access to health care, were the least served. Even though case fatality may be low, the severity and chronic nature of the disease often leads to devastating consequences of BU for the individual, their families and the affected communities. When diagnosed and treated early, BU can be a well-manageable disease. However, at the time, the majority of patients did not seek biomedical treatment until the condition was in later stages. Massive extension of BU lesions and delayed wound healing often causes irreversible physical disability due to deformities on arms, legs, trunks and in the face. Alone the painful changing of wound dressings already makes the disease a human tragedy for all affected. Stigma, discrimination, and the need for long-term hospital stays far away from home aggravated the social and economic conditions for affected households. The cost burden due to long-term treatments far from home and productivity loss of BU patients, and of parents accompanying children during hospital stays, can amount to catastrophic levels for the household economy, not seldom leading to abandonment of patients by the family [4].

A human-rights based perspective justified a global response to BU, but attention failed to increase.

2.4 Reason #4: The Pioneer Effect

While the needs-driven perspective to reduce inequity, social injustice and human suffering was convincing for philanthropic support, there was also the argument of likelihood for success. Philanthropic support tends to be open to taking risks, but only if these are calculated. Investment into BU research was likely to show a positive benefit/risk balance.

BU, coined *the mysterious disease*, was characterized by a great vacuum of knowledge and little understanding by researchers, implementers, and affected persons themselves. It was a disease where nearly nothing was known regarding transmission and epidemiology, where clinical presentation was difficult to interpret even by experts, where diagnostic procedures were complicated, costly and often leading to false results, where wide surgical excision of the affected tissue was the only recommended treatment option, where efficacy of antibiotic therapies was debated, where patients often had no access to treatment and where it was not clear how to reduce the risk of acquiring the disease.

To start to bring light into such nearly total darkness is a way to become a pioneer in an area where very few others are engaged but may join the bandwagon after they see success. The opportunity to be at the forefront of research and disease control, and to act as a catalyst by convening a variety of stakeholders and partners, was one of the most convincing arguments for the UBS-OF to engage in a multi-disciplinary BU initiative.

This pioneering effect in combination with the prospect that relatively small investments can make a real difference for many patients and their families was a powerful argument. The reason why little money could lead to big impact was also related to the fact that there was no need for big investments into new infrastructure, since much of the activities could piggyback on existing and well-established research and health care structures.

2.5 Reason #5: Existing Critical Mass of Researchers and Implementing Agencies

Research groups working on BU already existed, although the primary foci of researchers in the field were usually other mycobacterial diseases, like tuberculosis and leprosy, and not BU. Researchers with suitable expertise were therefore available and at the time when the Stop Buruli Initiative was launched, it was not the absence of scientific know-how or interest, but the lack of funding and coordination that impeded scientific progress.

A network analysis of institutions active in BU research at the time showed how small yet fragmented the scientific community was. Organizations active in BU research spanned all over the world, over four continents, with very diverse disciplines, expertise and experience.

The existence and diversity of this critical mass of researchers and institutions willing to engage in BU research was essential for rapid progress and potential for impact.

2.6 Reason #6: Scalability and Uptake

Despite fragmentation, a network of collaborations between institutions from different affected countries and national disease control programs already existed. The network analysis of institutions and publications concluded that most BU research was quite recent, mainly starting in the 1990s, but fast-growing especially in the early years of 2000 when the WHO Global Buruli Disease Initiative was established. By the year 2007, nearly a dozen experts had published more than 50 articles, and roughly 20 institutions worldwide were involved in research, with a high density of inter-institutional collaborations. Research activities were on-going primarily in four countries with BU case reports: Australia, Benin, Cameroon and Ghana.

Existing longstanding institutional partnerships between these four countries and researchers in the USA, Belgium, and Switzerland were a key element for setting up the new consortium. The organizational anchoring and partnership-based collaborations with national Ministries of Health and National BU control programmes were crucial. This permitted scientists to connect with practitioners and to consolidate and translate research efforts into action. From the beginning, favourable conditions existed for access to patients and specimens, as well as a high likelihood of integrating new findings into policy guidelines and practice. The WHO Global Buruli Ulcer Initiative, with its annual meetings that brought together all relevant stakeholders from across the globe and the Technical Advisory Group that included many of the researchers, enabled translation of key findings into normative guidelines at the global level.

2.7 Stop Buruli Initiative

In sum, the decision to invest in BU research and control followed a set of principles and reflections. An initiative was defined by the UBS-OF as "a relevant and neglected field, with a potential for change, where the foundation can take a leading role." BU matched the four elements contained in the definition. It is a *relevant* disease, as emphasized by international declarations and a World Health Assembly adopted resolution in 2004; it is *neglected*, i.e. not in the spotlight and receiving scarce resources due to both market and policy failures; there is a *potential for change* by helping to develop standard know-how and tools for diagnostics and therapies that are being implemented throughout African and other affected areas; and the UBS-OF can take a *leading role* through convening key stakeholders, creating cohesion among them, and attracting new funders to get aligned in a common effort to address the problem.

3 The Model: Transdisciplinary Research and Implementation

In 2007, the UBS-OF invited a core group of institutions to develop a consortium proposal with an overall research agenda, under the guiding principle to lead from innovation over validation of the findings and interventions to public health application. The invitation specified the major lines by mentioning transmission, ecological, social and economic, as well as clinical aspects, and by considering a focus on developing applicable and affordable tools; an integrated multidisciplinary approach; research capacity strengthening, including technology transfer and training of researchers from BU endemic countries; strengthening of North-South and South-South partnerships.

The consortium partners initially included seven institutions, with key expertise in different fields, including the Swiss Tropical and Public Health Institute (SwissTPH), Switzerland (research coordinator); Fairmed (former Aide Aux Lépreux Emmaüs-Suisse (ALES)), Cameroon; Institute of Tropical Medicine (ITM), Belgium; Noguchi Memorial Institute for Medical Research (NMIMR), Ghana; Programme national de lutte contre l'ulcère de Buruli et la lèpre, Benin; University of Melbourne, Australia; and University of Tennessee, USA.

This was the start of a model for a transdisciplinary approach, linking research to implementation, building local capacity across disciplines and through co-creating knowledge leading to action and, ultimately, impact for patients, their families, and countries. The role of the UBS-OF was to bring the diverse stakeholders together and to point the way to generate the tools and solutions that were needed, but also to advocate for and to ensure the taking up of those tools under real life conditions.

4 Key Elements for Advancing Transdisciplinarity

4.1 Transdisciplinarity Does Not 'Just Happen'

Transdisciplinary research agendas are often specified as a goal in the context of research for development promoted by development agencies or philanthropic grants. But experience shows how difficult it is to work across disciplines, and to consider policy and practice as an integral part of the research agenda. This requires patience, persistence and willingness to adapt the course based on factors internal and external to the initiative.

One key element is to convene the right disciplines and partners, and to jointly discuss and develop the aims and actions. But multi-disciplinary consortia are, by definition, complex constructs embracing not only different disciplines, but also different institutional cultures. Such a research consortium is of paramount complexity, since the element of North-South partnerships with all its socio-politically rooted facets adds to the already complex nature of a multi-stakeholder collaboration. A research consortium does not develop in a neutral space. Wanted or not, explicit or implicit, power relations among different partners may lead to unequal decision-making and imbalanced resource allocations.

To bring all disciplines and different partners into alignment and to foster an effective and efficient collaboration requires an active orchestration of the partnership. Beyond the financial resources, it is the convening power and the agile management of an engaged, unbiased donor that brings added value to the functioning of a complex consortium.

The development of the social sciences within the Stop Buruli consortium provides an illustrative example for the relevance of such an active and agile management approach. The consortium embraced research and action along four axes: disease transmission, diagnosis, treatment, and socio-economic and cultural aspects. The UBS-OF convened representatives from all different disciplines right from the beginning, and all partners showed great willingness to develop a joint effort along the four axes. But the social science axis started to lag behind, unable to integrate their research questions into the overall consortium agenda. The first social science proposal suggested a classic information, education, and communication approach with the aim to educate the population about signs and symptoms of BU, but ignored to consider the research priorities of the other axes. The misalignment of the social science research risked developing into a parallel direction, leading to a fragmentation of workstreams instead of the concerted action that had been envisioned.

4.2 Governance Model

The key to redirecting the social sciences towards a path of convergence with the other disciplines was not to request the researchers to rephrase and resubmit proposals, but to adapt the consortium's governance structure. The consortium had constituted a steering committee with representatives from the seven different partner organizations, but lacked a representative from a strong social science partner. In other words, the social scientists were within the research teams but lacked a voice at the higher level of decision-making, i.e. at the level of the steering committee.

The solution to the problem was to seek and include an eighth partner, and to add a senior-level social science representative to the steering committee. This is how, upon request of the UBS-OF, the Anthropology Unit of the University of Arizona joined the consortium. That way, the research directions discussed and decided by the steering committee gained in quality concerning transdisciplinarity.

The governance model of the Stop Buruli Consortium then consisted of eight members of the steering committee and for reasons of agility, a day to day management team of three rotating members, overseeing the different work packages along the four axes. The close relation of the steering committee with the national BU control programs in the three BU endemic African countries involved (Ghana, Benin, Cameroon) and with the WHO Technical Advisory Group for BU permitted flexibility for the consortium activities, but at the same time close interaction with the Ministries of Health for collecting data and developing recommendations and guidelines, nationally and globally. The regular WHO Meetings on BU control and research served as an important platform for exchange with other stakeholders, which enabled results from the research activities to be brought into the broader context and influence policy discussions.

5 Results

The three funding cycles (2007–2018) powered by the UBS-OF brought important advancements to research and control of BU as well as educating a future generation of African scientists. The broad transdisciplinary focus of the consortium with the integrated action not only secured an efficient and effective use of the funds, but it also leveraged funding from other donors, which continue with relevant, concerted research and action to this day. As the planning for the third cycle began, it was recognized by the UBS-OF and the consortium members that important channels had been built across disciplines and new partnerships formed, including a product development partnership with the Foundation for Innovative New Diagnostics (FIND), but that a formal structure to facilitate that cross-talk was no longer needed.

The research output of the consortium led to more than 60 publications, training of 25 African graduate students, more than 9000 BU sufferers having received improved treatment and education, as well as three hands-on, field workshops in Africa that allowed for professional and practical training of young scientists by senior experts in laboratory-based and field research. The consolidated approach achieved considerable success along the four axes. The following sections outline some of the major achievements and future perspectives:

5.1 Diagnosis

Diagnosis is a major challenge in BU control. District hospitals often rely on clinical diagnosis only, which is very often inaccurate and complicated by other disease conditions with similar presentation as BU. There is over/under treatment as a result of misdiagnosis and poor surveillance of the disease due to absence of effective diagnostic services. It is therefore important that clinical diagnosis is supported by laboratory confirmation.

The diagnostic field recorded impressive progress. The consortium optimized the method of fine needle aspiration for improved collection of specimens from closed lesions [5]. An illustrative and open access video-guide of how-to-do-it and WHO supported teaching and training activities improved the accuracy of diagnosis undertaken by qualified health staff in BU affected countries. In addition, optimization of PCR-based diagnosis by ITM improved laboratory confirmation of clinical cases [6, 7].

However, PCR technology based diagnosis is expensive, requires time and experienced health staff and technology, which are only available at specialized BU reference laboratories, which are far away from the affected patients and communities. Therefore, high priority was given to the development of rapid, low-tech, sensitive and specific diagnostic tests for use at district hospital and at community levels.

The identification of a *M. ulcerans* protein suitable for the development of an antigen capture assay [8] allowed to develop a simple ELISA executable at district hospital level. In collaboration with the FIND and Abbott/Standard Diagnostics

(SD), the monoclonal antibodies generated by SwissTPH are now used in further research aimed at the development of a point-of-care test. First prototypes of rapid diagnostic tests (RDTs) are under laboratory validation; if successful, these will pave the way for surveillance and early diagnosis of BU at the community and primary health care levels.

Another promising avenue for diagnostic test development is a technology known as loop-mediated isothermal amplification (LAMP) of pathogen DNA. This approach has recently been exploited in development of simple molecular tests for many infectious diseases, among them tuberculosis, sleeping sickness, leishmaniasis, malaria and Chagas disease. The Noguchi Memorial Institute for Medical Research (NMIMR) in Ghana and the Department for Infectious Diseases and Tropical Medicine (DITM) in Germany, in collaboration with FIND and other partners, are advancing the development of a Buruli LAMP test, which is highly accurate, and does require less specialised training or equipment than PCR.

The macrolide exotoxin mycolactone of *M. ulcerans* plays a crucial role in the pathogenesis of BU. Therefore major efforts have been made to raise mycolactone-specific antibodies. However, in spite of numerous attempts by several expert laboratories, generation of antibodies against mycolactone had failed for a long time. Based on quantitative structure-activity relationship analysis with synthetic mycolactones [9] SwissTPH has successfully developed a new immunization strategy, using a protein carrier conjugate of a synthetic truncated non-toxic mycolactone derivative. It has turned out that this formulation elicits mycolactone specific antibodies that neutralize the toxin [10]. This provides the basis for a new approach for BU vaccine design supported by the Medicor Foundation (see below). Furthermore, the *M. ulcerans* specific mycolactone may represent an ideal target for the development of a specific diagnostic test and the successful generation of mycolactone-specific antibodies has opened new options for diagnostic test development, an approach followed by SwissTPH in collaboration with FIND.

5.2 Treatment

The Stop Buruli consortium has also achieved highly relevant findings for BU treatment. Concerns that secondary lesions after streptomycin/rifampicin combination therapy given for 8 weeks were linked to relapses or failures of treatments could be discarded, supporting the WHO recommendations in the treatment guidelines for health workers.

A specific focus was given to wound care and healing. One of the findings showed that delay in wound healing may be related either to clinical deterioration, known as paradoxical reactions, or to secondary bacterial infections. It could also be shown by the NMIMR in collaboration with SwissTPH and the University of Melbourne that BU wounds are frequently super-infected with hospital-acquired multi-resistant bacteria [11–13].

The early pathogenesis of BU and the molecular mechanism of tissue necrosis caused by the toxin mycolactone produced by *M. ulcerans* were investigated [14–16] providing important insights relevant for diagnosis and treatment.

Wound management practices at primary and secondary health care facilities in Ghana were investigated and recommendations for improvements made [17]. In Benin a decentralized community-based treatment program for BU was implemented and assessed by the national control program supported by the University of Arizona [18]. The impact evaluation found that community confidence in decentralized BU care was greatly enhanced by clinic staff who came to be seen as having expertise in the care of most chronic wounds. The pilot program further demonstrates the added value of integrated wound management for skin NTD control. The model developed in this pilot study may serve as a foundation and proof of concept for a larger community-based decentralized wound care agenda.

A major conclusion of the consortium was that BU control should be integrated into horizontal programs for the integrated diagnosis and treatment of tropical skin diseases. WHO advocates for this approach and the UBS-OF continues to support a pilot project in Benin and Ivory Coast that includes several Stop Buruli members as well as the NGO Anesvad.

5.3 Transmission

Despite intensive environmental studies carried out by the consortium, the exact mode of transmission of *M. ulcerans* still remains unknown. However, studies brought new insights which contribute to the better understanding of the puzzle of transmission. For example, certain correlations between environmental factors and BU incidence could be shown by researchers from the NMIMR, which confirmed behavioural risk factors for contracting the disease, such as farming in swampy areas or application of leaves on wounds [19]. Sero-epidemiological studies showed that exposure to *M. ulcerans* increases at an age of about 4 years. However, many exposed individuals do not develop clinical disease [20]. Findings in *M. ulcerans* infected possums in Australia led researchers from the Melbourne University to propose that humans might harbour *M. ulcerans* in their gastrointestinal tract and shed the bacterium in their faeces. A study with faecal samples of BU patients from Ghana concluded however, that the human gastrointestinal tract is unlikely to be a significant reservoir of *M. ulcerans.*

M. ulcerans DNA has been detected in many environmental, in particular aquatic sources, but how bacteria enter the skin tissue is not clear. The hypothesis of bacteria entering through open wounds was tested using a guinea pig model by researchers of the University of Tennessee. While injection of *M. ulcerans* into the skin led to the development of lesions, the researchers were unable to produce infection via topical application of *M. ulcerans* on open abrasions. These results suggest that BU is not likely to be due to passive entry of bacteria into deeper skin tissue via superficial skin abrasions [21].

Evidence also showed that *M. ulcerans* is growing in amoeba, but this does not seem to represent an important environmental reservoir. In Australia, possums were found to develop clinical BU, and to serve as an animal reservoir in the Buruli endemic region of Victoria, but no comparable reservoir was found in Africa. While *M. ulcerans* DNA was found in the faeces of BU-affected possums in Australia, no such DNA could be demonstrated in faecal samples of human patients in Ghana [22].

A breakthrough was achieved through whole genome sequencing analyses which provided insights into the diversity, population structure and evolutionary history of *M. ulcerans*. The identification of local clonal complexes revealed that transmission over large distances is rare [23–26].

All in all, transmission of BU remains an enigma despite many insights into details relevant for transmission.

5.4 Socio-Cultural Aspects

The social science axis of the consortium investigated the cultural perceptions of BU and socio-economic factors influencing BU health-seeking and decision making. Using a formative research approach, interventions to improve early access of BU patients to diagnosis and treatment were designed and validated. In Cameroon, a BU community of practice involving multiple community stakeholders from clinic staff and former patients to community health workers and traditional healers resulted in improvement of case identification, reduction in health care seeking delay and an increase in treatment adherence [23, 27]. The process of social science intervention and major lessons learned from the Bankim case study in Cameroon are described in detail in chapter “Social Science Contributions to BU Focused Health Service Research in West-Africa” of this book. In Benin, establishing a therapeutic community in an in-patient ward resulted in positive impact on health staff and patient relations [28]. In Ghana, enhanced capacity of health staff in antibiotic treatment and wound care and community-based surveillance, patients screening and referrals to medical treatment improved early detection of disease [29]. In all three countries, pilot studies showed how patient support, such as transport or food offers improved treatment-seeking and adherence. The active engagement of different stakeholders from the communities, including former patients and traditional healers, addressing real life problems of the affected and their families contributed to improved early treatment and management of BU cases. The social science contributions and outcomes of the community-based studies and interventions are described in detail in the aforementioned “Social Science” chapter of this book.

6 Lessons Learnt and Conclusions

Transdisciplinary research and action approaches are promising and currently much promoted by many institutions. The scope is multidimensional, and the aim is to create impact based on research and evidence that is directly applicable to real-life conditions. The first important lesson to learn from the Stop Buruli Initiative is the key role of the funder as a convening partner of a diverse group of partners, bringing the right expertise to the table and facilitating co-development of research and intervention.

The second lesson, related to the complexity of multidimensionality, is the relevance of time. Collaborating across disciplines is challenging, and requires sufficient space and time for exchange and mutual learning. From the beginning, the Stop Buruli Initiative had a time horizon for support of 9–10 years, with 3-year cycles, and regular meetings. The long-term commitment permitted the consortium to adapt its priorities and approaches based on interim findings as well as to strengthen research capacity and to pave the way for sustainable solutions.

The organization of the overall activities into different work packages certainly helped to structure the research along clear topics, with defined milestones and deliverables for all work streams. However, the third important lesson that can be drawn from the consortium is the importance of good governance and an agile management of the donor. To develop, test and validate research results in the field, considering the political, social and economic conditions, is not a direct input-output-outcome equation, but requires dialogue and continuous adjustments. Such dialogue requires an institutionally neutral agency, which has the legitimacy, is able and sufficiently flexible to enable shared learning and to change course of action if needed. The experience of the social sciences' initial lack of integration illustrates how important the governance model with equal representation of different disciplines and partners is, and how flexible, responsive and patient a donor needs to be.

The fourth lesson to learn from the model is the weight of communication and advocacy for leverage. By actively reaching out to other potential donors, offering well-established and reliable platforms for collaboration, lines of research and action could be further expanded. Support by additional donors allowed to peruse promising outcomes and partnerships from the consortium leading to further transdisciplinary research projects.

The Medicor Foundation, a philanthropic foundation based in Triesen, Liechtenstein, supported over a period of 10 years multidisciplinary research on BU. In a first phase, biomedical and field research activities in Cameroon and Benin, aiming at the improvement of diagnosis and treatment of BU were facilitated. Motivated by promising results, Medicor supported in a second phase diagnostics and vaccine research focussed on the key virulence factor of *M. ulcerans*, the macrolide toxin mycolactone. Within a 6-year cooperation, early product development activities have generated concept validation and candidate products. These lead candidates are now handed over to other funders, such as FIND (see below), for full product development. In a third phase, the Medicor Foundation is currently supporting a project aiming at the demonstration that thermotherapy of BU can be successfully implemented at health post level.

To consider funding, it was important for the Medicor Foundation, (1) to work with committed, professional, trusted and experienced partners and (2) that specific research projects were defined that were expected to generate with limited funding results within a period of 2–3 years. An important conclusion of the 10-year experience with research support for BU is that on the other hand only a long-term commitment covering several project phases can generate tangible results.

The Foundation for Innovative New Diagnostics (FIND) has supported development of diagnostics for BU since 2014, with funding from various sources, including UBS-OF, the Swiss Agency for Development Cooperation (SDC), the German Federal Ministry of Education and Research through KfW, and recently Anesvad.

The network of partnerships established through the Stop Buruli Consortium contributed immensely in identification of the unmet diagnostic needs for BU, development of the FIND strategy for the disease, and access to clinical samples for use in test validation.

In conclusion, the coordinated effort advanced through the model of the Stop Buruli consortium led to a substantial push in BU research and control. While the big philanthropic organizations are more prone to address big societal problems, smaller foundations are well positioned to address 'niche' problems that remain otherwise unaddressed, but where they can have a big impact, if well-managed and with the right partners in place, both in their niche and also serving as a model.

References

1. WHO (2018) History of the GBUI. http://www.who.int/buruli/gbui/en/. Accessed 08 Jun 2018
2. Resolution WHA57.1 (2004) Surveillance and control of *Mycobacterium ulcerans* disease (Buruli ulcer). In: Fifty-seventh world health assembly, Geneva, 17–22 May 2004. Resolutions and decisions (WHA57/2004/REC/1). World Health Organization, Geneva, pp 1–2
3. Röltgen K, Pluschke G (2015) Epidemiology and disease burden of Buruli ulcer: a review. Res Rep Trop Med 6:59–73. https://doi.org/10.2147/RRTM.S62026
4. Peeters Grietens K, Um Boock A, Peeters H, Hausmann-Muela S, Toomer E, Muela Ribera J (2008) "It is me who endures but my family that suffers": social isolation as a consequence of the household cost burden of Buruli ulcer free of charge hospital treatment. PLoS Negl Trop Dis 2(10):e321. https://doi.org/10.1371/journal.pntd.0000321
5. Eddyani M, Fraga AG, Schmitt F, Uwizeye C, Fissette K, Johnson C, Aguiar J, Sopoh G, Barogui Y, Meyers WM, Pedrosa J, Portaels F (2009) Fine-needle aspiration, an efficient sampling technique for bacteriological diagnosis of nonulcerative Buruli ulcer. J Clin Microbiol 47:1700–1704. https://doi.org/10.1128/JCM.00197-09
6. Affolabi D, Sanoussi N, Vandelannoote K, Odoun M, Faïhun F, Sopoh G, Anagonou S, Portaels F, Eddyani M (2012) Effects of decontamination, DNA extraction, and amplification procedures on the molecular diagnosis of *Mycobacterium ulcerans* disease (Buruli ulcer). J Clin Microbiol 50:1195–1198. https://doi.org/10.1128/JCM.05592-11
7. Eddyani M, Lavender C, de Rijk WB, Bomans P, Fyfe J, de Jong B, Portaels F (2014) Multicenter external quality assessment program for PCR detection of *Mycobacterium ulcerans* in clinical and environmental specimens. PLoS One 9:e89407. https://doi.org/10.1371/journal.pone.0089407

8. Dreyer A, Röltgen K, Dangy JP, Ruf MT, Scherr N, Bolz M, Tobias NJ, Moes C, Vettiger A, Stinear TP, Pluschke G (2015) Identification of the *Mycobacterium ulcerans* protein MUL_3720 as a promising target for the development of a diagnostic test for Buruli ulcer. PLoS Negl Trop Dis 9:e0003477. https://doi.org/10.1371/journal.pntd.0003477
9. Scherr N, Gersbach P, Dangy J-P, Bomio C, Li J, Altmann K-H, Pluschke G (2013) Structure-activity relationship studies on the macrolide exotoxin mycolactone of *Mycobacterium ulcerans*. PLoS Negl Trop Dis 7:e2143. https://doi.org/10.1371/journal.pntd.0002143
10. Dangy J-P, Scherr N, Gersbach P, Hug MN, Bieri R, Bomio C, Li J, Huber S, Altmann K-H, Pluschke G (2016) Antibody-mediated neutralization of the exotoxin mycolactone, the main virulence factor produced by *Mycobacterium ulcerans*. PLoS Negl Trop Dis 10:e0004808. https://doi.org/10.1371/journal.pntd.0004808
11. Kpeli G, Buultjens AH, Giulieri S, Owusu-Mireku E, Aboagye SY, Baines SL, Seemann T, Bulach D, Gonçalves da Silva A, Monk IR, Howden BP, Pluschke G, Yeboah-Manu D, Stinear T (2017) Genomic analysis of ST88 community-acquired methicillin resistant *Staphylococcus aureus* in Ghana. PeerJ 5:e3047. https://doi.org/10.7717/peerj.3047
12. Kpeli G, Darko Otchere I, Lamelas A, Buultjens AL, Bulach D, Baines SL, Seemann T, Giulieri S, Nakobu Z, Aboagye SY, Owusu-Mireku E, Pluschke G, Stinear TP, Yeboah-Manu D (2016) Possible healthcare-associated transmission as a cause of secondary infection and population structure of *Staphylococcus aureus* isolates from two wound treatment centres in Ghana. New Microbes New Infect 13:92–101. https://doi.org/10.1016/j.nmni.2016.07.001
13. Yeboah-Manu D, Kpeli GS, Ruf M-T, Asan-Ampah K, Quenin-Fosu K, Owusu-Mireku E, Paintsil A, Lamptey I, Anku B, Kwakye-Maclean C, Newman M, Pluschke G (2013) Secondary bacterial infections of Buruli ulcer lesions before and after chemotherapy with streptomycin and rifampicin. PLoS Negl Trop Dis 7:e2191. https://doi.org/10.1371/journal.pntd.0002191
14. Bieri R, Scherr N, Ruf M-T, Dangy J-P, Gersbach P, Gehringer M, Altmann K-H, Pluschke G (2017) The macrolide toxin mycolactone promotes bim-dependent apoptosis in Buruli ulcer through inhibition of mTOR. ACS Chem Biol 12(5):1297–1307. https://doi.org/10.1021/acschembio.7b00053
15. Bolz M, Ruggli N, Ruf M-T, Ricklin ME, Zimmer G, Pluschke G (2014) Experimental infection of the pig with *Mycobacterium ulcerans*: a novel model for studying the pathogenesis of Buruli ulcer disease. PLoS Negl Trop Dis 8:e2968. https://doi.org/10.1371/journal.pntd.0002968
16. Ruf M-T, Bolz M, Vogel M, Bayi PF, Bratschi MW, Sopho GE, Yeboah-Manu D, Um Boock A, Junghanss T, Pluschke G (2016) Spatial distribution of *Mycobacterium ulcerans* in Buruli ulcer lesions: implications for laboratory diagnosis. PLoS Negl Trop Dis 10:e0004767. https://doi.org/10.1371/journal.pntd.0004767
17. Addison NO, Pfau S, Koka E, Aboagye SY, Kpeli G, Pluschke G, Yeboah-Manu D, Junghanss T (2017) Assessing and managing wounds of Buruli ulcer patients at the primary and secondary health care levels in Ghana. PLoS Negl Trop Dis 11:e0005331. https://doi.org/10.1371/journal.pntd.0005331
18. Amoussouhoui AS, Sopoh GE, Wadagni AC, Johnson RC, Aoulou P, Agbo IE, Houezo J-G, Boyer M, Nichter M (2018) Implementation of a decentralized community-based treatment program to improve the management of Buruli ulcer in the Ouinhi district of Benin, West Africa. PLoS Negl Trop Dis 12:e0006291. https://doi.org/10.1371/journal.pntd.0006291
19. Aboagye SY, Asare P, Otchere ID, Koka E, Mensah GE, Yirenya-Tawiah D, Yeboah-Manu D (2017) Environmental and behavioral drivers of Buruli ulcer disease in selected communities along the Densu River basin of Ghana: a case-control study. Am J Trop Med Hyg 96:1076–1083. https://doi.org/10.4269/ajtmh.16-0749
20. Ampah KA, Nickel B, Asare P, Ross A, De-Graft D, Kerber S, Spallek R, Singh M, Pluschke G, Yeboah-Manu D, Röltgen K (2016) A sero-epidemiological approach to explore transmission of *Mycobacterium ulcerans*. PLoS Negl Trop Dis 10:e0004387. https://doi.org/10.1371/journal.pntd.0004387

21. Williamson HR, Mosi L, Donnell R, Aqqad M, Merritt RW, Small PLC (2014) *Mycobacterium ulcerans* fails to infect through skin abrasions in a guinea pig infection model: implications for transmission. PLoS Negl Trop Dis 8:e2770. https://doi.org/10.1371/journal.pntd.0002770
22. Sarfo FS, Lavender CJ, Fyfe JAM, Johnson PDR, Stinear TP, Phillips RO (2011) *Mycobacterium ulcerans* DNA not detected in faecal samples from Buruli ulcer patients: results of a pilot study. PLoS One 6:e19611. https://doi.org/10.1371/journal.pone.0019611
23. Bolz M, Bratschi MW, Kerber S, Minyem JC, Um Boock A, Vogel M, Bayi PF, Junghanss T, Brites D, Harris SR, Parkhill J, Pluschke G, Lamelas Cabello A (2015) Locally confined clonal complexes of *Mycobacterium ulcerans* in two Buruli ulcer endemic regions of Cameroon. PLoS Negl Trop Dis 9:e0003802. https://doi.org/10.1371/journal.pntd.0003802
24. Doig KD, Holt KE, Fyfe JAM, Lavender CJ, Eddyani M, Portaels F, Yeboah-Manu D, Pluschke G, Seemann T, Stinear TP (2012) On the origin of *Mycobacterium ulcerans*, the causative agent of Buruli ulcer. BMC Genomics 13:258. https://doi.org/10.1186/1471-2164-13-258
25. Eddyani M, Vandelannoote K, Meehan CJ, Bhuju S, Porter JL, Aguiar J, Seemann T, Jarek M, Singh M, Portaels F, Stinear TP, de Jong BC (2015) A genomic approach to resolving relapse versus reinfection among four cases of Buruli ulcer. PLoS Negl Trop Dis 9:e0004158. https://doi.org/10.1371/journal.pntd.0004158
26. Vandelannoote K, Meehan CJ, Eddyani M, Affolabi D, Phanzu DM, Eyangoh S, Jordaens K, Portaels F, Mangas K, Seemann T, Marsollier L, Marion E, Chauty A, Landier J, Fontanet A, Leirs H, Stinear TP, de Jong BC (2017) Multiple introductions and recent spread of the emerging human pathogen *Mycobacterium ulcerans* across Africa. Genome Biol Evol 9:414–426. https://doi.org/10.1093/gbe/evx003
27. Awah PK, Boock AU, Mou F, Koin JT, Anye EM, Noumen D, Nichter M, Stop Buruli Consortium (2018) Developing a Buruli ulcer community of practice in Bankim, Cameroon: a model for Buruli ulcer outreach in Africa. PLoS Negl Trop Dis 12:e0006238. https://doi.org/10.1371/journal.pntd.0006238
28. Amoussouhoui AS, Johnson RC, Sopoh GE, Agbo IE, Aoulou P, Houezo J-G, Tingbe-Azalou A, Boyer M, Nichter M (2016) Steps toward creating a therapeutic community for inpatients suffering from chronic ulcers: lessons from Allada Buruli Ulcer Treatment Hospital in Benin. PLoS Negl Trop Dis 10:e0004602. https://doi.org/10.1371/journal.pntd.0004602
29. Ahorlu CK, Koka E, Yeboah-Manu D, Lamptey I, Ampadu E (2013) Enhancing Buruli ulcer control in Ghana through social interventions: a case study from the Obom sub-district. BMC Public Health 13:59. https://doi.org/10.1186/1471-2458-13-59

www.ingramcontent.com/pod-product-compliance
Lightning Source LLC
LaVergne TN
LVHW082119200926
848484LV00008B/1981